Thriving in the EdTech Revolution

A Practical Guide to Technology Integration in Teacher Education

Edited by

Harshith B. Nair

Regional Institute of Education (NCERT), Mysuru

G Viswanathappa

Regional Institute of Education (NCERT), Nellore

Vasileios Paliktzoglou

Bahrain Polytechnic, Isa Town, Bahrain

Series Editor
Vasileios Paliktzoglou

Series in Emerging Technologies in Education

VERNON PRESS

www.vernonpress.com

In the Americas:
Vernon Press
1000 N West Street, Suite 1200,
Wilmington, Delaware 19801
United States

In the rest of the world:
Vernon Press
C/Sancti Espiritu 17,
Malaga, 29006
Spain

Series in Emerging Technologies in Education

ISBN: 979-8-8819-0371-8

Cover design by Vernon Press with elements from Freepik.

Contents

List of Figures

List of Tables

Foreword

Beyond Translation: Technology Integration as a Creative Act

Punya Mishra

Director Innovative Learning Futures, Learning Engineering Institute
Mary Lou Fulton College, Arizona State University, USA

Lindsey McCaleb

Mary Lou Fulton College, Arizona State University, USA

Translation is that which transforms everything so that nothing changes.
Günter Grass

Creativity can be described as letting go of certainties.
Gail Sheeley

The French Lieutenant's Woman (1969) is a complex novel that works at multiple levels. Published in 1969 by John Fowles, it is at one level a story about a woman shunned by Victorian society and at another level a look back at the Victorian age from today's more permissive vantage point. What makes it particularly fascinating is how it works as a piece of meta-fiction: The author steps into the story multiple times to comment on the happenings, openly muses on how the story is to end, and finally chooses to give readers a choice of endings. Reading the last hundred pages is like being caught in a fictional labyrinth of possible endings, each forcing us to rethink everything that has gone before.

This complex structure was long considered impossible to film. As director John Frankenheimer (one of the many directors who tried and failed to adapt the novel to film) said, "There is no way you can film the book. You can tell the same story in a movie, of course, but not in the same way." But the film was made, and it stands to this day as one of the best adaptations of a novel ever put to screen. The secret to its success? The creative team, led by screenwriter Harold Pinter, didn't try to translate the book's meta-fictional devices (something more suited to print). Instead, they reimagined it completely, creating a film within a film where the Victorian love story is paralleled by a modern-day affair between the actors playing the roles.

Different medium, different approach, same deep impact.

The challenge of adapting a complex, multi-layered novel like *The French Lieutenant's Woman* mirrors the creative transformation expected of educators today. We're living through a revolutionary moment in educational technology. ChatGPT and other GenAI tools have burst onto the scene, VR and AR are becoming more accessible, and the digital transformation of education seems to be accelerating at a dizzying pace. Many educators feel like they're being asked to perform their own adaptation, transforming traditional teaching and learning for a rapidly transforming digital age. But just as with film adaptation, the real opportunity isn't in translation, it's in reinvention.

Here's the thing about creative acts: They require both courage and understanding. When Pinter and his team approached *The French Lieutenant's Woman*, they didn't just understand storytelling; they understood cinema as a creative medium. Similarly, effective technology integration isn't just about knowing how to use digital tools or deeply understanding standard pedagogical approaches. On the other hand, it *is* about understanding their creative potential, their limitations, and most importantly, how they might enable entirely new forms of teaching and learning.

This creative mindset gives us permission to "wreak havoc with the structure" while remaining true to the essence. The film version of *The French Lieutenant's Woman* barely resembles the novel in structure, yet it captures its themes and emotional impact perfectly. In education, this might mean completely reimagining a lesson or unit, moving away from traditional formats while keeping core learning objectives intact. It's not about digitising worksheets or putting lectures online, it's about seeing technology integration as an opportunity for creative reinvention.

This creative reinterpretation of technology by educators isn't just an abstract concept; it's actually formalised in the TPACK (Technological Pedagogical Content Knowledge) framework (Mishra & Koehler, 2006). Just as a successful film adaptation requires creative synthesis of cinematic techniques and narrative understanding, effective technology integration demands active integration of technological knowledge, content expertise, and pedagogical craft. These aren't separate considerations — they're the raw materials that educators weave together to create something new. What matters is the thoughtful, creative decision-making that goes into integrating these tools into the learning experience. Sometimes, as with any creative work, knowing what to leave out is as important as knowing what to include.

And just as film adaptation is inherently collaborative — involving writers, directors, actors, and countless others — technology integration thrives on creative collaboration. Teachers working with technology specialists, curriculum designers, and, yes, even students, can create more robust and effective learning experiences. In today's context, this collaboration might even extend to GenAI tools, which can serve as creative partners in brainstorming, content creation, and instructional design.

This brings us to why we are excited to introduce this book: *Thriving in the EdTech Revolution: A Practical Guide to Technology Integration in Teacher Education.* Throughout these chapters, you'll find voices of educators and researchers who approach technology integration not as a technical challenge, but as a creative opportunity. They're sharing stories of creative success and instructive failure, offering frameworks for thinking about technology integration, and most importantly, demonstrating how to maintain pedagogical integrity while embracing technological innovation.

The chapters ahead explore everything from foundational frameworks to cutting-edge applications, from global perspectives to local challenges. But running through all of them is this central idea: Technology integration is fundamentally a creative act. It requires understanding both the medium and the message, the tool and the task, the technology and the teaching, and then creating something new from these elements.

We're living in an age where the pace of technological change can feel overwhelming. Every day seems to bring new tools, new possibilities, and new challenges. But perhaps that's exactly why approaching technology integration as a creative act is so valuable. Creative acts don't succeed by slavishly copying what came before or by blindly embracing every new technique. They succeed by understanding the essence of what they're trying to achieve and finding innovative ways to realise that vision.

As you read through this book, we invite you to approach technology integration as an act of creative translation, preserving the core of effective teaching while boldly reimagining how it can be expressed through new digital mediums. Explore how different technologies might not just replace but reinvent educational experiences. And remember that, like any creative act, technology integration isn't about replication, it's about transformation.

Because it is only when we approach technology integration as a creative act that we can achieve what seemed paradoxical in our opening quotes, we can transform our teaching practices while keeping our educational values intact, just as Günter Grass suggested about translation. And like Gail Sheeley's view of creativity, this transformation requires us to let go of our certainties about traditional teaching methods, not in surrender but in purposeful exploration.

This creative tension – between transformation and preservation, between letting go and holding on – is where the real magic happens. In education, this means having the courage to radically reimagine our teaching methods while maintaining our commitment to deep learning. After all, the magic happens not in the digital tools themselves, but in the creative acts of teachers who, like skilled translators, know how to transform their practice while preserving its essential purpose, just as the best adaptations capture not the letter but the spirit of their source.

Introduction
Future-Making Educators

Harshith B Nair

Regional Institute of Education (NCERT), Mysuru, India

G Viswanathappa

Regional Institute of Education (NCERT), Nellore

Vasileios Paliktzoglou

Bahrain Polytechnic, Isa Town, Bahrain

Education is at a turning point where tradition and change coexist. To meet the challenges of the twenty-first century, we must reconsider what we teach and how and why we teach it. Education must be both a stabilising and a transformative force as society struggles with rapid technological breakthroughs, environmental challenges, and changing cultural landscapes. Educators find their fundamental purpose in the constantly changing relationship with these forces, not just as knowledge facilitators but also as scaffolds of a sustainable and inclusive future.

The great potential of technology, in particular artificial intelligence (AI), resides at the centre of this change. Previously unthinkable levels of personalisation are now possible because of AI technologies like data analytics, virtual tutors, and adaptive learning systems. These innovations minimise constraints that have long impeded equity and access and allow educators to tailor learning experiences to each student's requirements. However, there are also disadvantages, including the potential increase of already-existing inequalities, algorithmic bias, and ethical challenges around data privacy. Educators must embrace these technologies critically, ensuring they serve the majority rather than just a few and preserving the fundamental principles of education, inclusion, equity, and integrity.

The Role of Educators as Future-Makers

Educators today stand at the forefront of shaping a future defined by unprecedented challenges and opportunities. Far beyond the traditional classroom, their role now encompasses the fostering of critical thinking, creativity, and a lifelong thirst for learning are qualities essential to thriving in an interconnected, ever-evolving world. In this expanded role as "future-makers," teachers cultivate not just students but also the leaders, innovators, and global citizens who will steer humanity's course.

This shift demands a revolutionary approach to teacher preparation, one that goes beyond conventional content delivery. Educators must master digital literacy, ethical technology use, and global citizenship to nurture a generation capable of navigating technological advances with a moral compass. Drawing lessons from trailblazing efforts worldwide or grassroots innovations in adaptive learning closer to home, we can see the transformative potential of educators who embrace change. It is within their classrooms that the seeds of progress are sown, making teachers not only knowledge providers but architects of a brighter, more inclusive tomorrow.

A Vision for Educator Preparation

The future demands educators who are as dynamic as the challenges they will face, individuals capable of inspiring and empowering their students in a world where change is the only constant. To achieve this, teacher preparation must undergo a radical evolution, becoming a crucible for developing digital fluency, creative pedagogical approaches, and a capacity for systemic leadership. This is not merely an academic exercise but a moral imperative for the societies we wish to build. Imagine programs that don't just teach educators how to use technology but challenge them to wield it as a tool for equity, innovation, and transformative learning. Picture a system where teaching excellence is measured not by rote outcomes but by the ability to ignite curiosity, foster critical inquiry, and create lifelong learners. Such a vision is not only necessary but urgent.

Educators must be equipped to bridge the digital divide, confront systemic inequities, and inspire students to become active architects of the future. The call to action is clear: to rethink, reimagine, and revolutionise education for a world that demands nothing less than our very best.

Chapter Overview

The book begins by laying the foundation for understanding the transformative role of technology in education. It introduces the reader to the historical context of digital integration, showing how traditional teaching methods have evolved into digital-first approaches. The narrative emphasises the potential of digital tools to revolutionise learning, using case studies to illustrate early innovations that paved the way for today's digital classrooms. This context builds a strong base, preparing educators and learners to grasp the underlying principles of educational technology. Building on this foundation, the book delves into the powerful role of artificial intelligence and machine learning in classrooms. These technologies offer unprecedented personalisation, adapting to the needs of individual learners while enhancing engagement. Ethical considerations, such as data privacy and algorithmic fairness, are explored, providing readers with a balanced view of the promise and challenges of AI. By connecting these discussions to the global context, the narrative broadens to explore how educational technology is being adopted and adapted across diverse socio-economic and cultural landscapes.

As the book transitions into practical applications, it introduces the concept of blended learning. This model harmonises in-person and digital teaching methods, offering a dynamic approach to education. By examining frameworks like TPACK and SAMR, educators are guided on how to seamlessly integrate digital tools while maintaining inclusivity. The discussion naturally leads into the potential of cutting-edge technologies such as virtual reality and gamification, which transform classrooms into immersive environments, fostering deeper student engagement and retention.

The narrative takes a thematic turn, exploring the digital humanities and their impact on history education. By integrating tools like interactive mapping and digital storytelling, students are not merely passive learners but active participants in shaping historical narratives. This theme of empowerment continues with a focus on strategies for seamless technology integration, emphasising the importance of collaborative apps, personalised feedback systems, and multimedia presentations to enrich the learning experience.

Addressing the pressing issue of inequality, the book shines a light on the digital divide. It underscores how access to technology shapes learning outcomes and proposes innovative solutions to bridge this gap. This naturally flows into a discussion on pedagogical innovation, highlighting how emerging technologies can complement traditional teaching methods to create adaptive, student-centred learning experiences. The role of artificial intelligence in advanced curricula is particularly notable here, showcasing its ability to tailor educational pathways and enhance assessment techniques. Ethical considerations return as the narrative evaluates responsible technology integration, discussing issues like algorithmic bias and data security. These discussions are paired with examples of global collaborative learning projects that use technology to foster creativity, cultural exchange, and global citizenship. The focus then shifts to educators themselves, emphasising the importance of continuous professional development. By adopting frameworks like TPACK and participating in lifelong learning, teachers can confidently navigate digital transformations and inspire their students.

The book highlights technologies such as augmented reality and blockchain, predicting their impact on education in the years to come. This forward-looking discussion transitions seamlessly into an exploration of sustainability, linking digital learning to global goals such as climate education. The narrative ends with a deep dive into the application of digital tools in specialised fields like medical and veterinary education, showcasing advanced simulations and ethical considerations that prepare students for real-world challenges. Through each chapter, the book weaves a cohesive narrative that underscores the transformative potential of technology in education. It equips readers with the theoretical understanding, practical strategies, and ethical frameworks needed to thrive in the digital age, advocating for the importance of adapting and innovating in a rapidly evolving landscape. This makes it an invaluable resource for educators, learners, and stakeholders committed to shaping the future of education.

AI-Use Declaration Statement

The chapters of this volume and the content of the chapters are original contributions of the authors to the best of our knowledge. LLMs and Proprietary software like Grammarly AI have been used for syntactical refinement and expression.

PART I: THE FOUNDATIONS OF EDTECH AND AI IN EDUCATION

Chapter One
Assessing the Digital Readiness of Pre-Service Teachers

Bhaskar Gurramkonda

Central University of Kerala, India

Vanitha C

Central University of Kerala, India

Abstract

In the fast-changing digital world, pre-service teachers (PSTs) need to have the required digital skills to effectively incorporate technology into their teaching. This research examines the digital preparedness of pre-service teachers in India, with emphasis on their capacity to utilise different digital tools and technologies required for contemporary education. A quantitative method was used, employing the Pre-Service Teachers Digital Readiness Scale (PSTDRS) to evaluate competencies along five dimensions: Basic Digital Skills for Teaching (BDST), Ethical Knowledge (EK), Techno-pedagogical Knowledge (TPK), Socio-digital Practice (SDP), and Critical Content Refinement (CCR). The results show that while PSTs register moderate digital readiness, there are important gaps in their capacity to employ sophisticated digital tools and critically evaluate online content. The research also compares these findings with global best practices, pointing out successful approaches in Estonia and South Korea. The study underlines the necessity for better teacher training programmes that address higher-order digital capabilities, ethical use of technology, and ongoing professional development. Suggestions for increasing digital competence in teacher education are offered.

Keywords: digital readiness, pre-service teachers, teacher education, technology integration.

***

Introduction

The fast evolution of digital technologies has widely impacted the field of education, with teachers needing to incorporate digital tools and online platforms into pedagogy. As future teaching staff, pre-service teachers need to be suitably prepared to negotiate the digital change, so assessing whether they are digitally ready is an important research topic. Digital preparedness includes not only the technological skills necessary for using digital technologies but also having the capacity for self-directed learning in digital platforms (Fuchs et al., 2022). Research indicates that it is critical to cultivate digital preparedness to support innovative teaching practice and increase the satisfaction of the learners (Huang, 2022).

Although most of the previous research has been on in-service teachers, there is a rising interest in knowing the digital preparedness of pre-service teachers, who are being trained at the moment to cope with twenty-first-century education needs (Milawati & Sholeh, 2020; Ng et al., 2024). Digital preparedness also goes beyond having technical skills and encompasses the psychological capital and self-efficacy to accommodate digital learning environments (Al-Awidi & Aldhafeeri, 2017). Testing pre-service teacher preparedness is crucial in facilitating their future effectiveness in executing digital curricula and embracing technology-influenced instruction methods (Scherer et al., 2021). In spite of such developments, very significant gaps still exist in the literature on the pertinent factors affecting the digital readiness of pre-service teachers.

While Scherer et al. (2021) investigate in-service teachers' preparedness for online teaching, there is less focus on the preparedness of pre-service teachers, particularly regarding self-regulated learning (Fuchs et al., 2022) and the use of digital preparedness to foster entrepreneurial competence and innovation (Huang, 2022). Moreover, studies with pre-service teachers in non-Western settings, including Kuwait (Al-Awidi et al., 2017) and (Milawati & Sholeh, 2020), have emphasised the importance of localised studies that examine cultural and contextual determinants of digital readiness. Qualitative and quantitative methods combined through mixed-methods research have worked well to document these nuances (Kleih et al., 2022; Venkatesh et al., 2016). Yet there remains a need for further exploration in this area. In light of the COVID-19 pandemic, which has accelerated the transition to online learning, digital readiness has become a focal point of educational research (Pagente & Selecios, 2023).

Pre-service teachers not only need to be proficient in the use of digital tools but also in applying them to pedagogical practice in a manner that supports efficient learning outcomes (Quaicoe & Pata, 2015; Taggart & Roulston, 2024).

Unlike other research works that concentrate on the digital preparedness of students, institutes, and teachers, this study aims to determine the level of digital readiness of pre-service teachers, who carry a lot of potential to reform themselves during their teaching period. Therefore, the current study was primarily concerned with evaluating the readiness of the pre-service teachers in the area of five dimensions namely Basic Digital Skills for teaching (BDST), Ethical Knowledge use in teaching with technology (EK), Techno-pedagogy knowledge (TPK), Socio-digital practice (SDP), and Critical Content Refinement (CCR), these dimensions lay greater emphasis on the teaching component compared to the individual academic participation as stated in Hong and Kim (2018). These dimensions are used by the researcher to investigate the impact of gender, locality, age, qualification, pedagogy subject, type of institute they study, learner generation, academic grades of students, exposure to technology, and the devices they own. The following research questions were developed for analysing the above impacts.

Research Questions

1. What is the current level of digital readiness among pre-service teachers?
2. What role does the institutional use of technology play in pre-service teachers' digital readiness?
3. What cultural and contextual factors influence the digital readiness of pre-service teachers in non-Western settings?
4. To what extent does digital readiness correlate with pre-service teachers' grades?

By addressing these research questions, this study aims to contribute to the existing body of knowledge by filling critical gaps in the literature, particularly regarding the digital readiness of pre-service teachers and the factors influencing their preparedness for the evolving demands of education in the twenty-first century.

Objectives of the Study

Based on the above research questions, the objectives of this study are:

1. To determine the current level of digital readiness among pre-service teaches across five key dimensions.
2. To examine the influence of factors such as time of first technology exposure and device ownership on pre-service teachers' digital readiness.
3. To explore the influence of demographic and contextual factors on the digital readiness of pre-service teachers in Kerala.
4. To determine the correlation between pre-service teachers' digital readiness and their academic grades.

Research Hypotheses

- **H01:** There is no significant correlation between the mean scores of the digital readiness of the pre-service teachers in terms of their gender, age, locality, exposure to technology, and the device they use.
- **H02:** There is no significant correlation between digital readiness to the academic achievement of the pre-service teachers.

Method

This study used a quantitative research design utilising a normative survey method to collect data on the Digital Readiness of Pre-service teachers. The sample consisted of 151 pre-service teachers across various teacher training institutes from the randomly selected districts of Kerala.

Tool

This quantitative study used a Pre-Service Teachers' Digital Readiness Scale (PSTDRS) tool developed by the researchers with the guidance of Guzmán-Simón et al. (2017) for theoretical understanding and Hong & Kim (2018) for the preparation of the tool's dimensions. Further, items were taken from available sources and arranged in the following dimensions: Basic Digital Skills for teaching (BDST), Ethical Knowledge use in teaching with technology (EK), Techno-pedagogy knowledge (TPK), Socio-digital practice (SDP), and Critical Content Refinement (CCR). Milawati and Sholeh (2020), Pagente and Selecios (2023), and Taggart and Roulston (2024) served as references for the item formation. The scale consists of 24 items with a three-point Likert-type scale with options Yes, Need Help, and No. The reliability of the scale using Cronbach's alpha was found to be 0.781.

Data Analysis

Descriptive and correlation analyses were used to describe the data quantitatively through SPSS software and Excel. The dependent variable, Digital Readiness, was evaluated using scores from the PSTDRS scale. With a significance level of 0.05, t-test, ANOVA, and Pearson correlation were analysed. Data representation was done with the help of Datawrapper.

Results

The population used for the study is pre-service teachers from Kerala, India. These 151 samples were gathered via stratified random sampling from six teacher training institutes from four different districts of Kerala, not limited to one specific region of the state. The use of a stratified sample ensures that the sample is representative based on demographic aspects. As revealed in Table 1, the sample comprises 35.76% (54) male and 64.23% (97) female pre-service teachers. In terms of locality, 44.37 % (67) of the sample are rural, 31.78% (48) semi-urban, and 23.84% (36) urban. The figures differed in their percentage with regard to age, with the biggest percentage of the age range under consideration being 22-24 with 50.33% (76), and the 25-30 age group occupying 35.09% (53) with the remaining two being 5.96% (9), 8.60% (13) for 18-21 and 30+ age ranges respectively. For the topic of Pedagogy, the distribution of data is as follows: 17.8% (27), 37.74% (57), 36.42% (55), and 7.94% (12) for the Arts, Science, Language, and others, respectively. For the Exposure to Technology, the respondents who received exposure to the technology from their SSLC are 36.42% (55), from their +2 (intermediate) 19.86% (30), from their UG 29.80% (45), from their 5.96% (9), and from the post-COVID period are 7.94% (12). Data in the context of availability of devices such as mobile, laptop, Wi-Fi connection, etc., indicates 28.47% (43) have only one device (Mobile), 35.76% (54) have two devices (Mobile+Laptop), 27.81% (42) have three devices (Mobile+Laptop+Wi-Fi), and 7.94% (12) have more than three devices.

Table 1.1 Demographic data.

Sl.No.	Group	Category	Sample	Percentage
1.	Gender	Male	54	35.76
		Female	97	64.23
2.	Locality	Rural	67	44.37
		Semi-Urban	48	31.78
		Urban	36	23.84
3.	Age	18-21	9	5.96
		22-24	76	50.33
		25-30	53	35.09
		30+	13	8.60
4	Pedagogy subject	Arts	27	17.8
		Science	57	37.74
		Language	55	36.42
		Other	12	7.94
5.		SSLC	55	36.42
		+2	30	19.86

	Exposure to technology	UG	45	29.80
		PG	9	5.96
		After covid only	12	7.94
6.	Device they posses	Only one device (mobile)	43	28.47
		Two devices (mobile + laptop)	54	35.761
		Three devices	42	27.81
		More than 3 devices	12	7.94

The responses made by the pre-service teachers to the digital readiness scale were examined through descriptive statistics, percentage analysis, and an item-based assessment. To find the differences among attitudes based on subsamples, inferential statistics like the independent sample t-test and one-way ANOVA with post-hoc method were applied. The statistical significance level was set at 0.05. The descriptives of the data showed a positive skewness, with the questions being answered more on the positive side. This skewness was later reduced by removing the outliers and using the statistical calculation of log10(total) through SPSS, and the normality is assumed through the Q-Q plot and box plot. Log10 values are wieldy used for the skewed distribution to avoid serious misinterpretation (Feng et al., 2014).

Figure 1.1 Showing the Assumed Normality of the Data.

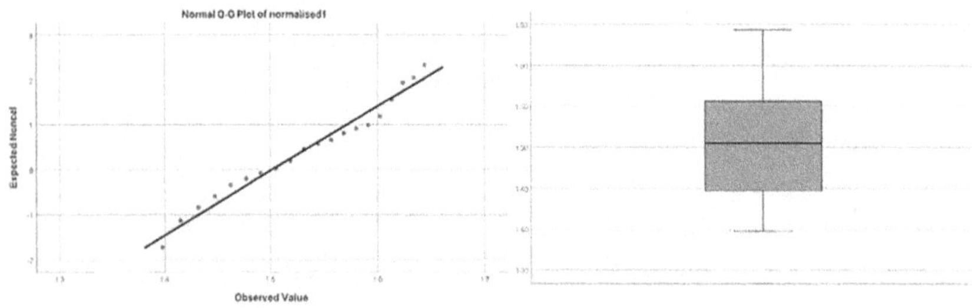

Hereafter, the normalised data, which is log10 (total), is the data used throughout the analysis in place of raw data. The variation of the values is seen in Table 1.2.

Table 1.2 Variation of Data between Raw and log10 (normalised data).

Descriptives	**Raw data**	**Log10 data**
N	151	151
Mean	32.13	1.5013
Median	32	1.5051
Mode	28	1.45
Standard Deviation	5.235	0.067
Variance	27.409	0.005
Skewness	0.426	0.214
Kurtosis	-0.915	-1.090

Pre-service teachers' digital readiness was examined through an item-based assessment. The item-based assessment was carried out on five dimensions: Basic Digital Skills for Teaching (BDST), Ethical Knowledge (EK), Techno-pedagogical Knowledge (TPK), Socio-digital Practice (SDP), and Critical Content Refinement (CCR). These items according to the dimensions are presented in Table 4 to Table 8. Their response levels were divided into three categories based on their mean and standard deviation. The overall mean for all the items is 1.33, and the standard deviation is 0.21. Based on the mean and standard deviation, the three categories of the data are divided as follows: Item mean, which is less than M-1SD=1.33-0.21=1.12, indicates that the respondents had a high level of digital readiness. Similarly, the item mean, which is more than M+1SD= 1.33+0.21=1.54, indicates that the respondents had a low level of digital readiness. Further, the item mean, which lies between 1.12 and 1.54, is considered moderate. However, from Table 1.3, it is quite evident that the overall digital readiness level of pre-service teachers is moderate in level, with 58.94% at this level, whereas it is also a point of concern that the digital readiness of pre-service teachers at the low level is more than that of the high level.

Table 1.3 Level of Digital Readiness in Pre-Service Teachers.

Level	**Frequency**	**Percentage**
Low	34	22.51
Moderate	89	58.94
High	28	18.57

The above table shows the overall results, but an in-depth understanding of the data can be obtained through the item-wise analysis.

Table 1.4 Responses Based on Dimension 1 (Basic Digital Skills).

Sl. No.	Statements	Mean	SD	% respondents at the lower end	Result in terms of readiness
1	I can upload and download media, including online photos, files, video files, and sound files.	1.092105	0.30011	6.849315	High
2	I can manage software or apps from a computer or mobile device	1.236842	0.461402	15.06849	Moderate
3	I can take any type of print for my teaching and learning purposes.	1.282895	0.484278	18.26484	Moderate
4	I can download and attach files from email.	1.105263	0.374932	5.022831	High
5	I can convert the materials into different formats.	1.223684	0.495899	11.41553	Moderate
6	I can use basic computer tools like PowerPoint, Word, and Excel for teaching and learning.	1.092105	0.30011	6.849315	High

From the above table, it is clear that both moderate and low levels are equally distributed; however, the percentage of pre-service teachers who can perform small computer tools like PowerPoint and Excel is considerably high, and a considerably low percentage of the sample is weak at this skill. From the above item-based responses analysis, the pre-service teachers are well-versed in the basic skill set required for teaching and learning.

Table 1.5 Responses Based on Dimension 2 (Ethical Knowledge).

Sl. No.	Statements	Mean	SD	% respondents at the lower end	Result in terms of readiness
1	I can set up and change security options in web browsers.	1.546053	0.595939	31.05023	Low
2	I can browse safely without falling prey to cyberattacks.	1.453947	0.563026	27.39726	Moderate

3	I can give restricted access to people while sharing educational content.	1.414474	0.593712	21.91781	Moderate
4	I know about intellectual property rights while using content for teaching and learning.	1.322368	0.525634	19.17808	Moderate

The readiness of the pre-service teacher is high and moderate in this particular domain. From the above table, it is quite evident that a considerable amount (31.05%) of pre-service teachers are not very good at using the internet securely, which leads to security threats and cyberattacks, although readiness regarding cyberattacks falls into the moderate category (Mean = 1.453). Readiness concerning the security options in web browsers low threshold (1.54).

Table 1.6 Responses Based on Dimension 3 (Techno-pedagogy).

Sl. No.	Statements	Mean	SD	% respondents at the lower end	Result in terms of readiness
1	I know how to use smart boards and projectors in the classroom	1.375	0.630988	15.52511	Moderate
2	I know how to use different LMS software	2.019737	0.696823	35.61644	Low
3	I can create assessment tools for my class	1.460526	0.661392	19.6347	Moderate
4	I can use cloud spaces to store and share educational materials	1.605263	0.710501	5.936073	Low
5	I can effectively use technology to enhance my teaching methods.	1.197368	0.466446	10.50228	High

Table 6 provides clear evidence that the overall readiness of the sample in this techno-pedagogy dimension is moderate and lower. While a high proportion (about 90%) of the sample believed that they can effectively use technology in teaching and learning, the items on other aspects, like coping with the real and trending techno tools, showed that more than 35% of the sample are at the low level when it comes to maintaining the LMS and cloud spaces. Even the use of variant assessment tools is only at the moderate level on average.

Table 1.7 Responses Based on Dimension 4 (Socio-digital Practice).

Sl. No.	Statements	Mean	SD	% respondents at the lower end	Result in terms of readiness
1	I can interact with students and colleagues using video conferencing tools.	1.092105	0.341662	5.022831	High
2	I can share my opinions on educational topics using online platforms.	1.269737	0.623527	6.392694	Moderate
3	I can use chatting apps for learning and content sharing with students.	1.25	0.534959	11.41553	Moderate
4	I can create teachable content from social networking apps and sites.	1.335526	0.589309	15.52511	Moderate
5	I can use chat apps for classroom discussions.	1.157895	0.468052	5.936073	Moderate

This dimension, shown above through Table 7, revealed that the sample pre-service teachers are at the moderate to high level of readiness. Pre-service teachers are very well-versed in the use of chat apps and interactive applications. With a range of a minimum of 5.02% to a maximum of 15.52%, the sample is at the lower end of the readiness level in this dimension.

Table 1.8 Responses Based on Dimension 5 (Critical content refinement).

Sl. No.	Statements	Mean	SD	% respondents at the lower end	Result in terms of readiness
1	I can recognise bias or misinformation in digital educational content.	1.493421	0.575896	29.22374	Moderate
2	I can critically interpret digital media content relevant to teaching and learning.	1.473684	0.598379	26.0274	Moderate
3	I can generate educational content using AI tools.	1.210526	0.527814	7.762557	Moderate
4	I can inform others about effective ways to search for teaching content online.	1.210526	0.460348	12.32877	Moderate

Table 8 shows a clear distribution of data into a moderate level of readiness. While readiness with AI tools is moderate, only 7.76% of the sample pre-service teachers are at the lower end. Yet, nearly 30%, which is 29.22%, which is a notable percentage of pre-service teachers, are not in a position to recognise the biased misinformation, and 26.02% struggle to critically interpret the information on the internet.

The item-based assessment provides concise information about the data distribution, and the percentage analysis helps to identify the actual difficulties of pre-service teachers. However, for a deeper understanding of the actual variables, these data have to be studied with inferential and descriptive statistics. For statistical analysis of the data, independent variables, namely, gender, locality, age, exposure to the technology, and devices they possess, were considered, and these were evaluated through t-tests and one-way ANOVA. Further, the relation between academic achievement and digital readiness was analysed by Pearson's product-moment correlation.

Table 1.9 Mean Comparison of Gender for Digital Readiness Using t-test.

Gender	N	Mean	Std. Dev.	t-value	p-value	Result
Male	54	1.430	0.063	1.095	0.275	Not significant
Female	97	1.5060	0.729			

Table 9 discloses the t-test results, with 54 male and 97 female pre-service teachers. The mean log10 readiness score for males is 1.430, and the standard deviation is 0.063; for females, the mean is 1.506, and the standard deviation is 0.072. The p-value for significance is 0.275, which is greater than 0.05. It is observed that the calculated t-value is less than the critical value of 1.96 at the 0.05 level. Hence, the null hypothesis was not rejected.

The data were analysed using one-way ANOVA to examine the digital readiness of the pre-service teachers based on their locality, age, exposure to devices, and the devices they have.

Table 1.10 Mean Comparison of Locality, Age, Exposure to Devices, and Devices They Have for Digital Readiness Through One-Way ANOVA.

Group	Measure	Sum of Squares	Df	Mean Squares	F ratio	p-value	Result
Locality	Between group	0.008	2	0.04	0.800	0.4511	Not significant
	Within group	0.721	148	0.05			

Age	Between Group	0.036	3	0.012	2.557	0.058	Not significant
	Within group	0.693	147	0.005			
Exposure to devices	Between group	0.180	4	0.045	11.955	0.00	Significant
	Within group	0.721	147	0.004			
Devices they have	Between group	0.076	3	0.025	5.728	0.001	Significant
	Within group	0.653	147	0.004			

Table 10 showed that locality, age, exposure to devices, and devices they have only exposure to devices and devices they have variables showed a significant difference with p-values of 0.00 and 0.001, respectively. These two p-values are way less than the significance value of 0.05, resulting in the null hypothesis.

Table 1.11 Mean and Standard Deviation of Exposure to Devices for Digital Readiness.

Variable	**N**	**Mean**	**Std. Dev.**
SSLC	55	1.4681	0.06057
+2	30	1.4092	0.0655
UG	45	1.544	0.0491
PG	9	1.4689	0.07806
After COVID only	12	1.5445	0.08045

Table 1.11 presents the descriptive statistics of pre-service teachers' digital readiness. This table reveals the mean readiness of pre-service teachers who were exposed to technology in different stages: SSLC, +2, UG, PG, and After COVID-19 are 1.4681, 1.4092, 1.544, 1.4689, and 1.5445, respectively. The Scheffe test is used for the post hoc analysis of the data. The test analysed the group variables SSLC, +2, UG, PG, and After COVID-19. The significant difference between the mean scores can be seen in Table 1.12.

Table 1.12 Post-Hoc Comparison of Variables in Time of Exposure to Technology.

Variables compared	Mean difference	p-value	Result
SSLC and +2	-0.2218	0.639	Not significant
+2 and UG	-0.5415	0.009	Significant
UG and PG	0.07550	0.026	Significant
PG and after COVID-19	-0.07567	0.104	Not significant
SSLC and UG	-0.07632	0.00	Significant
SSLC and PG	-0.00082	1.00	Not significant
SSLC and After COVID-19	-0.7649	0.005	Significant
+2 and PG	0.02136	0.933	Not significant
+2 and after COVID-19	-0.0543	0.158	Not significant
UG and after COVID-19	-0.00017	1	Not significant

The post hoc results using the Scheffe test revealed that there is a significant difference in the mean scores of readiness of the pre-service teachers observed in comparison with exposure to the technology at the UG level, compared with the +2 level, with a p value of 0.009. The mean difference reveals that the exposure to technology at the UG level comparatively shows improved readiness with exposure at the +2 level.

Similar results could be seen in the case of UG and SSLC, and after COVID-19 and SSLC, which showed significant differences with p values of 0.00 and 0.005, respectively.

Table 1.13 Post-Hoc Table of Devices They Possess.

Variable	N	Mean	Std. Dev.
1 device (mobile) (A)	43	1.5328	0.07045
Two devices (mobile + laptop) (B)	54	1.5008	0.06892
Wi-Fi at home (C)	42	1.4798	0.06022
Additional devices (D)	12	1.4668	0.06315

Table 1.13 reveals the mean readiness of pre-service teachers who had devices like mobiles, laptops, Wi-Fi, and other devices. These devices are coded as 1 device (mobile) (A), two devices (Mobile+Laptop) as B, Wi-Fi at home, along with two other devices (C), and additional technological devices (D) for

simplicity in the analysis and interpretation process. The mean scores of A, B, C, and D are 1.5328, 1.5008, 1.4798, and 1.4668, respectively.

Table 1.14 Post-Hoc Comparison Readiness in Terms of Devices They Possess.

Variables compared	Mean difference	p-value	Result
A and B	0.03201	0.142	Not significant
B and C	0.02098	0.507	Not significant
C and D	0.01295	0.950	Not significant
A and C	0.05299	0.005	Significant
A and D	0.06593	0.030	Significant
B and D	0.03392	0.470	Not significant

The post hoc results from Table 1.14 using the Scheffe test revealed a significant difference in the mean scores of readiness of the pre-service teachers observed in comparing A with C and A with D, having p-values 0.005 and 0.030, respectively. The mean difference revealed that pre-service teachers with only one device, mobile (A), have a significant difference compared to the pre-service teachers having a laptop, Wi-Fi, and any other devices.

Table 1.15 Correlation Analysis of Five Dimensions with Academic Achievements.

Digital Readiness	Academic Achievement		
	Correlation Coefficient	p-value	Interpretation
Basic digital skills	0.067	0.410	Not significant
Ethical skills	-0.032	0.695	Not significant
Techno pedagogy	-0.44	0.594	Not significant
Socio-digital	-0.048	0.555	Not significant
Critical content refinement	-0.038	0.643	Not Significant

The results of the Pearson moment correlation study are shown in Table 1.15, with digital readiness showing a significant effect on the academic achievement of pre-service teachers, though the considered dimensions in this study are different from the dimensions of Hong and Kim (2018), Pagente and Selecios (2023) which aligned with the same results of not having significant difference with the academic achievements of pre-service teachers.

Unlike the results of Kim et al. (2019), which showed a significant correlation between digital readiness with academic scores, our study's correlation between the dimensions and academic achievement did not show any significant results. An attempt was made to compare the influence of teachers using technology for teaching on the pre-service teachers' readiness.

However, the influence of teachers on the students has not been studied here; instead, a mere attempt at comparison is made.

Figure 1.2 Showing the Comparison of Tools Teachers Use and the Purpose for Which They Use Them.

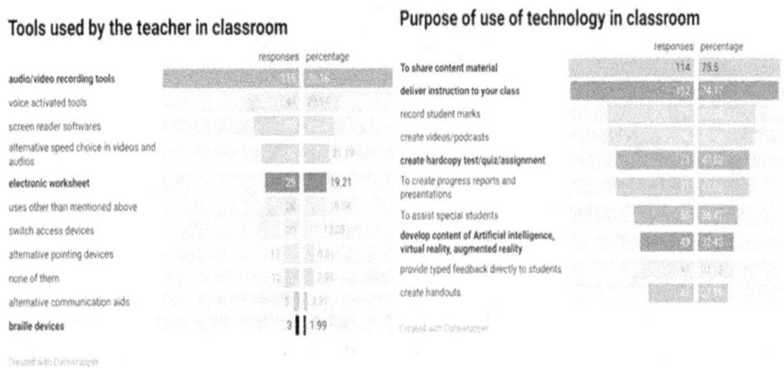

Figure 1.2 shows that almost 76.16% of teachers use audio/video recordings frequently in the classroom, followed by 29.14% using voice-activated tools, which are AI-integrated. Electronic worksheets, which are a part of e-assessment (innovative assessments were not of much focus), had 19.2%, and among all the tools, the least used tool is Braille devices, with only 1.99%.

Further, the tools which are used by the teachers are mainly focused on content sharing, that is, e-learning, with 75.5%, and delivering instruction (with almost the same share percentage: 74.17%). Less than 50% (47.02%) used tools for assessment; the percentage is the same for the use of progress reports, that is, recording their assessment marks. For developing the content through artificial intelligence, however, the result is 32.45%.

There are more aspects like the above-mentioned variables, which can vary the perception of the pre-service teachers who have not been studied specifically here. When we compare Figure 3 and Figure 2 with the pre-service teachers' perception, it is seen that the use of basic technical tools in the classroom did form some understanding of the digital skills of the students. Of the pre-service teachers, 67.55% felt that they should be trained thoroughly with the ed tech. The pre-service teachers agreed that ed tech makes them more efficient as teachers. However, it is also a point of concern that there are not enough ed tech resources in their institutes.

Figure 1.3 Students' Perception of the Use of Technology in Teaching and Learning.

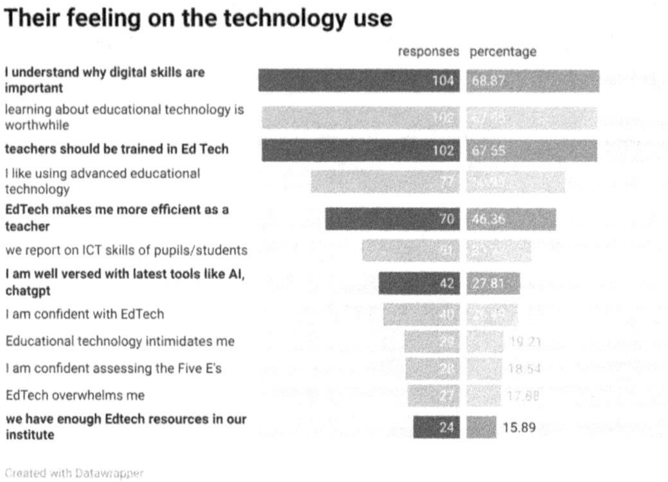

From the pictures, it can be inferred that pre-service teachers widely acknowledge the importance of technology in education, but there are gaps in training and resources that need to be addressed. It also observed that pre-service teachers are less confident, about 18.54%, and are only confident about assessing the five E's of the lesson.

Discussion

This study presents an overall evaluation of the digital readiness of pre-service teachers and highlights the imperative necessity of interventions aimed at enhancing their capacities, especially regarding advanced technology and digital literacy. Comparing our results with corresponding international research helps to contextualise these findings and provide useful strategies for increasing digital readiness in teacher education programmes.

Actionable Steps for Teacher Educators and Institutions

Our findings confirm that, although pre-service teachers score moderately on primary tools, such as PowerPoint and Excel, numerous pre-service teachers are challenged when it comes to sophisticated tools (Cuhadar, 2018; Liza & Andriyanti, 2020), i.e., Learning Management Systems (LMS) and cloud technology (low readiness in Table 6). This result is consistent with Fuchs et al.'s (2022) research, which investigated pre-service teachers' online self-regulated learning and concluded that digital preparedness is central to their capacity to handle sophisticated educational technologies. Both studies emphasise the need for teacher education programmes to go beyond the instruction of fundamental digital competencies, including explicit, experiential training with sophisticated tools. Based on the suggestions from Fuchs et al. (2022) and these results, institutions can enhance the integration of foundational digital literacy into their curriculum. In addition to emphasising structured, hands-on workshops that enable pre-service teachers to work with these tools in actual classroom settings (Rodrigues, 2020). Additionally, curriculum reform needs to prioritise emphasising developing skills in AI and Learning Management Systems (LMS). Tomczyk (2024) argues that without such comprehensive changes, pre-service teachers will continue to face challenges in integrating technology into their classrooms. Given that the demographic data indicated, most of the sample falls in the 21-25 age category. For these individuals, peer-assisted learning, as indicated by Huang (2022), might encourage mutual skill development and confidence in employing digital tools. Institutions may implement these peer-assisted lessons.

Milawati and Sholeh (2020) highlighted the significance of teacher preparedness in employing digital technologies in Indonesian schools, especially after COVID-19. Their study corroborates our findings, indicating that institutions need to introduce immediate interventions like digital literacy audits and micro-credential programmes to establish a strong foundation for pre-service teachers' tech skills.

Comparative Analysis with Global Best Practices

Comparing our study with those conducted in other countries offers valuable insights into how digital readiness is being addressed across different contexts. Guzmán-Simón et al. (2017) found that Spanish undergraduate students viewed digital competence as essential to their academic success, yet many still lacked the necessary skills to critically engage with digital platforms. This resonates with our findings, where pre-service teachers struggled to critically interpret online information and recognise biased data. Similarly, Scherer et al. (2021) profiled teachers' readiness for online teaching and found significant variability in digital competencies, much like the variability we observed in our study between pre-service teachers with different levels of exposure to technology.

Best practices from countries such as South Korea, highlighted by Hong & Kim (2018), emphasise the importance of national initiatives to foster digital readiness in higher education. In contrast to this study and Cuhadar (2018), which found a considerable portion of pre-service teachers underprepared to handle advanced tools, South Korean institutions integrate digital readiness into their curricula at a more comprehensive level, ensuring that pre-service teachers are exposed to a wide range of digital competencies. Following this model, local teacher education programs could enhance their global relevance by incorporating more foreign best practices, particularly those that focus on continuous digital engagement and upskilling.

One of the key findings of the study is the intervention of technology at particular levels and at particular times, which could be quoted from Instefjord and Munthe (2016) as an underrepresentation of digital competency in teacher education. As Tomczyk (2024) mentioned, countries such as Estonia and South Korea give immense importance to digital competence in teacher education follow an interesting model of training from basics to advanced and, beginning from elementary and continuing throughout the teacher training, which is quite the opposite to our intervention of technology in a particular time and at a particular level, as noted in our study.

Ethical Considerations and Equitable Access

A major concern emerging from our study is the lack of knowledge about secure internet use, with over 30% of pre-service teachers expressing uncertainty in this area. This is a critical issue, especially considering the increasing reliance on digital platforms for educational purposes. Ethical concerns such as data privacy, cybersecurity, and equitable access to technology were also highlighted in research by Livingstone et al. (2021), who stressed the importance of addressing these issues within teacher training programs. Our findings mirror these concerns, revealing gaps in pre-service teachers' understanding of online security measures and ethical tech use.

To address this, we recommend that teacher education programs introduce specific courses on digital ethics and online security. Serezhkina (2021) also underscored the importance of equipping teachers with digital skills that prioritise ethical considerations, particularly in protecting student data and ensuring equitable access to technology. In regions where technology resources are limited, institutions must also focus on providing the necessary infrastructure to ensure all pre-service teachers have access to the tools they need to succeed.

Linking Digital Readiness to Student Outcomes

One of the key findings of this study is the moderate level of digital readiness among pre-service teachers, which has implications for student outcomes. Research by McKnight et al. (2016) found that teachers with higher digital readiness are more effective at promoting emotional well-being (Händel et al., 2022), engaging students and enhancing their learning outcomes. Similarly, Huang (2022) highlighted how digital readiness in educational settings, particularly when combined with innovative teaching methods, positively influences student satisfaction and academic achievement

However, our study revealed that many pre-service teachers lack critical thinking skills when it comes to interpreting digital content. This is particularly concerning because teachers who cannot critically engage with digital platforms are unlikely to foster these skills in their students. Thus, enhancing pre-service teachers' digital readiness is not only crucial for their own development but also for improving student outcomes. Future research should further explore this link, providing more concrete evidence on how improving digital competencies in teachers can directly benefit their students.

Recommendations for Ongoing Professional Development

Ongoing professional development is essential to maintaining and enhancing digital competencies among teachers. Pagente and Selecios (2023) stress the importance of continuous professional development to keep educators updated on emerging technologies. Their findings align with ours, showing that while pre-service teachers may feel confident with basic technologies, they often fall short when it comes to advanced tools and innovative teaching methods (Cuhadar, 2018).

To address this, we recommend that teacher education programs incorporate continuous professional development opportunities. These could include workshops on emerging technologies, peer mentoring, and participation in professional learning communities focused on EdTech. Quaicoe and Pata (2015) suggested that institutional support enhances digital readiness. Thus, by offering structured opportunities for lifelong learning, institutions can ensure that their educators remain digitally literate throughout their careers, a necessity in the ever-evolving field of education.

Conclusion

This study highlights the moderate level of digital readiness among pre-service teachers, with notable gaps in their ability to use advanced technologies and critically interpret online information. By comparing our findings with global research, we see the potential to adopt best practices from countries that have successfully integrated digital readiness into their education systems. Furthermore, addressing ethical concerns and ensuring equitable access to technology must be priorities for teacher education programs. Continuous professional development is essential for maintaining digital competencies, and the link between teacher readiness and student outcomes should be explored further. By implementing these strategies, teacher education programs can better prepare pre-service teachers to thrive in an increasingly digital educational landscape.

References

Cuhadar, C. (2018). Investigation of pre-service teachers' levels of readiness to technology integration in education. *Contemporary Educational Technology, 9*(1). https://doi.org/10.30935/cedtech/6211

Feng, C., Wang, H., Lu, N., Chen, T., He, H., Lu, Y., & Tu, X. M. (2014). Log-transformation and its implications for data analysis. *Shanghai Archives of Psychiatry,* 26(2), 105–109. https://doi.org/10.3969/j.issn.1002-0829.2014.02.009

Fuchs, K., Pösse, L., Bedenlier, S., Gläser-Zikuda, M., Kammerl, R., Kopp, B., Ziegler, A., & Händel, M. (2022). Preservice teachers' online self-regulated learning: does digital readiness matter? *Education Sciences, 12*(4), 272. https://doi.org/10.3390/educsci12040272

Guzmán-Simón, F., García-Jiménez, E., & López-Cobo, I. (2017). Undergraduate students' perspectives on digital competence and academic literacy in a Spanish university. *Computers in Human Behavior, 74,* 196–204. https://doi.org/10.1016/j.chb.2017.04.040

Händel, M., Stephan, M., Gläser-Zikuda, M., Kopp, B., Bedenlier, S., & Ziegler, A. (2022). Digital readiness and its effects on higher education students' socio-emotional perceptions in the context of the covid-19 pandemic. *Journal of Research on Technology in Education, 54*(2), 267–280. https://doi.org/10.1080/15391523.2020.1846147

Hong, A. J., & Kim, H. J. (2018). College students' digital readiness for academic engagement (drae) scale: scale development and validation. *The Asia-Pacific Education Researcher, 27*(4), 303–312. https://doi.org/10.1007/s40299-018-0387-0

Huang, Y. (2022). The role of digital readiness innovative teaching methods in music art e-learning students' satisfaction with entrepreneur psychological capital as a mediator: evidence from music entrepreneur training institutes. *Frontiers in Psychology,* 13. https://doi.org/10.3389/fpsyg.2022.979628

Instefjord, E., & Munthe, E. (2016). Preparing pre-service teachers to integrate technology: an analysis of the emphasis on digital competence in teacher education curricula. *European Journal of Teacher Education,* 39(1), 77–93. https://doi.org/10.1080/02619768.2015.1100602

Kim, H. J., Hong, A. J., & Song, H.-D. (2019). The roles of academic engagement and digital readiness in students' achievements in university e-learning environments. *International Journal of Educational Technology in Higher Education,* 16(1), 21. https://doi.org/10.1186/s41239-019-0152-3

Kleih, A.-K., Lehberger, M., & Sparke, K. (2022). Using photographs in mixed methods research: an illustration integrating quantitative and qualitative analyses from consumer research. *Journal of Mixed Methods Research, 16*(4), 438–457. https://doi.org/10.1177/15586898211056748

Livingstone, S., Stoilova, M., & Nandagiri, R. (2021). Data and privacy literacy: The role of the school in educating children in a datafied society. *Aufwachsen in überwachten Umgebungen,* 219–236. https://doi.org/10.5771/9783748921639-219

Liza, K., & Andriyanti, E. (2020). Digital literacy scale of english pre-service teachers and their perceived readiness toward the application of digital technologies. *Journal of Education and Learning (EduLearn),* 14(1), 74–79. https://doi.org/10.11591/edulearn.v14i1.13925

McKnight, K., O'Malley, K., Ruzic, R., Horsley, M. K., Franey, J. J., & Bassett, K. (2016). Teaching in a digital age: how educators use technology to improve student learning. *Journal of Research on Technology in Education,* 48(3), 194–211. https://doi.org/10.1080/15391523.2016.1175856

Milawati, & Sholeh, M. (2020). Teacher's readiness in using digital technology for learning in samarinda city high school. *Proceedings of the 2nd International Conference on Social Science and Character Educations* (ICoSSCE 2019), Yogyakarta, Indonesia. https://doi.org/10.2991/assehr.k.200130.064

Mubarak Al-Awidi, H., & M Aldhafeeri, F. (2017). Teachers' readiness to implement digital curriculum in kuwaiti schools. *Journal of Information Technology Education: Research,* 16, 105–126. https://doi.org/10.28945/3685

Ng, D. T. K., Su, J., & Chu, S. K. W. (2024). Fostering secondary school students' ai literacy through making ai-driven recycling bins. *Education and Information Technologies,* 29(8), 9715–9746. https://doi.org/10.1007/s10639-023-12183-9

Pagente, D. P., & Selecios, A. C. (2023). Digital readiness for online learning and academic achievement of teacher education students during the covid-19 pandemic. *International Journal of Science and Research Technology,* 8(6).

Quaicoe, J., & Pata, K. (2015). Modelling factors influencing digital readiness: a case of teachers in ghana's basic school system. https://doi.org/10.1007/978-3-319-25515-6

Rodrigues, A. L. (2020). Digital technologies integration in teacher education: the active teacher training model. *Journal of E-Learning and Knowledge Society,* 16, 24–33. https://doi.org/10.20368/1971-8829/1135273

Scherer, R., Howard, S. K., Tondeur, J., & Siddiq, F. (2021). Profiling teachers' readiness for online teaching and learning in higher education: Who's ready? *Computers in Human Behavior, 118,* 106675. https://doi.org/10.1016/j.chb.2020.106675

Serezhkina, A. (2021). Digital skills of teachers. *E3S Web of Conferences,* 258, 07083. https://doi.org/10.1051/e3sconf/202125807083

Taggart, S., & Roulston, S. (2024). Educational technologies in schools in northern ireland: Teachers' digital competencies in the goldilocks zone. *Journal of Digital Learning in Teacher Education,* 1–19. https://doi.org/10.1080/21532974.2024.2393575

Tomczyk, Ł. (2024). Digital competence among pre-service teachers: A global perspective on curriculum change as viewed by experts from 33 countries. *Evaluation and Program Planning,* 105, 102449. https://doi.org/10.1016/j.evalprogplan.2024.102449

Venkatesh, V., Brown, S., University of Arizona, Sullivtable 1an, Y., & State University of New York, Binghamton. (2016). Guidelines for conducting mixed-methods research: an extension and illustration. *Journal of the Association for Information Systems,* 17(7), 435–494. https://doi.org/10.17705/1jais.00433

Chapter Two

Empowering Teachers and Students in the Digital Age: Help Educators Cultivate Their Own Digital Competence and Confidence

Shraddha Verma

Kalinga University, Chhattisgarh, India

Vijayalakshmi A

CHRIST (Deemed to be University), Bangalore, India

Abstract

This chapter examines the changing roles and identities of teachers in the digital age, stressing the importance of digital skills and confidence to navigate technology in education. As technology is integrated into classrooms, teachers must develop proficiency with digital tools to create engaging, student-centred environments. This shift repositions teachers from traditional knowledge providers to facilitators who foster students' critical thinking, collaboration, and lifelong learning. Digital competence involves a range of skills, knowledge, and attitudes for effective tech use. Professional development tailored to digital literacy, like that suggested by Tondeur et al. (2017), is essential for building these competencies. Teachers should merge technological and pedagogical skills, guided by the TPACK model, to integrate technology in ways that boost student engagement. To enhance critical thinking, educators should use interactive digital resources that align with learning goals (Scherer et al., 2019). Collaboration, supported by digital tools like Google Classroom and Microsoft Teams, is also crucial for modern learning, reflecting today's team-oriented workforce (Vygotsky, 1978). A commitment to lifelong learning is vital, as Voogt et al. (2013) emphasise. Teachers who continually develop their digital skills model this mindset, inspiring adaptable, curious learners. In sum, empowering educators requires ongoing development, adaptive roles, and thoughtful technology use to prepare students for the digital age.

Keywords: digital competence, professional development, life-long learning.

***

Introduction

Education has undergone a significant transformation as a result of the digital revolution, which requires educators to not only adopt new technology but also change their responsibilities from those of traditional teachers to those of facilitators of dynamic, individualised learning experiences. By offering useful tactics, theoretical underpinnings, global perspectives, case studies, and an analysis of the difficulties encountered during this transformative journey, this chapter delves into the vital significance of giving educators the digital competence and confidence they need.

A wide range of abilities, know-how, and dispositions necessary for the efficient and critical application of digital technology in educational settings is included in the concept of digital competence. To effectively incorporate technology into their teaching methods, teachers need to have pedagogical insights in addition to technical expertise (Tondeur et al., 2017). Programs for professional development that are adapted to the various requirements of educators are essential because they promote a culture of ongoing learning and adjustment.

The theoretical foundation for incorporating technology into teaching methods is provided by the Technological Pedagogical Content Knowledge (TPACK) framework. TPACK, which was created by Mishra and Koehler (2006), guides teachers in utilising technology to improve learning outcomes and student engagement by highlighting the interaction between pedagogical techniques, subject matter expertise, and technological understanding.

Effective Techniques for Teacher Preparation: Professional Development Programs

1. Practical workshops: recurring seminars on new teaching approaches and digital tools
2. Mentorship programs: assigning teachers to one another to share knowledge and assist with integrating technology
3. Online learning communities are places for cooperation, resource exchange, and continuous professional discussion.

Integration of Digital Literacy into Teacher Education

Pre-service and in-service teacher training should embed digital literacy competencies aligned with TPACK principles. Curricula should emphasise practical applications of technology to enhance teaching effectiveness and student engagement.

Improving Critical Thinking with Technology

In the digital age, when students must analyse, assess, and solve problems in complex and dynamic situations, critical thinking abilities are essential. Higher-order thinking abilities can be scaffolded by digital tools, including simulations, multimedia materials, and online platforms, which promote profound conceptual understanding and application (Scherer et al., 2019).

Promoting Cooperation in Online Learning Environments

In today's professional and educational settings, teamwork is essential. Social networking tools, Microsoft Teams, Google Classroom, and other digital platforms enable collaborative learning experiences, equipping students with the communication and teamwork skills they need in their future careers (Vygotsky, 1978).

Effective Techniques for Collaborative Education

1. Co-teaching models: These involve teachers working together to plan and carry out lessons utilising digital resources.
2. Global classroom initiatives: Intercultural partnerships that extend students' horizons and promote global citizenship.
3. Digital feedback mechanisms: To improve learning results, peer review and real-time feedback procedures are conducted using digital platforms.

Developing a Mindset of Lifelong Learning

Teachers must engage in ongoing professional development to stay up to date with pedagogical advances and technology advancements. Promoting self-directed learning and reflective practice among educators sets an example for students to follow throughout their lives, equipping them to succeed in a fast-changing digital environment (Voogt et al., 2013).

Techniques for Continuous Education

1. Ongoing professional development programs: Organised courses for continuing education in educational technology and digital pedagogy.
2. Development of digital portfolios: Reflective practice using digital portfolios that record accomplishments and professional development.
3. Learning Management System integration: Using an LMS to efficiently track progress and provide materials for professional development.

In this chapter, we will discuss a few of the worldwide situations and case studies that are well accepted for their teaching methods.

Finland's Empowerment Model for Teachers

Continuous professional development in digital pedagogy and teacher autonomy are given top priority in Finland's educational system. Teachers receive extensive training and materials, which encourage creativity and help them adjust to new technology.

Singapore's Programs for Professional Development

Comprehensive professional development programs aimed at incorporating ICT into teaching methods have been put in place in Singapore. Significant gains in teacher satisfaction and educational outcomes have resulted from this strategy, underscoring the need to provide educators with organised assistance.

Difficulties with Implementation

Notwithstanding the advantages, obstacles to digital empowerment include gaps in the infrastructure, opposition to change, and disparities in access to training and technology. To overcome these challenges, stakeholders must work together to make infrastructure investments, offer thorough training, and foster an innovative and supportive culture.

Teaching Methods

Empowering teachers in this era of the digital world involves various teaching methods that educators use to help convey knowledge to students. When they use teaching methods, the objective is to improve their learning outcomes and engage students in the classroom during the session, which can ensure that students understand the concepts. Various teaching methods contribute to different subjects and classroom dynamics. Here is a list of various teaching methods that are used in classrooms.

Lecture Method

This is the widely used classroom teaching strategy in education. In this method, the instructor delivers the content in a very structured manner with a preplanned course plan. This method is a teacher-centred approach, and it limits the interaction of students in the classroom. There are various advantages of this method, like the instructor can guide the students with complex topics. It is very much suited if the group of students is large. At the same time, this method has various disadvantages, like teachers not being able to give individual attention to students in the classroom, possibly because of the high number of students. This may lead to a lack of interaction between the student and the teacher.

In many of the scenarios of a classroom, a lecture-based method is useful if it is prepared in a well-structured way and delivered such that students can understand the concept. If the student group in a classroom is large, lectures can be an efficient method to deliver the content to everyone in the classroom. This method ensures a consistent learning experience among students. This method also provides a better experience to students when a teacher has to introduce a new topic. This could help the students to understand the key concepts before understanding the topic in depth. Generally, all the classroom hours are time-bound. In such situations, lecture-based methods might come in handy for a timely delivery.

Discussion-Based learning

This is a student-centric approach that encourages active participation of students in the teaching environment. In this case, a discussion is facilitated by the teacher, and the students participate by voicing their opinions and sharing their ideas. Students are encouraged to think critically and solve problems. It can promote active engagement of students in a discussion in the classroom and improve the communication skills of students. The major advantage of this method is that it provides collaborative learning and support from peers. At the same time, it is important to note that this method has some flaws, like being time-consuming and limiting participation in a larger group of students. A very careful planning of each session is required to make sure that the discussion is productive. Discussion-based teaching methods are recommended when the class size is small and every member can participate in the discussion. Workshops and tutorial classes are some examples where we can follow this method efficiently. For subjects or topics that have multiple viewpoints, discussion-based methods will be useful as students can voice out their opinions and debate on the topic. This could engage all the students to understand multiple viewpoints on that topic. In a discussion-based classroom, students can learn from their peers. When each one voices their opinion, others understand a different perspective on a topic. This might help the students broaden their understanding and get new insights into the topic.

Problem-Based Learning

The problem-based method is where students in a classroom are involved in learning by solving real-world problems. This method gives very much importance to critical thinking and creativity of a student. The role of the instructor in this method is to serve as a facilitator, wherein they guide the students as they find solutions to problems. This method is most commonly used in fields like medicine and engineering. There are various advantages to learning using this method: Students will have a deeper knowledge of the subject. It will contribute to solving a problem and making the appropriate decisions. The major problem the instructors will face in this method is that the assessment will be difficult to quantify. When there is a need to connect theoretical understanding of a concept to real-world problems, it helps in understanding the concept better. Problem-based learning includes teamwork, which can help students work in groups and solve complex problems through discussions and the exchange of ideas.

Experiential Learning

This method gives importance to how students can be taught with hands-on experience. It believes that when students experience the concept they learn, it can be an encouragement to their creativity and initiative. There are various methods that are followed to learn, like experiments or projects, and fieldwork. This method is commonly used in science education. The advantage of this method is that it helps in learning a concept through real-world applications. To implement this method of teaching, there is a requirement for various resource materials and a good amount of time. When a topic includes understanding through practical experiments, like laboratory experiments, field work, etc., this method helps students to gain hands-on learning experience. Various skill-based learning activities, like role-plays and workshop activities, can help in acquiring skills like leadership, problem-solving, and other technical skills. When students have to understand and solve complex problems, experiential learning is one of the best methods to identify possible solutions and learn from the outcomes.

Flipped Classroom

This is a teaching method where the usual method of teaching is reversed. In this case, students learn content by using some of the recorded lectures and other reading materials, and they engage the class. The classroom will be engaged in various discussions and activities. This helps in promoting active learning among students. The most important concept here is that students will be confident enough to take up the responsibility of their own learning and understanding of the concept. The learning of the students is at their own pace, and hence helps students to learn at their own pace.

Cooperative Learning

In this method, groups are made of students, and they share their ideas to learn a concept. When students work in groups, each member in the group shares their knowledge. This helps in each member contributing, and it encourages collaborative learning and support for peers. This method of learning improves teamwork and interpersonal communication. It is also possible that each member understands the content thoroughly.

Even though there are various teaching methods that can be adopted by teachers, there is no one specific method that can fit all modes of teaching. How effective a teaching method is for a specific set of students depends on various factors like the subject that is being taught, the learning environment, the teaching style of the teacher, and the needs of the student group. The most effective way is to use a combined method that can include lectures, discussion among the students, and learning through experiments.

Teaching Aids

Educators follow various teaching methods to improve the delivery of class content; it is necessary to discuss the tools that support these teaching methods. These tools help in making the classroom more engaging and improve the effectiveness of teaching. In this section, various teaching tools are explained:

Table 2.1 Teaching Aids as per Teaching Method.

Teaching Method	Teaching Aid	Advantages	Disadvantages
Lecture Method	Various presentation software, like PowerPoint and Google Slides, can be used for demonstration.	As these slides are prepared before the class, they will have structured content.	Limited interaction
	Interactive whiteboards/ smartboards	Effective for large groups	Less individual attention
	Lecture capture software (e.g., Panopto)	Easy to review content later	Real-time feedback
	Clickers/response systems		
Discussion-Based Learning	Online discussion platforms (e.g., Padlet, Google Classroom)	Encourages critical thinking and peer collaboration	Time-consuming
	Collaborative documents (e.g., Google Docs, Microsoft Teams)	Enhances communication skills	Limited participation in large groups
	Polling and survey tools (e.g., Mentimeter, Poll Everywhere)	Engages even shy students	Requires careful session planning
Problem-Based Learning	Case study repositories (e.g., Harvard Business Review Case Studies)	Promotes creativity and critical thinking	Difficult to assess
	Simulations and Software (e.g., MATLAB, Simulink)	Provides real-world problem-solving experience	Time-intensive

	Project management tools (e.g., Trello, Asana)	Fosters collaboration	
	Brainstorming tools (e.g., MindMeister, Miro)	Encourages deeper learning	
Experiential Learning	Laboratory equipment	Hands-on experience deepens understanding	Requires resources
	Virtual labs (e.g., Labster, PhET Simulations)	Real-world applications of concepts	
	Project-based learning platforms (e.g., PBLWorks)	Encourages practical knowledge	
Flipped Classroom	Video creation tools (e.g., Screencast-O-Matic, Loom)	Encourages self-paced learning	Requires student discipline
	Learning Management Systems (e.g., Moodle, Blackboard)	More interactive classroom experience	
	Interactive video platforms (e.g., Edpuzzle, PlayPosit)	Promotes active learning	
	Quizzing tools (e.g., Kahoot, Quizizz)	Immediate feedback for students and teachers	
	Collaborative tools (e.g., Padlet, Flipgrid)	Engages students in pre-class activities	
Cooperative Learning	Group collaboration tools (e.g., Slack, Microsoft Teams)	Improves teamwork and communication	Dominant students may overshadow others.
	Shared documents (e.g., Google Docs, Canva)	Encourages peer learning	Potential for unequal participation
	Peer assessment tools (e.g., Peergrade, Turnitin PeerMark)	Fosters accountability and deeper understanding	
	Gamification tools (e.g., Classcraft, Kahoot)	Adds motivation and fun to the learning process	

Empowering Digital Competence in Students

In this world, which is driven by technology, digital competence is a very important skill with which each student has to be equipped. Digital competence is the ability of students to use various digital tools and technologies effectively, following the ethics of their usage. Students must be proficient in using these tools to succeed in this technology-driven world.

Digital competence ranges from basic computer literacy to writing code and analysing data. As students, this involves using various digital tools like word processors, software for presentation and being able to identify various online learning platforms that use them for increasing their knowledge. More importantly, students must be aware of ethical standards and be cyber safe. Students must be aware of how to use the content ethically and protect themselves in the cyber world.

Digital Tools in the Classroom

In order to empower students with digital knowledge, one of the most important methods is to incorporate various digital tools in the classroom during teaching. There are various platforms of learning management systems, namely, Google Classroom and Moodle, which permit interaction between teachers and students digitally. Teachers can create a classroom and post various activities like materials, assignments, questions, quizzes, etc. This allows a discussion and interaction between the two, encouraging students to access the materials and assignments even beyond the boundaries of a classroom. For each of the activities in these classrooms, there are various properties associated with it that can be set by the teacher to ensure that the students submit the work on time. These learning platforms that provide flexibility in access encourage students to get familiarised with the digital environments.

The assignments provided in class should help students develop critical digital literacy. Since there is a lot of information in the digital world, students should develop the capability to critically identify reliable sources from which data and information can be accessed. Teachers can introduce students to various fact-checking websites like snopes.com and factcheck.org that can help in identifying the reliability of information.

There are various other tools, like Google Docs and Microsoft Teams, where students in a group can work and collaborate, and where teachers can monitor the progress of the work. In this case, students and teachers can contribute to the team regardless of their physical location. Tools like this are essential in a digitally connected world. If teachers can introduce these tools to students, digital communication skills can be developed, and faster completion of work can be expected. In this world of fast-changing technology, empowering students with digital competence is essential. By integrating various digital tools in the classroom, we can foster critical thinking, collaborative work and digital ethics in students. As technology is an integral part of the present and future, students with digital literacy will be prepared to contribute to the fast-changing society.

Empowering Digital Competence in Teachers

As technology is showcasing its power in all fields, it is important to equip teachers with digital competence, as students are being equipped with digital competence. Digital competence among teachers will help them acquire the necessary skills that they need to integrate teaching with digital tools in the classroom effectively. This enhances learning and interest to learn among students.

Quality improvement programs (QIP) have to be conducted to enhance the digital skills of every teacher in this tech-driven educational environment. Various QIPs that give training in digital tools and teaching platforms can help teachers in integrating them into their curriculum and help improve teaching methodology. QIPs will also help teachers keep updated with the ever-changing technological landscape. When they are updated with the latest tools, their confidence in using the tool increases, and it creates a positive impact in the classroom while engaging students. Faculty Development Programs (FDPs) that focus on enhancing digital skills will be added to the life of a teacher, where they enhance their skills with the latest technology. Most importantly, teachers learn how to use technology so that they can make the classroom interactive and engaging. Teachers can be introduced to various teaching methodologies that are discussed in this chapter, like flipped classrooms and cooperative learning, which can cater to the diverse learning needs of students. Programs like FDPs can boost the confidence level of teachers and promote lifelong learning. It also helps in sharing the experience and exchanging knowledge among their peers. FDPs enhance the digital skills of a teacher that are necessary for classroom teaching in this era of technology. This helps them incorporate technology in classroom teaching efficiently and effectively. Various teaching strategies can be introduced in FDPs that can be used by teachers to support their subjects. FDPs contribute to professional growth in this technology-driven modern era.

Various programs, like workshops and training, are very important for the professional growth of a teacher in an institution. These programs can help teachers get introduced to new tools and techniques in the current era. And as per the requirements of the topic, it can be used by teachers in their teaching.

Conclusion

Empowering teachers in this age of the digital world demands very constructive strategic methods in an individual's professional development. By training teachers with various technological tools and teaching methodologies, it can be ensured that all are prepared for the transformative world of technology. This method of equipping teachers with methods that can be incorporated in classrooms will help learners to actively engage in the classroom and bring out better learning outcomes. This ultimately prepares students for success in the future in this digitally equipped world.

References

Mishra, P., & Koehler, M. J. (2006). Technological pedagogical content knowledge: A framework for teacher knowledge. *Teachers College Record,* 108(6), 1017-1054. https://doi.org/10.1111/j.1467-9620.2006.00684.x

Scherer, R., Siddiq, F., & Tondeur, J. (2019). The technology acceptance model (TAM): A meta-analytic structural equation modeling approach to explaining teachers' adoption of digital technology in education. *Computers & Education,* 128, 13-35. https://doi.org/10.1016/j.compedu.2018.09.009

Tondeur, J., van Braak, J., Ertmer, P. A., & Ottenbreit-Leftwich, A. (2017). Understanding the relationship between teachers' pedagogical beliefs and technology use in education: A systematic review of qualitative evidence. *Educational Technology Research and Development,* 65(3), 555-575. https://doi.org/10.1007/s11423-016-9481-2

Voogt, J., Erstad, O., Dede, C., & Mishra, P. (2013). Challenges to learning and schooling in the digital networked world of the 21st century. *Journal of Computer Assisted Learning,* 29(5), 403-413. https://doi.org/10.1111/jcal.12029

Vygotsky, L. S. (1978). *Mind in society: The development of higher psychological processes.* Harvard University Press.

Chapter Three
EdTech Frameworks for Effective Technology Integration in Teacher Education

Yashi Goyal

Central Institute of Education, University of Delhi, New-Delhi, India

Abstract

This chapter explores the transformative role of educational technology (EdTech) in modern classrooms and presents various frameworks for its effective integration. It argues that EdTech, when used thoughtfully, can revolutionise learning environments by shifting from passive to active student engagement. Key frameworks, such as TPACK, SAMR, and T3, provide educators with structured approaches to embedding technology in pedagogy, emphasising adaptation, student-centred practices, and collaboration among educational stakeholders. Frameworks for broader applications, including ESRM, ICT4E, and SABER-ICT, address policy-level decision-making to ensure equity, sustainability, and accessibility in EdTech initiatives. These frameworks offer pathways for assessing and enhancing technology use, guiding educators, administrators, and policymakers toward creating inclusive, technology-driven educational ecosystems. The chapter advocates for continuous professional development and adaptable implementation to keep pace with technological advancements, envisioning a future where EdTech facilitates personalised learning and fosters critical thinking and lifelong learning in students.

Keywords: educational technology, student-centred learning, professional development.

***

Introduction

In today's classrooms, interactive smartboards have replaced the old blackboards. Instead of passively absorbing information, students are now actively constructing knowledge through virtual simulations. This shift is part of the growing impact of EdTech, which is revolutionising the education landscape at a rapid pace. The conventional classroom image of rows of desks facing a teacher at the front of the room is being replaced by a more dynamic, learner-centred environment, thanks to the tidal wave of technological advancements collectively known as Educational Technology or EdTech. From immersive virtual reality experiences that transport students to ancient civilisations, to adaptive learning platforms that personalise instruction, EdTech brings innovation to classrooms.

Integrating technology in education is about more than just adding fancy gadgets. At its core, effective EdTech aims to enhance the learning process. For example, a well-designed EdTech program can provide interactive visualisations to help students struggling with complex scientific concepts. Alternatively, a different EdTech platform can connect a student who thrives in a collaborative environment with peers worldwide for a shared project, fostering a sense of global community. The possibilities are diverse and tailored to the student population itself. However, careful consideration is needed when integrating EdTech. Merely throwing technology at education problems is not a guaranteed recipe for success. The key lies in effective integration. The technology should complement rather than replace traditional teaching methods. This is where the concept of EdTech frameworks comes into play.

An EdTech framework acts as a blueprint for integrating technology meaningfully. These frameworks provide educators, administrators, and developers with guiding principles. They do not dictate a one-size-fits-all approach but offer a roadmap for making informed decisions about EdTech selection, implementation, and assessment. For instance, imagine a teacher planning their lesson or a school considering investing in a new online learning platform. An EdTech framework could help them address essential questions, such as does this platform align with their curriculum goals? How will it be integrated into existing teaching practices? How will they ensure equitable access for all students? By addressing these questions proactively, the teacher and the school board can make a more informed decision about whether this particular EdTech solution is the right fit.

The benefits of utilising EdTech frameworks extend beyond individual classrooms. These frameworks can inform policy decisions at the district and national levels, ensuring that EdTech initiatives are part of a cohesive strategy for educational improvement, not fragmented efforts. Additionally, EdTech frameworks can guide curriculum development, ensuring that technology is used to enhance, not replace, the core content students need to master. However, EdTech frameworks are not static entities. The field of education is constantly evolving, and the tools and technologies available to educators are continually changing. Therefore, these frameworks must be adaptable and capable of incorporating new advancements while staying true to their core principles of effective technology integration. They should also be readily accessible and understandable by all stakeholders, from educators on the frontlines to policymakers shaping the educational landscape.

In the following subheadings, we will delve deeper into different EdTech frameworks, each offering a unique lens for integrating technology thoughtfully and exploring frameworks that guide educators in selecting the right tools, emphasising effective classroom practices, and ensuring equitable access for all learners. By demystifying these frameworks, educators and leaders will be equipped with the knowledge and confidence to navigate the exciting yet complex world of EdTech. It is not just about integrating technology into classrooms; it is about transforming the very essence of learning. The potential of EdTech is limitless, and with the right frameworks as our compass, we can ensure that this revolution in education empowers students to become lifelong learners, critical thinkers, and the architects of a brighter future.

Foundational Frameworks for Teachers for Effective Technology Integration in Classrooms

TPACK Framework

The Technological Pedagogical and Content Knowledge (TPACK) framework is a cornerstone for effective technology integration in education. The foundation for this framework was the PCK model introduced by Lee Shulman in the mid-1980s. Shulman indicated that teachers have a unique kind of knowledge called Pedagogical Content Knowledge or PCK. This knowledge involves selecting each lesson's most effective strategies and practices based on the content (Shulman, 1986). Mishra and Koehler (2006) developed the TPACK model, emphasising the interplay between three key knowledge domains for teachers.

Technological knowledge (TK) is the understanding and proficiency in using various technologies for educational purposes. Pedagogical knowledge (PK) encompasses teaching strategies and instructional design expertise. Content knowledge (CK) refers to a deep understanding of the subject matter that the teacher needs to teach. TPACK goes beyond simply knowing how to use technology. It is about understanding how technology can enhance specific content delivery and learning experiences. It also encourages collaboration among technology experts, curriculum specialists, and instructional leaders. This framework is essential in teacher education, equipping future educators with the skills to navigate the ever-evolving technological landscape.

A science teacher can enhance students' understanding of complex biological systems by using virtual reality simulations to immerse them in the human body. Students can also use online graphing tools to visualise and analyse mathematical functions, which helps foster a deeper conceptual understanding. Additionally, a class can collaborate on a digital timeline using an online platform to promote research and communication skills while exploring historical events.

Figure 3.1 TPACK Framework.

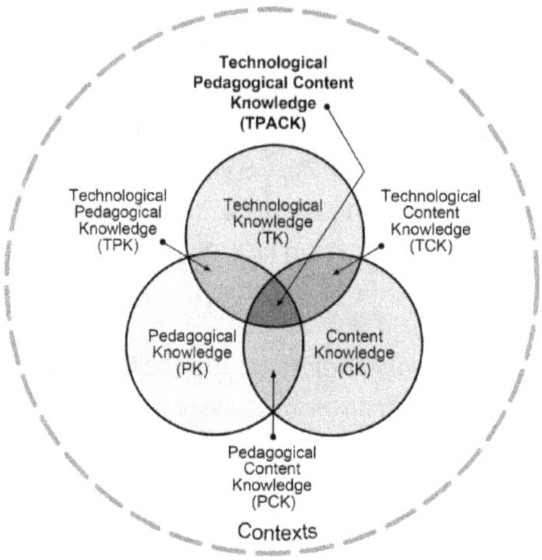

Reproduced by permission of the publisher, © 2012 by tpack.org.

However, the successful implementation of TPACK can vary across regions due to limited access to technology and reliable internet connectivity. Comprehensive TPACK-based training programs for educators may be uneven across regions. These issues can be addressed by integrating TPACK principles in teacher education programs through dedicated courses and workshops or embedding them within existing curricula. It is crucial to align curriculum requirements with technology integration strategies. Encouraging creativity using readily available technology, such as simple presentation tools or educational apps accessible on mobile devices, can help tackle these challenges. Facilitating knowledge sharing among teachers through online communities or regional workshops is also essential. Technology integration strategies should be tailored to address regional needs and cultural contexts.

SAMR Model

The SAMR model, designed by Dr Ruben Puentedura, provides a valuable framework for teacher education programs to guide instructors in effectively integrating technology into their curriculum. It categorises four levels of technology integration, progressing from basic substitution to a complete transformation of learning experiences (Puentedura, 2010).

Figure 3.2 SAMR Model.

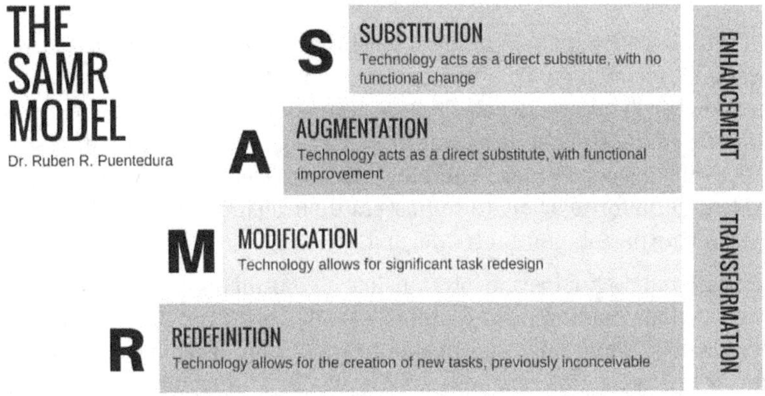

Reproduced from Wikimedia Commons (CC BY 4.0).

Substitution is the most basic level, where technology replaces traditional tools. For example, a teacher might use a digital whiteboard instead of a traditional one, or students might complete quizzes online instead of on paper. While this offers convenience, it does not fundamentally change the learning experience. Augmentation is the level at which technology enhances existing activities. A teacher could use a presentation tool with animations to explain a concept, or students might use online research tools to supplement textbook learning. Technology adds value, but the core task remains the same.

Modification signifies a more significant shift. Technology allows for tasks to be redesigned to improve learning outcomes. For example, a teacher might create a flipped classroom where students watch video lectures for homework and use class time for interactive activities. Students could collaborate on projects using online platforms, fostering communication and teamwork skills.

Redefinition is the highest level, where technology transforms learning experiences entirely. Imagine a simulation game where students can practice decision-making or a virtual reality field trip that allows them to explore remote locations. Technology creates new learning opportunities that would not be possible without it.

Teacher educators can utilise the SAMR model as a fundamental framework to instruct pre-service teachers on effectively selecting and integrating technology. During micro-teaching sessions, students can use technology at different SAMR levels. Instructors can assess their current courses and pinpoint opportunities to progress from substitution and augmentation to modification and redefinition.

T3 Framework

The T3 framework enhances modern teaching and learning practices through educational technology. This framework divides technology integration into three domains: T1 (Translational), T2 (Transformational), and T3 (Transcendental). It aims to emphasise the importance of technology in education, assist in decision-making concerning technology use in the learning environment, and provide criteria for teachers to self-assess their technology use while setting and achieving meaningful goals (Magana, 2017).

Translational technology involves using digital methods to perform tasks previously done manually. For example, Google Forms can be used instead of paper surveys. While this brings value to traditional teaching, the impact is relatively low as the task remains unchanged. For T1 integration, teachers can be equipped with basic digital literacy skills for classroom management, communication, and content delivery (e.g., using online learning platforms and creating presentations). Technology can be integrated into existing lesson plans to enhance traditional methods (e.g., online simulations for science experiments).

Transformational technology, on the other hand, is that which changes the nature of the task, role of the individual entrusted with the task, or the impact of the task upon those involved (Disruptive Classroom Technologies – CEESA, 2022). This stage encourages students to engage in new thought processes and create new content, leading to a shift from the role of a student to that of a teacher. For T2 integration, educators can be encouraged to redesign lessons using technology to personalise learning experiences (e.g., adaptive learning software and digital portfolios). Collaboration and communication should be fostered through technology tools (e.g., online discussion forums and project management platforms).

The transcendent stage represents the highest form of added value for educational technologies. It empowers students to find their pathways in the learning process, solve complex problems, and initiate meaningful change, leveraging technology to empower them. For T3 integration, emerging technologies like Virtual Reality (VR) and Augmented Reality (AR) for immersive learning experiences can be explored. Students can be involved in creating technology-based solutions to real-world problems (e.g., coding games to teach math concepts). Student agency should be developed by empowering them to use technology for self-directed learning and knowledge creation.

Another example could be a teacher education program partnering with a local museum to create a T3 project. Student teachers can design a VR tour of the museum exhibits, incorporating historical documents, interactive quizzes, and student-created audio guides. This project requires collaboration, technology skills, and innovative thinking, aligning with the T2 and T3 aspects of the framework. The T3 framework could be adapted to diverse contexts. In case of limited access to technology, low-tech or offline alternatives such as educational radio programs and television programs can be explored. By utilising these strategies and the T3 Framework, teacher education programs can ensure that future educators are well-equipped to harness the power of technology for transformative learning experiences that benefit all students.

Frameworks for School-Level EdTech Integration

E-Learning Stakeholders' Responsibility Matrix

E-learning Stakeholders' Responsibility Matrix (ESRM) harmonises the efforts put in by e-learning and promotes collaborative relations among the stakeholders. Prepared for higher educational institutions for understanding, integration, and adaptation of educational technology programs, the matrix outlines the core roles of all the stakeholders by describing the reasons and concerns of all of them (Wagner et al., 2008). Stakeholders involved in the process of e-learning for teacher training are policymakers, teacher training institutions, faculty members, student teachers, and developers of technology. Policymakers determine the national strategic planning, allocation of the funds, and the general standards that need to be established in relation to the quality that such e-learning programs require to maintain. Institutions for the education of teachers need to ensure the design of more functional e-learning modules, along with proper training for all faculty members, so they can access the use of technology on the part of the students. The lecturers also participate in e-learning by incorporating it into their teaching, providing support to the students, and assessing the effectiveness of e-learning. The student teachers take an active role in e-learning activities, acquire digital literacy skills, and provide feedback. The technology providers are involved in developing user-friendly e-learning platforms, providing technical support, and ensuring data security.

To illustrate how this framework can work, let us consider a hypothetical example. A teacher education program could use ESRM to create a micro-credentialing program that integrates technology into the classroom. The Ministry of Education could provide funding, teacher education institutes could design online modules, faculty could receive training, and students could use the provided tablets. Technology providers could offer a user-friendly platform with offline access to address regional connectivity challenges. ESRM provides a valuable framework for successful e-learning implementation in teacher education. Teacher education programs can use technology to prepare future educators for the digital age by clearly defining stakeholder roles, addressing regional challenges, and measuring impact.

ICT for Education

ICT4E is conceptualised as a frame of action that supports the design of activities which will provoke long-lasting changes in pedagogical practice in schools. It focuses on the integration of technology in teaching and learning (Rodríguez et al., 2012). The framework has four components: implementation, intervention, transference and total cost. Implementation involves applying ICT in the educational context, considering infrastructure, resources, and teacher training. Intervention refers to the pedagogical approaches and strategies that leverage ICT for enhanced learning outcomes. The transference component focuses on the sustainability and scalability of the ICT initiative, ensuring its long-term impact. Total cost encompasses the economic, social, and environmental costs associated with the ICT project.

Some practical strategies have been discussed below that can be utilised to apply this framework. In terms of implementation, there can be an overhaul in infrastructure, resources and teacher training. Teacher education institutions can be provided with high-speed internet access and modern computing facilities. Teacher trainees can use personal devices or tablets for learning and practice. Open-source educational resources and platforms should be developed to reduce costs. Teacher training, ICT pedagogy and digital literacy courses should be incorporated into the teacher education curriculum. Online and blended learning options should be offered for teacher training. Professional development opportunities should be provided so that teachers stay updated with technological advancements. Intervention can include pedagogical approaches such as training teachers to use ICT for differentiated instruction, personalised learning, and collaborative learning, encouraging online platforms for student assessment and feedback and integrating ICT into lesson planning and classroom management.

During the transference component, sustainability and scalability should be kept in mind. ICT support centres should be established in schools to provide technical assistance to teachers. Online communities can be created for teachers to share best practices and resources. ICT needs to be integrated into school improvement plans to ensure long-term commitment. Open-source educational materials and platforms should be such that they can be adapted to different contexts. In other words, government and other non-governmental organisations will be needed to collaborate so that the use of information and communication technologies in teaching is promoted. When quantifying the overall costs involved, the cost-effectiveness of various solutions that can solve the problems must be reflected. Alternative options, such as open source, can be provided to cut costs. Further ways of sourcing funding for the implementation of the use of these technologies must be sought. The impact of the increase in teacher workload and a decline in teacher well-being should be estimated to form part of the social costs.

There must be efforts to bridge the digital divide for fair access to technology. Environmental costs can't be ignored, so energy-efficient devices and sustainable ICT practices need to be promoted. ICT4E frameworks are implemented differently in various regions due to different socio-economic, cultural, and technological contexts. Solutions should therefore be customised to meet these different needs. Local expertise and resources to support ICT integration should be utilised. Through a holistic and systematic approach based on the ICT4E framework, educational institutions can use technology effectively to improve teaching and learning outcomes while addressing regional challenges and being accountable.

Frameworks for Policymakers

SABER-ICT Framework

The SABER (Systems Approach for Better Education Results) ICT Framework is a comprehensive tool designed to assist policymakers in making informed decisions about the effective use of Information and Communication Technologies (ICTs) in education. Developed by the World Bank, it offers a structured approach to analysing, benchmarking, and improving ICT policies. The SABER-ICT Framework is mainly for policymakers and governments to help design and evaluate critical policies related to ICT use in K-12 education (Trucano, 2016). The framework is built upon six key domains. These include policy and strategy, which outline the overall vision, goals, and strategies for ICT integration in education. The institutional arrangements focus on the organisational structures and human resources required for effective ICT implementation. The infrastructure component examines ICT hardware, software, and connectivity availability and quality. The teacher education and development aspect addresses teachers' preparation and ongoing professional development to use ICTs effectively. The content and pedagogy part deals with developing, adapting, and using ICT-supported learning materials and teaching methods. The assessment and evaluation emphasise the importance of measuring the impact of ICT integration and using data to inform decision-making.

Regarding teacher education and development, it talks about incorporating ICT pedagogy, digital literacy, and technology integration into teacher education programs. It provides opportunities for pre-service teachers to develop practical skills through internships, micro-teaching, and project-based learning and equips teachers with skills to create engaging digital learning materials. It also suggests offering ongoing training and support to help teachers stay updated with technological advancements. The successful implementation of the SABER ICT Framework depends on addressing regional challenges and leveraging unique opportunities. Developing countries should focus on building basic infrastructure, providing affordable access to technology, and developing relevant digital content. They should emphasise innovation, personalised learning, and the ethical use of technology. For instance, India's Digital India initiative aims to bridge the digital divide by providing affordable internet access and digital literacy programs.

PISA ICT Framework

The PISA ICT Framework gives an overview of how students use and access technology and its impact on learning. In addition, the framework discusses schools and teachers' use of technology in the classroom. This data helps policymakers understand how system-level factors have an effect on the usage of technology by students and schools. It further gives education institutions and nations the possibility to analyse their situation and compare themselves with others (OECD, 2023). The critical components of the PISA ICT framework include access, utility, outcomes, and contextual factors. It talks about the availability and affordability of ICT devices and internet connectivity. It seeks to understand how students employ ICT for various purposes, including information-seeking, communication, creation, and problem-solving. It addresses the impact of ICT use on students' cognitive performance, well-being, and ICT skills. It also considers contextual factors such as socio-economic status, school resources, teacher ICT competence, and educational policies.

The PISA ICT Framework underscores the critical role of teachers in fostering students' digital competence. Teacher education programs must incorporate the following measures to prepare teachers for this role. Teachers should be equipped to seamlessly integrate ICT into teaching core subjects, enhancing student engagement and learning outcomes. Teacher education programs must address the diverse needs of students, including those with disabilities, by providing strategies for inclusive ICT use. Teachers must develop students' ability to evaluate digital information and become responsible digital citizens, critically. Teacher education should prepare educators to design and implement effective blended learning models that combine face-to-face and online instruction.

Teachers should be proficient in using digital tools for formative and summative assessment to provide timely feedback to students. A culture of continuous learning and adaptation to emerging technologies must be promoted in teacher education. Experienced teachers will be the ones to guide new teachers on how to use ICT in their teaching settings. The availability of digital resources for the teachers, opportunities for professional development, and technical support are all highly essential. Many regions of the world experience a gap, popularly known as a digital divide, characterised by an imbalance in the accessibility of ICT. In addressing such a gap, the issue of teaching under minimal resource conditions must also be integrated into teacher education. The cultural and linguistic contexts of students must be taken into consideration for the teacher preparation programs regarding ICT integration. Policies will be supportive, and there will be a strong base of ICT infrastructure to ensure successful integration. Thus, regional and specific challenges can be met through the strategies that can be taken following the efficient teacher preparation programs for proper utilisation of ICT in student learning, along with superior results.

Additional EdTech Frameworks

TIM Framework

The Technology Integration Matrix (TIM) is a comprehensive framework developed by the Florida Center for Instructional Technology (FCIT) at the University of South Florida. It provides a structured approach to evaluating and enhancing technology integration in instructional settings. The TIM is grounded in constructivist learning theory and emphasises creating meaningful learning environments. There are five characteristics of meaningful learning environments. They are active, collaborative, constructive, authentic, and goal-directed. In active learning environments, students learn through exploration and discovery. In collaborative learning environments, students work together to construct knowledge. In constructive learning, students build understanding through creating and producing. Learning experiences are relevant and connected to real-world contexts in authentic learning environments. In goal-directed learning, students have clear learning objectives and understand the purpose of their activities. These characteristics are related to five stages of technology implementation, which include entry, adoption, adaptation, infusion, and transformation (Maclemale, n.d.).

At the entry level, technology takes over the other traditional tools. Adoption involves technology in enhancing the practices already in use. The adaptation level involves the use of technology in modifying already existing practices. Infusion means introducing technology in the learning process in a way that facilitates new kinds of learning. Lastly, technology fundamentally changes the learning environment at the transformation level and creates new possibilities.

The TIM creates a matrix of 25 cells, representing the intersection of the five characteristics and five levels. This matrix provides a visual representation of technology integration and helps educators identify areas for improvement. TIM can be incorporated into the curriculum to provide pre-service teachers with a foundational understanding of technology integration. The TIM can be adapted to various educational contexts and applied across different subject areas and grade levels. Educators can effectively leverage technology to enhance student learning and achieve desired outcomes by systematically implementing the TIM and utilising appropriate measurement and accountability strategies.

Triple E Framework

Dr Liz Kolb developed the Triple E Framework in 2011, and it is intended to aid teachers in evaluating whether they use technology tools in meaningful ways to facilitate their instruction, and gives educators avenues to increase use (Staff, 2024). It puts emphasis on the process of student-centred learning with assurance that technology supports learning by acting as an instrument of facilitation, rather than controlling the act of learning. Technology should be used to support the engagement of students and then enhance and extend their learning while engaged (Staff, 2024).

It comprises three main stages of the framework. In the stage of engagement, the learning content captures the attention and interest of the students because of the technology. In the enhancement stage, technology is used further for the mastery of subject matter by including some elements, such as simulations, data analysis, and interactives, for the students. During the extension stage, technology enables them to solve real-world problems or construct new knowledge using what is already known while working collaboratively. The Triple E Framework is a four-step procedure (Kolb, 2017). Firstly, determine goals for learning, followed by the selection of a relevant tool. Next, the tool should engage students to be active and social learners. Lastly, students should relate what they are learning to real-world tasks and contexts. The use of technology tools in class connects what students are learning to real-world tasks and contexts.

To make the above framework clearer, let us take an example. A teacher might use technology to prepare printed materials for students. There is no engagement in this case. However, in the case of the Triple E framework, there should be co-use for engagement to exist. Kolb defines co-use as the social use of a digital tool between people. This approach shifts learners from being passive to becoming active, allowing them to focus on their learning goals while using a device. Examples of engagement are that students play a Quizlet Live game, brainstorm on Padlet, or play a Kahoot game. Enhancement is the development of critical thinking, analysis, and evaluation skills, and to support the same, the teacher can encourage the students to create digital crosswords and quizzes using platforms like these. To further support enhancement, the teacher should create scaffolds that would help the students to understand ideas and concepts more quickly.

In other words, teachers should support differentiated learning since every classroom has all kinds of students at different levels. Finally, to achieve enhancement in an optimal way, students should be able to show their understanding better than with the traditional tools in the application of technology. Assessment using a learning management system can be done by having a brainstorming session or group discussion among students. In addition, teachers can use project-based learning and give students choices of tools to demonstrate their learning. Lastly, extension is a result of students learning outside the class. However, many teachers mistake extensions for homework. Homework that does not extend into the student's everyday lives is not helpful. Instead of asking students to write words to practice the vocabulary, they can be asked to use applications to discover the vocabulary that is present around them in their real lives.

Kolb (2017) has developed a rubric that is very straightforward in determining how well educators integrate technology into each of the lessons. For any yes, there are two points, for every somewhat, one point, and for every no, it is zero points. The rubric uses a traffic light system. The red zone is the lowest level of integration, ranging from zero to six points. Activities that fall into the red zone are not using original engagement. The yellow zone meets two of the three stages, ranging from seven to twelve points. It usually indicates that the lesson has either engagement and enhancement or engagement and extension. The green zone has the value from thirteen to eighteen, which represents excellent lessons to students; these make learning more relevant and interconnected to life in real life outside the class. They make the students more enthusiastic. The teachers should strive for these kinds of lessons.

Implementing EdTech Frameworks

Many EdTech frameworks are available, each of which has its unique focus and target audience. One must choose depending on the context. An EdTech framework provides a holistic approach; it gives educators, administrators, and policymakers a roadmap for strategically implementing technology in their contexts. Such a framework helps at a classroom level; the same could be at a bigger scale, with a view towards SAMR in such classrooms. The TPACK framework helps teachers seamlessly embed technology into teaching practice. Frameworks at the school leader level must attend to issues beyond those related to teacher preparation. Scaling EdTech across the district requires frameworks that deal with scalability and sustainability.

Once a framework is selected, the real work begins: integrating its principles into daily practice. The primary focus of most frameworks is establishing a clear vision for how technology will support learning goals. All stakeholders should share this vision. Collaborative workshops should be conducted to define specific and measurable outcomes for EdTech implementation. Successful technology integration depends on educators' preparedness and comfort level. Investment in professional development opportunities aligned with the chosen framework is crucial. This could include workshops on integrating specific EdTech tools, developing technology-based lesson plans, and fostering student digital literacy. Adequate technical infrastructure plays a crucial role in EdTech implementation. Current infrastructure, such as hardware, software, and internet connectivity, should be assessed, and any gaps should be addressed.

Additionally, a dedicated support team should be established to troubleshoot technical issues and provide ongoing technical assistance to educators. Technology should be seamlessly integrated with the curriculum rather than being an afterthought. Therefore, specific areas within the curriculum need to be identified where technology can enhance learning, and lessons that leverage the affordances of specific EdTech tools need to be designed. Open educational resources (OERs) and other digital content that complement the curriculum and align with the chosen framework must be explored. It is essential to ensure equitable access to technology for all students. This may involve providing students with the necessary devices and ensuring diverse representation in the chosen EdTech tools and learning materials. Different approaches should be explored to cater to students with various learning styles and abilities. Despite the valuable guidance frameworks provide, student voices and preferences should be considered. Opportunities for students to explore different EdTech tools should be incorporated, and they should be encouraged to provide feedback on their experiences. This fosters student agency and buy-in for technology use in the classroom.

Successful EdTech implementation is an iterative process. Hence, flexible, adaptive strategies based on specific contexts should be adopted and continuously monitored to evaluate the effectiveness of EdTech implementation. Progress should be regularly evaluated against the established goals. A mixed-method approach to evaluation should be used, employing both quantitative data (e.g., student achievement scores) and qualitative data (e.g., teacher observations and student feedback). Based on the evaluation findings, EdTech implementation strategies should be refined, and the chosen framework can be adapted if needed. This continuous cycle of evaluation and improvement ensures that EdTech investments yield the most significant benefit for students' learning.

Recommendations for Ongoing Professional Development Aligned with EdTech Frameworks

To truly benefit from these frameworks, teacher training must be dynamic, adapting to teachers' evolving technological, pedagogical, and subject-specific knowledge. Effective professional development is pivotal in creating opportunities for both pre-service and in-service teachers to move beyond theoretical understanding toward practical, impactful classroom integration. Below, we present specific recommendations for effective professional development aligned with prominent EdTech frameworks, tailored to different stages of teacher development and diverse instructional contexts.

The TPACK framework stresses the need for teachers to possess a sound, integrated understanding of technology, pedagogy, and content knowledge. To apply TPACK effectively, teachers must be trained in specific digital tools, their pedagogical applications, and content relevance. For pre-service teachers, introduce the TPACK framework early in teacher education programs through courses that integrate technology with curriculum planning and instructional strategies. Simulated teaching experiences, where pre-service teachers can design and deliver technology-integrated lessons, help build initial confidence. Micro-teaching sessions give room to experiment with various tools and receive feedback on their technological, pedagogical, and content choices. Collaborations with mentor teachers, where students co-design and reflect on lessons through a TPACK lens, further deepen understanding. For in-service teachers, professional development should be continuous and job-embedded, allowing teachers to reflect regularly on and refine their use of TPACK in real-time. Establishing learning communities where teachers can discuss successes and challenges with technology integration enriches practice.

The SAMR model helps teachers evaluate the sophistication of their technology use, from simple substitution to transformative redefinition. Effective professional development for SAMR involves assisting teachers to understand each level and gradually move toward the transformative use of technology. First, it familiarises pre-service teachers with SAMR through structured assignments that involve creating lessons at each level of the model. For instance, assign projects where they design a lesson at the "substitution" level and then explore how to augment or modify the same lesson with additional technology applications. Encouraging them to observe experienced educators who integrate SAMR concepts during their practicum also reinforces real-world applications. For in-service teachers, professional development workshops or coaching sessions can be organised where teachers analyse their current practices in the context of SAMR. Case studies or sample lessons showcasing the different SAMR stages help teachers visualise progression in their own classrooms. Professional development can also include peer observations. Teachers can benefit from collaborative lesson redesign, aiming to elevate a lesson through the SAMR levels by incorporating more sophisticated technology applications.

The T3 framework is focused on how technology can actively engage students in higher-order thinking. For pre-service teachers, integrate the T3 framework within instructional planning courses to encourage active student engagement with technology. Assignments should involve creating lessons that encourage student collaboration, creativity, and problem-solving. They should be allowed to experiment with tools that support active learning, such as project-based learning platforms or interactive simulations. Structured reflection on these experiences can help them to understand the transformative and transcendent impact of technology on learning. In-service professional development should focus on realigning technology use with engagement goals through workshops and practice sessions where teachers analyse student engagement. Tools like student surveys and engagement analytics could be added to measure the impact of technology on learning. Professional development can involve sessions on strategies for moving beyond the translational phase, such as encouraging students to lead inquiry projects using digital tools, thereby improving students' agency and ownership over their learning process.

For ICT4E and Saber-ICT, bringing prospective teachers' exposure to cheap, but very powerful digital tools or methods with huge impacts has potential inputs to diverse types of education settings. By this method of role-playing simulation or experiencing various teaching approaches and practices, it prepares student-teachers to think innovatively in solving problems at schools or centres. Courses on inclusive education that integrate ICT4E principles can help pre-service teachers understand the importance of digital equity. For teachers already in the field, professional development should include sessions on identifying and implementing accessible, affordable technologies to improve inclusivity. Peer-sharing sessions where teachers discuss successful strategies in ICT-constrained settings can be particularly valuable. Furthermore, collaborating with community organisations and NGOs to share resources and insights on overcoming digital divides is a powerful way for teachers to apply Saber-ICT principles.

PISA-ICT and TIM (Technology Integration Matrix) frameworks use technology to foster students' critical thinking and real-world problem-solving skills. Effective professional development in these frameworks emphasises helping teachers incorporate technology in ways that are relevant to students' everyday lives. Training can include developing project-based assessments integrating technology skills with real-world applications, such as data analysis or digital storytelling. Courses could emphasise assessing and building students' critical thinking through technology-based projects, fostering readiness for PISA-ICT style evaluations. Professional development for in-service teachers could include workshops on creating interdisciplinary, project-based learning experiences that engage students with real-world issues. Teachers can also benefit from exposure to digital tools that simulate professional environments, allowing them to incorporate technology as a bridge between classroom learning and real-world applications.

The Triple E framework is centred on using technology to extend, enhance, and engage students in learning activities beyond the traditional classroom scope. Professional development aligned with the Triple E framework helps teachers make targeted technology choices that support learning objectives. Introduce the Triple E framework through reflective exercises where they analyse and modify lesson plans better to support learning engagement, enhancement, and extension. Assign pre-service teachers to create lesson plans that focus on real-world applications of content knowledge, promoting technology use beyond classroom walls. Encourage in-service teachers to work collaboratively on lesson plans that intentionally align with the Triple E framework. PD sessions could include opportunities for teachers to evaluate their own or others' lesson plans against the Triple E principles, receiving constructive feedback on ways to align technology to better use with learning goals.

The effective application of EdTech frameworks is, therefore, inextricably linked to professional development. In fact, each of these frameworks provides a different path for meaningful integration of technology. Professional development activities can be designed to tailor and combine theoretical understanding with real-world application to help teachers from all career stages make their best use of technology, thereby improving student learning outcomes, increasing student engagement, and closing educational gaps. Meaningful professional development, therefore, enables teachers to develop an effective reflective practice and responsiveness to the shift in the landscape of technology as a leverage for innovative learning experiences, as well as being highly inclusive.

Future Trends in Educational Technology and the Evolution of EdTech Frameworks in Teacher Education

Education technology has rapidly changed and continues to do so more sharply than ever in recent history. Emerging trends such as AI, personalised learning, virtual reality, and data-driven instruction are promising to enhance teacher education and student outcomes. Integration of these developments requires the alteration of traditional frameworks in educational technology. Each of these frameworks is uniquely designed to guide and support technology integration, thus providing critical scaffolding for the educator looking to do so effectively. As new technologies are adopted and their educational potential becomes better understood, such frameworks will likely evolve in response to new pedagogical challenges, ethical considerations, and instructional opportunities.

Artificial intelligence may transform the classroom landscape to a great extent, from automated grading tools to intelligent tutoring systems. It can even provide predictive analytics or personal learning paths for students. In other words, AI can give a bouquet of tools to support learning and make it effective, more customised, and supportive. AI-driven tools can further support the teacher by automatically handling most repetitive tasks. Instant feedback is provided on students' performances; real-time data-driven insights are used to inform the pedagogical decisions in the process. Frameworks like TPACK and SAMR will have to expand the scope of their frames further to accommodate the complexities and ethical considerations that AI brings into education. For instance, TPACK might add another dimension that addresses AI knowledge, considering that teachers will need a basic understanding of AI's strengths and weaknesses and the ethical issues it raises. Similarly, the SAMR model could evolve to accommodate AI-specific applications, perhaps introducing a fifth tier that goes beyond "redefinition" to consider the implications of AI-powered adaptive learning environments. The TIM and Triple E frameworks may incorporate elements that help teachers evaluate AI tools based on educational outcomes rather than novelty. A data-informed layer could be added to these frameworks, assisting educators in interpreting analytics provided by AI systems and making more informed instructional choices.

Personalised learning tries to adjust the learning experiences to the learners' individual needs, strengths, and interests. Adaptive learning software, along with data analytics, can support teachers in creating individual learning pathways, pacing lessons based on the progress of students and emphasising specific skill gaps. As a result, the transition towards personalisation is considered part of the new education movement that shifts more toward competency-based models, away from traditional one-size-fits-all curricula. Concerning the ESRM and the ICT4E, where information and communication technologies are seen as wholly and well-entrenched parts of the education learning process, some more comprehensive overarching tenets will have to guide efforts in personalisation with interventions in learning. How this kind of technology affects and directs learning outcomes should become about how each child, through their effort in various activities, pursues individual goals and accomplishments with increased personalisation on competency progression. ICT4E, with its broad focus on technology in education, could serve as a foundation for developing tools that monitor and adjust learning experiences in real time. SAMR and T3 frameworks might also introduce new layers or criteria to evaluate the level of personalisation in technology use. For example, while SAMR currently focuses on moving from substitution to redefinition, an additional stage or parallel model might focus on the extent to which the technology enables individual learning pathways. For T3, transformation in a personalised learning context could involve leveraging technology to create adaptive assessments and provide individualised feedback.

The power to collect, analyse, and apply data from digital platforms provides powerful new ways to understand student performance and refine instructional strategies. Learning analytics can provide insights into engagement, comprehension, and skill acquisition, allowing educators to intervene more effectively and customise instructional support. The TIM and Pisa-ICT frameworks that inform ICT in classrooms and foster ICT literacy will likely need to adjust to include aspects of data literacy. Teachers will require competencies in interpreting complex data sets, data privacy implications, and effective use of insights informing instructional strategies. Data analysis should also be built into the assessment parts of these frameworks, making the teacher use data to track short and long-run effects of integrating technology into learning outcomes. This may add a separate element to TPACK data literacy to recognise a new form of knowledge acquired in mastering data analytics tools to enhance learning analytics. This would be integrated if it were found to align with the framework of developing ICT skills in education, allowing teachers to use data to inform their instructional decisions.

Through VR and AR technologies, an immersive experience through hands-on engagement, field trips with a lab simulation, the actual historic reenactments, and the more accurate models in 3D are possible, thereby making abstract topics easily accessible, leading to the engagement of many minds for experiential learning not easily achieved via normal procedures. The T3 and Triple E frameworks, which focus on engagement and enhancement in the use of technology, can be extended to include specific criteria that can be used to measure the educational value of immersive technologies. For instance, T3 could have an immersive dimension that would determine whether VR and AR experiences enhance learning or are more of a form of entertainment. The Triple E Framework could also involve even more granular pieces to look at whether VR and AR tools support active learning and engagement with content; hence, teachers can differentiate between shallow uses of VR/AR and applications that enhance learning outcomes. SAMR's redefinition level could also become more nuanced, distinguishing between different types of redefinitions based on the immersion level of the experience. A separate layer might help educators assess whether VR and AR tools offer new ways to interact with content beyond traditional media's limits.

By embracing these changes, the educational technology frameworks can support teachers in navigating this dynamic landscape of digital tools, thereby promoting a more effective, responsible, and student-centred approach to integrating technology into education. With these anticipated evolutions, educators and policymakers will be better placed to support meaningful technology integration, adapting to the needs of teachers and students alike in this ever-changing digital world.

Conclusion

The journey into the realm of EdTech has unveiled a landscape rich with potential and complexity. Integrating technology into education is no longer a luxury but a necessity, reshaping the dynamics of teaching and learning. This chapter has delved into a range of EdTech frameworks, each representing a unique lens through which one might view and practice technology in the classroom and beyond. From the foundational TPACK framework, where technology, pedagogy, and content knowledge all intertwine in very subtle ways, to the progressive stages of integration set forth by the SAMR model, educators have an important toolkit for navigating this technological landscape effectively. The T3 model extends the boundaries even further by encouraging educators to think of technology as a transforming agent in changing how learning occurs.

It is essential to derive insights from ESRM, ICT4E, SABER-ICT, PISA-ICT, and TIM frameworks, other than just classroom concentration. Those that point to the relevance of collaboration between stakeholders, equal access, and information-based decision-making are central in determining the future direction of education. Thus, Triple E is a practical guide for teachers to effectively integrate technology in the teaching-learning process in the light of engagement, enhancement, and extension.

The power of EdTech frameworks lies in their ability to bridge the gap between theory and practice. They provide a structured approach to integrating technology, ensuring that it is not merely a superficial addition but a catalyst for meaningful learning experiences. By understanding and applying these frameworks, educators can create dynamic, engaging, and effective learning environments that empower students to become critical thinkers, problem solvers, and lifelong learners.

However, the successful implementation of EdTech frameworks requires a holistic approach. It necessitates ongoing professional development, robust infrastructure, and unwavering support from policymakers and administrators. Recognising that technology is a tool, not a panacea, is essential. Its effectiveness depends on how it is used and integrated into the broader educational context.

As we look ahead, the future of EdTech holds immense promise. By embracing the insights offered by these frameworks and fostering collaboration among educators, policymakers, and technologists, we can harness the power of technology to create a more equitable, inclusive, and transformative educational landscape. The journey has just begun, and the destination is a future where technology empowers learners to reach their full potential.

References

Disruptive Classroom Technologies – CEESA. (2022, November 17). https://www.ceesa.org/elementor-8784/

Kolb, L. (2017). *Learning first, technology second: The Educator's Guide to Designing Authentic Lessons.* International Society for Technology in Education.

Maclemale, R. (n.d.). *Matrix.* TIM. https://fcit.usf.edu/matrix/matrix/

Magana, S. (2017). *Disruptive Classroom Technologies: a framework for innovation in education.* https://doi.org/10.4135/9781071872628

Mishra, P., & Koehler, M. J. (2006). Technological Pedagogical Content Knowledge: a framework for teacher knowledge. *Teachers College Record, 108*(6), 1017–1054. https://doi.org/10.1111/j.1467-9620.2006.00684.x

OECD. (2023). *PISA 2022 Assessment and Analytical Framework.* OECD Publishing. https://doi.org/10.1787/dfe0bf9c-en

Puentedura, R. (2010). *SAMR and TPCK: Intro to advanced practice.*

Rodríguez, P., Nussbaum, M., & Dombrovskaia, L. (2012). ICT for education: a conceptual framework for the sustainable adoption of technology-enhanced learning environments in schools. *Technology, Pedagogy and Education, 21*(3), 291–315. https://doi.org/10.1080/1475939x.2012.720415

Shulman, L. S. (1986). Those who understand: Knowledge growth in teaching. *Educational Researcher, 15*(2), 4–14. https://doi.org/10.3102/0013189x015002004

Staff, E. C. (2024, February 22). *The Triple E framework for more effective technology integration in adult education.* EdTech Center @World Education. https://edtech.worlded.org/the-triple-e-framework-for-more-effective-technology-integration-in-adult-education/

Trucano, M. (2016). SABER-ICT Framework Paper for Policy Analysis. In *World Bank, Washington, DC eBooks.* https://doi.org/10.1596/26107

Wagner, N. L., Hassanein, K., & Head, M. M. (2008). Who is responsible for E-Learning Success in Higher Education? A Stakeholders' Analysis. *Educational Technology & Society/Journal of Educational Technology & Society, 11*(3), 26–36.https://library.oum.edu.my/oumlib/sites/default/files/file_attachments/odl-resources/4332/who.pdf

PART II: DIGITAL TOOLS AND INNOVATIVE PEDAGOGICAL STRATEGIES

Chapter Four

Enhancing Pedagogy with Digital Tools: Strategies for Effective Technology Integration in Teacher Education

Amit Nitinrao Kolhe

Sanjivani Group of Institutes, Kopargaon, Maharashtra, India

Smita Tukaram Kumbhar

Sanjivani College of Pharmaceutical Education and Research (Autonomous), Kopargaon, Maharashtra, India

Dhananjay Arun Kumbhar

Sanjivani University, Kopargaon, Maharashtra, India

Abstract

In the modern educational landscape, integrating digital tools in teaching is essential. The chapter, Enhancing Pedagogy with Digital Tools: Strategies for Effective Technology Integration in Teacher Education, provides a practical approach for embedding technology in teacher training. It addresses current technology's benefits and barriers, emphasising frameworks like TPACK and SAMR to guide lesson planning and integration. Through case studies, it illustrates how these models translate theory into classroom practice. The chapter centres on three pillars: professional development, curriculum design, and pedagogical strategies, highlighting methods like peer-led workshops and collaborative communities for continuous teacher development. Examples, including flipped classrooms and gamified learning, demonstrate technology's potential to engage students. Equity issues are also addressed, with strategies to bridge the digital divide, ensuring all students benefit from technology. Looking ahead, the chapter explores emerging technologies like AI for personalised learning. It ultimately offers teacher educators a roadmap for integrating digital tools thoughtfully, ensuring they enhance rather than replace solid pedagogical practices.

Keywords: Blended Learning, Collaborative Learning, Personalised Education, Digital Pedagogy, Educational Innovation.

***

Introduction

Role of Technology in Modern Education

Technology has transformed nearly every aspect of modern society, and education is no exception. In today's educational landscape, the integration of digital tools plays a pivotal role in enhancing the learning process, fostering student engagement, and equipping learners with the essential skills needed for the digital age (Anderson, 2016; Bates, 2015; Greenhow et al., 2009). Digital tools such as computers, tablets, interactive whiteboards, and online learning platforms have reshaped traditional classrooms into dynamic, interactive environments. These tools not only enrich the delivery of educational content but also accommodate diverse learning styles and needs, making education more inclusive and accessible to all students (Darling-Hammond et al., 2014).

Importance of Integrating Digital Tools in Pedagogy

Figure 4.1 Flowchart of the Process of Integrating Technology in Teacher Education.

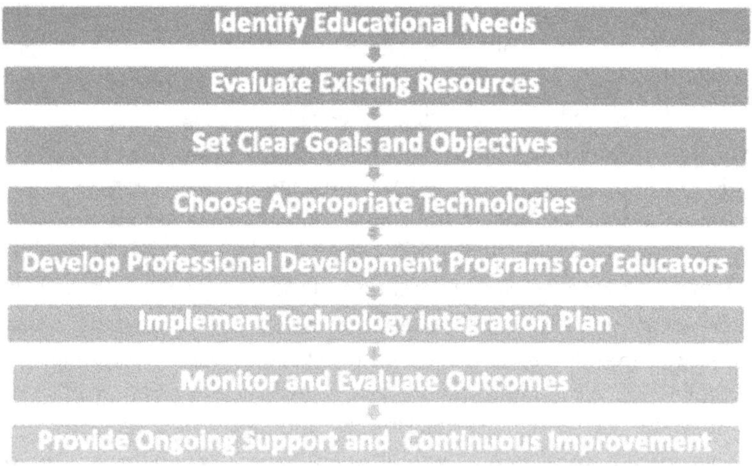

Integrating digital tools into pedagogy is crucial for preparing students for a future in which digital literacy and technological competence are indispensable (Anderson & Shattuck, 2012; Selwyn, 2011). When used effectively, technology can increase student engagement, promote collaborative learning, and offer personalised learning experiences that cater to individual student needs (Anderson, 2016; Ertmer et al., 2012). Moreover, digital tools enable educators to adopt data-driven approaches to track progress, assess student performance, and modify instruction as needed (Bates, 2015; Tamim et al., 2011). As educational methods continue to evolve, integrating technology ensures that teaching strategies remain relevant and adaptive in the face of rapid societal and technological changes (Fullan, 2013; Lawless & Pellegrino, 2007).

Theoretical Foundations

Technological Pedagogical Content Knowledge (TPACK)

The TPACK framework underscores the intersection of technology, pedagogy, and content knowledge, emphasising that effective teaching with technology requires more than just technical proficiency (Mishra & Koehler, 2006; Koehler & Mishra, 2009). It necessitates a deep understanding of how technology can complement teaching strategies and subject matter. Educators must integrate technology in ways that enhance learning while maintaining a balance between their pedagogical expertise and subject-specific content. TPACK provides a comprehensive roadmap for educators to navigate the complexities of integrating digital tools in their instructional practices, ensuring that technology serves as a meaningful extension of both content delivery and pedagogical methods (Harris & Hofer, 2009).

SAMR Model (Substitution, Augmentation, Modification, Redefinition)

Developed by Dr. Ruben Puentedura, the SAMR Model offers a progressive structure for integrating technology into educational practices (Puentedura, 2014). The model's four levels are:

- Substitution: Technology acts as a direct substitute for traditional tools without functional change (e.g., replacing paper-based worksheets with digital versions).
- Augmentation: Technology substitutes traditional tools with some enhancements (e.g., using Google Docs instead of paper to facilitate collaborative editing).
- Modification: Technology allows for significant redesign of tasks (e.g., creating interactive presentations using multimedia tools).
- Redefinition: Technology enables tasks that were previously inconceivable (e.g., conducting real-time data analysis with global collaboration tools).

By helping educators evaluate their technology integration practices, the SAMR model encourages movement from basic substitution toward transformative teaching approaches that reimagine learning experiences (Davies & West, 2014).

Constructivist Learning Theory

Constructivist learning theory posits that learners build knowledge through active engagement and reflection, a process that is greatly enhanced by technology (Garrison et al., 2000; Zhao & Frank, 2003). Digital tools such as virtual labs, interactive simulations, and educational games provide learners with hands-on experiences that foster deeper understanding and encourage critical thinking (Ching Hsu & Baldwin, 2018). These resources align seamlessly with constructivist principles, allowing students to experiment, explore, and construct their knowledge in a dynamic and interactive environment. Through technology, constructivist learning becomes more engaging, offering rich, immersive opportunities for active learning and problem-solving (Spires et al., 2012).

Heutagogy and Self-Determined Learning

Heutagogy, or self-determined learning, shifts the focus from teacher-centred instruction to learner autonomy, empowering students to take control of their own learning (Blaschke, 2012). In heutagogical learning environments, students set their learning goals, identify suitable resources, and tailor their learning strategies to meet their individual needs. Technology plays a crucial role in supporting heutagogy by providing access to a wide range of resources, fostering self-paced learning, and offering tools for tracking progress and reflecting on growth (Voogt & Roblin, 2012). This learner-centred approach, facilitated by digital tools, encourages flexibility, independence, and personalisation, preparing students for lifelong learning in an increasingly digital world (Bolliger & Halupa, 2012).

Strategies for Effective Technology Integration

Designing a Digital-Ready Curriculum

Creating a curriculum that is ready for digital integration involves more than simply adding technology to traditional teaching methods. It requires a strategic, thoughtful approach to curriculum design that ensures technology serves as a complement to the educational content and pedagogy, rather than as a distraction or a superficial add-on (Mouza & Lavigne, 2013). A digital-ready curriculum seamlessly incorporates technology into all aspects of learning, whether for delivering content, assessing progress, or facilitating collaboration and creativity. This approach ensures that digital tools are used to enhance the learning experience, making education more interactive, engaging, and responsive to the diverse needs of students.

Designing a digital-ready curriculum starts with understanding how digital tools can support different learning objectives across various subjects (Harris et al., 2009). It involves identifying key touchpoints where technology can transform a lesson into a more dynamic and interactive experience. This might mean integrating virtual labs in science classes, using interactive timelines in history lessons, or leveraging multimedia projects in language arts. The goal is to embed technology in ways that elevate content delivery and deepen student engagement, ensuring that it actively contributes to the development of essential skills.

Aligning Technology with Learning Objectives

One of the fundamental principles of effective technology integration is ensuring that digital tools are not just used for the sake of using technology, but are aligned with specific learning objectives (Ertmer & Ottenbreit-Leftwich, 2010). This alignment involves mapping out how different technological tools and platforms can help students achieve learning goals more effectively. For instance, in a math class, using graphing calculators or online platforms that visualise complex equations can significantly enhance understanding of mathematical concepts. Similarly, in a language arts class, digital storytelling tools can be employed to improve narrative writing and multimedia presentation skills.

When technology is closely tied to learning objectives, it ensures that each tool or platform used in the classroom has a clear purpose and measurable outcomes. This also enables educators to assess whether the technology is genuinely enhancing learning or merely functioning as a novelty (Kirkwood & Price, 2014). Moreover, this approach helps teachers avoid technology overload, where the presence of too many tools can become overwhelming for both teachers and students, detracting from the core educational goals.

For example, integrating virtual simulations in a biology lesson on cellular processes allows students to visualise and interact with the material in ways that would be impossible with traditional textbooks alone. Similarly, adaptive learning platforms in mathematics provide personalised exercises that adjust in difficulty based on the student's performance, helping each learner progress at their own pace. When aligned with learning objectives, these digital tools ensure that technology enhances understanding and supports the development of key competencies.

Incorporating Digital Literacy Skills

Incorporating digital literacy into the curriculum is essential for preparing students to navigate a world increasingly dominated by technology (Ribble,

2012). Digital literacy goes beyond basic computer skills; it encompasses the ability to find, evaluate, and create information online in a responsible and ethical manner. Educators must teach students how to critically assess the credibility of online sources, understand the implications of their digital footprint, and practice safe and responsible online behaviour. This includes instruction on privacy, data protection, and understanding algorithms that shape online content consumption (Ng, 2012).

Digital literacy also involves equipping students with the technical skills they need to effectively use a range of digital tools. Whether it is teaching students how to navigate online research databases, use productivity software like word processors and spreadsheets, or create multimedia presentations, these skills are critical for both academic success and future employability. By integrating digital literacy lessons into the curriculum, educators can help students become more confident and competent in their use of technology (Hobbs, 2010).

An effective way to incorporate digital literacy is through project-based learning that requires students to use digital tools to solve real-world problems (Larson & Miller, 2011). For instance, a social studies project could involve researching a global issue, creating an interactive presentation, and sharing their findings with a wider audience using digital platforms. This not only reinforces subject knowledge but also develops critical digital literacy skills such as evaluating sources, creating digital content, and understanding intellectual property and copyright issues.

Selecting and Evaluating Digital Tools

Criteria for Choosing Educational Technology

Selecting the right digital tools is a critical step in ensuring that technology enhances the learning experience without overwhelming students or teachers. With the vast array of educational technologies available today, it is essential to evaluate tools based on their educational value, usability, and how well they align with pedagogical goals (Mishra & Koehler, 2006). Here are the key criteria to consider:

- Relevance: The tool must align with curriculum objectives and support specific learning goals. Before adopting any technology, educators should ask whether it helps achieve the intended learning outcomes. For example, a tool that fosters critical thinking in science or problem-solving in mathematics is more valuable than one that merely entertains students without reinforcing core content (Harris et al., 2009). Relevance ensures that the technology serves a clear academic purpose, enhancing subject-specific knowledge or skills.

- Usability: Ease of use is a significant factor in determining the success of technology integration. A tool should have an intuitive interface that requires minimal training for both teachers and students. Complicated or clunky tools can lead to frustration and disengagement, detracting from the learning experience (Kirkwood & Price, 2014). Tools that offer user-friendly designs, clear instructions, and seamless integration with existing systems are more likely to be adopted successfully, making technology integration more fluid.
- Engagement: One of the primary benefits of digital tools is their ability to engage students through interactive and dynamic content. When selecting a tool, it is essential to consider how well it captivates students' attention and motivates them to participate actively in the learning process. Tools that promote interactivity, such as virtual simulations, gamification, or multimedia content, can enhance student engagement and create a more immersive learning environment (Prensky, 2001). This is particularly important for learners who may be less inclined to engage with traditional teaching methods.
- Accessibility: Technology should be inclusive and accessible to all learners, regardless of their abilities or learning preferences. This includes offering features that support diverse learners, such as text-to-speech, customizable font sizes, and closed captioning for videos (Kerr & Ritchhart, 2018). Accessibility also extends to ensuring that the tools are compatible with various devices, especially in classrooms where students may be using different types of technology (e.g., tablets, laptops, or smartphones). Selecting tools that are accessible promotes equity in education, ensuring that every student can participate fully in the learning process.
- Cost-effectiveness: Budget constraints are often a significant consideration when integrating technology in schools. Educators and administrators must weigh the cost of a tool against its potential educational benefits (Roblyer & Doering, 2013). Free or low-cost tools that offer substantial value are ideal, especially for institutions working with limited resources. However, it is also crucial to consider the long-term sustainability of using the tool, including subscription fees, maintenance, and updates. A cost-effective tool provides significant educational benefits while fitting within the institution's budgetary constraints.

Examples of Effective Digital Tools

Effective digital tools can significantly enhance the learning experience by aligning with pedagogical objectives, engaging students, and addressing specific subject needs. Here's a breakdown of some widely recognised tools, along with more detail on how they support learning outcomes across different educational levels and subjects:

Kahoot!

Subject Area: Multiple disciplines

Educational Level: K-12, Higher Education

Purpose: Game-based formative assessments

Kahoot! transforms quizzes into interactive, game-like experiences, making learning fun and engaging. It's especially effective for formative assessments, allowing teachers to quickly gauge student understanding and provide immediate feedback. The platform supports various content areas, from vocabulary quizzes in language arts to math and science reviews, and offers a competitive element that motivates students to participate actively (Morris, 2021). Its versatility and ability to promote active learning make it a popular choice across multiple disciplines.

Google Classroom

Subject Area: All disciplines

Educational Level: K-12, Higher Education

Purpose: Organising assignments, communication, and collaboration

Google Classroom serves as an all-in-one platform for managing assignments, distributing resources, and facilitating communication between teachers and students. It allows for seamless integration with other Google tools, such as Docs, Sheets, and Slides, and enables educators to streamline workflows by providing a centralised location for assignments and feedback (Gordon & Sun, 2018). The tool also supports collaboration by enabling students to work together on shared documents and projects, fostering an environment of digital teamwork.

Desmos

Subject Area: Mathematics

Educational Level: Middle School, High School, Higher Education

Purpose: Interactive graphing calculator and math activities

Desmos is an interactive tool primarily used for graphing and exploring mathematical concepts in real-time. Its intuitive interface allows students to visualise complex mathematical functions, from simple linear equations to advanced calculus topics. Desmos goes beyond being a calculator; it provides interactive math activities that promote inquiry-based learning. Teachers can create custom activities or use pre-made lessons to engage students in problem-solving, making abstract mathematical concepts more tangible and accessible (Desmos, 2019).

Nearpod

Subject Area: Multiple disciplines

Educational Level: K-12, Higher Education

Purpose: Interactive lesson delivery and formative assessments

Nearpod allows educators to create interactive lessons by incorporating multimedia content, quizzes, polls, and virtual reality experiences. The platform promotes student engagement by enabling teachers to deliver content in real-time, with options for both synchronous and asynchronous learning. Teachers can embed formative assessments throughout the lessons, instantly gathering data on student comprehension and adjusting instruction as needed (Khalil & Elkhider, 2020). The tool is effective across disciplines, offering flexibility for lessons in subjects ranging from social studies to STEM.

Figure 4.2 Bar Chart of Adoption Rate of Digital Tools in Teacher Education.

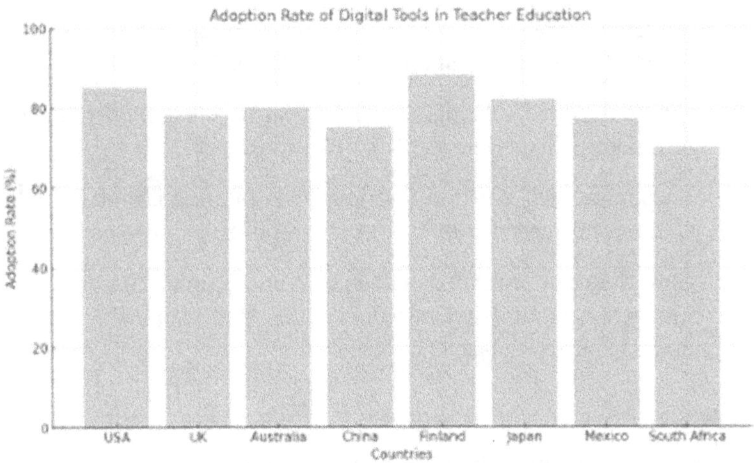

Implementing Blended Learning Models

Blended learning models integrate traditional face-to-face teaching with online learning, offering a flexible and personalised approach to education. By combining the strengths of both environments, these models create opportunities for deeper engagement, more individualised learning, and the development of critical digital literacy skills. Two key models of blended learning are the Flipped Classroom Approach and Hybrid Learning Environments, both of which leverage technology to enhance the learning experience (Garrison & Vaughan, 2008; Graham, 2006).

Flipped Classroom Approach

The flipped classroom approach rethinks the traditional lecture model by shifting the initial content delivery outside the classroom. Instead of using class time for direct instruction, students engage with instructional materials—such as recorded lectures, readings, or multimedia content—before coming to class. In the classroom, students then participate in hands-on, collaborative activities, discussions, and problem-solving sessions that build upon their understanding of the content (Bergmann & Sams, 2012).

How It Works: Instructors create or curate materials that students can access independently, such as videos or reading assignments. This allows students to learn at their own pace, reviewing materials as needed. Class time is then devoted to applying concepts through group work, experiments, debates, or other interactive activities that encourage critical thinking and collaboration.

Benefits:

- Students come to class prepared, allowing for deeper engagement with the material (Bergmann & Sams, 2012).
- Promotes active learning, as students participate in more meaningful, collaborative experiences (Bishop & Verleger, 2013).
- Encourages self-directed learning and time management skills as students take responsibility for their initial exposure to content.
- Class time can be more dynamic, with opportunities for personalised instruction, addressing misconceptions, and supporting individual learning needs (Jensen et al., 2014).

Challenges:

- Requires students to be motivated and disciplined to engage with pre-class materials (Bishop & Verleger, 2013).
- Instructors need to ensure that out-of-class content is accessible and engaging for all students (Jensen et al., 2014).
- Some students may lack the necessary technology or home environment to fully participate in the flipped model (Zhang, 2014).

Example: In a flipped science class, students might watch a video on the principles of thermodynamics at home, then engage in experiments or group discussions in class to apply the concepts. The teacher can facilitate real-time problem-solving and address any confusion students may have about the content.

Hybrid Learning Environments

Hybrid or blended learning environments combine in-person and online instruction to create a more flexible learning experience. This model allows students to engage with content both synchronously (in real-time during class sessions) and asynchronously (on their own schedule, often through online platforms). Hybrid learning can accommodate different learning preferences, support varied paces of learning, and allow educators to design more personalised and student-centred learning experiences (Garrison & Kanuka, 2004; Hrastinski, 2008).

How It Works: In a hybrid model, part of the instruction happens in a physical classroom, while another part takes place online, either through recorded lectures, discussion boards, or interactive online activities. Students might spend a few days a week in the classroom for face-to-face instruction and the rest of the time engaging in online coursework.

Benefits:

Hybrid or blended learning environments combine in-person and online instruction to create a more flexible learning experience. This model allows students to engage with content both synchronously (in real-time during class sessions) and asynchronously (on their own schedule, often through online platforms). Hybrid learning can accommodate different learning preferences, support varied paces of learning, and allow educators to design more personalised and student-centred learning experiences (Garrison & Kanuka, 2004; Hrastinski, 2008).

Challenges:

- Requires a well-planned balance between online and in-person activities to avoid redundancy or confusion (Hrastinski, 2008).
- Can lead to issues with student engagement or accountability in the online portion if not carefully monitored (Garrison & Vaughan, 2008).
- Technology access can be a barrier for some students, and instructors must ensure equitable access to online resources (Zhang, 2014).

Example: In a hybrid history course, students might attend a weekly in-person seminar for discussions and group activities, while completing the bulk of their content learning—such as watching lectures and engaging in online debates—through a learning management system like Canvas or Blackboard.

Enhancing Student Engagement and Interaction

Interactive Whiteboards and Smart Classrooms

Interactive whiteboards (IWBs) and smart classrooms are powerful tools that transform traditional teaching methods into more engaging and interactive experiences. These technologies provide a platform for educators to present multimedia content, facilitating various teaching and learning activities.

Key Features:

Multimedia Presentations: IWBs support the integration of videos, images, and interactive simulations, making lessons more visually appealing and engaging.

Real-Time Interaction: Students can participate directly in lessons by answering questions, manipulating digital objects, or collaborating on problems. This immediate feedback encourages active participation.

Collaborative Learning: Smart classrooms often include technology that allows for group work and collaborative projects, where students can share ideas and work together on tasks, fostering teamwork and communication skills.

Benefits:

- Enhances student engagement through dynamic, interactive lessons that cater to various learning styles.
- Supports active learning by allowing students to participate in discussions and activities, making the learning experience more relevant and memorable (Garrison & Vaughan, 2008).
- Encourages exploration and inquiry, as students can access a wealth of digital resources and tools that facilitate deeper understanding.

Challenges:

- Requires training for educators to effectively utilise these technologies and integrate them into their teaching practices.
- Potential technical issues can disrupt lessons and cause frustration if not addressed promptly.

Example: In a language arts class, an interactive whiteboard can be used to display a digital story map, allowing students to contribute their ideas and insights while discussing character development and themes. The teacher can incorporate multimedia elements, such as video clips or online articles, to enrich the discussion.

Gamification and Game-Based Learning

Gamification and game-based learning are innovative strategies that leverage the inherent interest students have in games to enhance engagement and motivation in the learning process.

Gamification: This approach incorporates game elements, such as points, badges, leaderboards, and challenges, into non-game contexts. By creating a competitive or rewarding environment, educators can motivate students to complete tasks and achieve learning goals.

Game-Based Learning: This method involves using actual games—video games, board games, or simulations—to teach concepts and skills. Game-based learning provides immersive experiences where students can apply their knowledge in a fun and engaging context.

Benefits:

- Increases motivation and engagement by making learning enjoyable and relevant. Students are more likely to invest time and effort when they see their learning as a game (Anderson & Krathwohl, 2001).
- Promotes critical thinking, problem-solving, and collaboration as students navigate challenges and interact with peers.
- Offers immediate feedback, allowing students to understand their progress and areas for improvement.

Challenges:

- Designing effective gamified experiences that align with curriculum objectives can be time-consuming and require creativity.
- Some students may struggle with competition, so it's essential to foster a supportive environment that encourages all learners.

Example: In a math class, educators might use a game-based learning platform like **Prodigy Math**, where students engage in a fantasy-themed game that reinforces math skills through quests and challenges. Students earn rewards and progress through levels, fostering a sense of achievement and encouraging consistent practice.

Personalised Learning with Technology

In today's educational landscape, personalised learning is increasingly recognised as a vital approach to cater to the diverse needs of students. Technology plays a crucial role in enabling personalised learning experiences that address individual strengths, weaknesses, and learning preferences. Two significant components of personalised learning are adaptive learning platforms and data-driven instruction.

Adaptive Learning Platforms

Adaptive learning platforms utilise advanced algorithms and artificial intelligence to tailor educational content to the individual needs of students. These platforms continuously assess student performance and adjust the learning experience accordingly, providing personalised pathways for each learner.

Key Features:

- **Individualised Learning Paths:** The platform adapts the curriculum based on students' unique abilities, ensuring they engage with material that is appropriately challenging. For example, if a student excels in a particular subject, the platform may present advanced topics sooner, while offering additional support to those struggling with foundational concepts.
- **Real-Time Feedback:** Students receive immediate feedback on their performance, allowing them to understand their strengths and weaknesses and adjust their learning strategies in real time.
- **Engaging Content:** Adaptive platforms often incorporate multimedia resources—such as videos, interactive simulations, and quizzes—to maintain student interest and cater to different learning styles.

Benefits:

- Enhances student engagement by providing relevant content tailored to individual learning needs.
- Supports mastery learning by allowing students to progress at their own pace, ensuring they have a solid understanding of each topic before moving forward.

- Reduces frustration for students who may feel overwhelmed by a one-size-fits-all approach to education.

Challenges:

- Implementation costs can be significant, especially for schools with limited budgets.
- Requires professional development for educators to effectively integrate these platforms into their teaching practices.

Example: Platforms like **DreamBox Learning** and **Knewton** offer adaptive learning solutions for mathematics and other subjects. These tools assess student responses and adjust the difficulty of problems in real time, ensuring that learners remain challenged yet supported.

Data-Driven Instruction

Data-driven instruction involves collecting and analysing various forms of data to inform teaching practices. This approach enables educators to make evidence-based decisions that enhance student learning outcomes.

Key Features:

- **Assessment Data Analysis:** Educators can use data from formative and summative assessments to identify trends and patterns in student performance. This information helps in recognising both strengths and areas for improvement.
- **Personalised Instruction:** By analysing data, educators can tailor their instruction to meet the specific needs of each student. For example, if a significant number of students struggle with a particular concept, the teacher can adjust the curriculum to reinforce that area.
- **Tracking Progress Over Time:** Continuous monitoring of student performance allows educators to track growth and adjust instructional strategies as needed.

Benefits:

- Empowers educators to make informed decisions about curriculum and instruction based on concrete evidence rather than intuition.
- Facilitates timely interventions for students who may be at risk of falling behind, ensuring that support is provided when it is most needed.

- Promotes accountability among educators, as data can be used to demonstrate the effectiveness of instructional strategies and identify areas for professional growth.

Challenges:

- The need for adequate training in data analysis tools and methods for teachers who may be unfamiliar with interpreting data.
- Data privacy concerns must be addressed, ensuring that student information is kept secure and confidential.

Example: Schools that use tools like **Edmodo** or **Google Classroom** can track student engagement and performance metrics, allowing teachers to adjust lesson plans based on data insights. Additionally, platforms like **iReady** provide diagnostic assessments and personalised learning recommendations based on student performance data.

Table 4.1: Global Strategies for Effective Technology Integration in Teacher Education.

Strategy	**Description**	**Example Tools**	**Global Examples & Expected Outcomes**
Blended Learning	Combining traditional classroom methods with online educational resources	LMS (e.g., Moodle, Blackboard), Google Classroom	**Global Examples:** The Open University (UK), INTEC (Malaysia) **Expected Outcomes:** Increased student engagement and flexibility in learning.
Flipped Classroom	Inverting traditional teaching by delivering instructional content online	YouTube, Khan Academy, Edpuzzle	**Global Examples:** University of Queensland (Australia), Shanghai Jiao Tong University (China) **Expected Outcomes:** More interactive and hands-on classroom activities.
Collaborative Learning	Encouraging students to work together on projects and problem-solving	Google Docs, Microsoft Teams, Padlet	**Global Examples:** Monterrey Institute of Technology (Mexico), University of Helsinki (Finland) **Expected Outcomes:** Enhanced communication, teamwork, and critical thinking skills.

Gamification	Using game design elements to motivate and engage students in learning	Kahoot!, Classcraft, Quizizz	**Global Examples:** European Schoolnet Academy (EU), Keio University (Japan) **Expected Outcomes:** Increased motivation and improved retention of knowledge.
Personalised Learning	Tailoring education to meet the individual needs of each student	Adaptive Learning Platforms (e.g., DreamBox, iReady)	**Global Examples:** Summit Learning Program (USA), AltSchool (USA) **Expected Outcomes:** Improved learning outcomes and student satisfaction.
Digital Storytelling	Using digital tools to tell stories and share experiences	Storybird, Adobe Spark, WeVideo	**Global Examples:** University of Cape Town (South Africa), National University of Singapore (Singapore) **Expected Outcomes:** Enhanced creativity and communication skills.
Virtual and Augmented Reality	Implementing VR/AR to create immersive learning experiences	Google Expeditions, Oculus Rift, AR Apps	**Global Examples:** Stanford University (USA), University of Barcelona (Spain) **Expected Outcomes:** Enhanced engagement and understanding of complex concepts.
Data-Driven Instruction	Using data analytics to inform and improve teaching practices	Learning Analytics Tools (e.g., Tableau, Power BI)	**Global Examples:** University of Edinburgh (UK), Indian Institute of Technology (India) **Expected Outcomes:** Informed decision-making and targeted interventions for students.

Professional Development for Educators

As technology continues to evolve rapidly, the role of educators must also adapt to meet the changing landscape of education. Ongoing professional development is essential for equipping educators with the skills and knowledge necessary to integrate technology effectively into their teaching practices (Ertmer & Ottenbreit-Leftwich, 2010; Harris & Hofer, 2009). This section explores the importance of continuous learning, various training opportunities, and strategies to foster a culture of professional growth among educators (Lawless & Pellegrino, 2007; Koehler & Mishra, 2009).

Importance of Ongoing Professional Development

Ongoing professional development is crucial for several reasons:

- **Staying Current with Technological Advancements:** As new educational technologies emerge, educators must familiarise themselves with the latest tools and trends. This understanding helps them select appropriate resources for their students and leverage technology effectively in the classroom.
- **Enhancing Instructional Strategies:** Professional development provides educators with the latest research and best practices in pedagogy, enabling them to improve their teaching methods. Continuous learning promotes the adoption of innovative instructional strategies that enhance student engagement and learning outcomes.
- **Adapting to Diverse Learning Needs:** Ongoing training equips educators with strategies to address the diverse needs of students. Professional development opportunities can focus on inclusive practices, differentiation, and personalised learning, ensuring that all students receive the support they need.
- **Fostering Confidence and Competence:** Regular professional development can help educators build confidence in using technology and instructional strategies. This confidence translates into more effective teaching practices and a willingness to experiment with new approaches.

Training Programs and Workshops

Training programs and workshops are critical components of professional development, offering educators practical, hands-on experiences with digital tools and strategies. Key characteristics of effective training include:

- **Relevance to Educators' Needs:** Training programs should be tailored to meet the specific needs of educators, addressing the unique challenges they face in their teaching contexts. For example, workshops focused on integrating specific tools like **Google Classroom** or **Kahoot!** can provide practical insights and skills for effective implementation.
- **Hands-On Experience:** Educators benefit most from training that allows them to engage directly with the technology. Hands-on workshops, where teachers can practice using tools in real-time and develop lesson plans that incorporate these technologies, foster deeper learning.

- **Alignment with Instructional Goals:** Professional development should align with the broader instructional goals of the school or district. By connecting training to specific educational objectives, educators can see the immediate relevance of the skills they acquire.
- **Follow-Up Support:** Effective training programs include follow-up support, such as coaching or mentoring, to help educators implement what they have learned in their classrooms. This ongoing support can enhance the sustainability of new practices.
- **Online Communities of Practice** provide invaluable opportunities for educators to connect, collaborate, and share experiences related to technology integration and instructional practices. Benefits of these communities include:
- **Resource Sharing:** Educators can exchange resources, lesson plans, and ideas for using technology effectively. Access to a wide array of materials enriches the teaching toolkit and encourages innovation.
- **Peer Support and Collaboration:** Online communities foster a sense of belonging among educators, enabling them to seek advice, share successes, and discuss challenges. This collaborative environment nurtures professional relationships and encourages collective problem-solving.
- **Continuous Learning:** Online forums and discussion groups allow educators to engage in ongoing discussions about best practices, research findings, and new technologies. This continuous dialogue supports lifelong learning and keeps educators informed about emerging trends in education.
- **Accessibility:** Online communities can break down geographical barriers, allowing educators from different regions or countries to connect and learn from one another. This diversity enriches the conversation and broadens perspectives.

Building a Culture of Continuous Improvement

Creating a culture of continuous improvement within educational institutions involves several key strategies:

- Encouraging Experimentation: Educators should be encouraged to experiment with new technologies and teaching methods. Providing a safe space for trial and error allows educators to explore innovative practices without fear of failure.

- Reflective Practices: Promoting reflective practices encourages educators to assess their teaching methods critically. Regular reflection on what works and what does not fosters a mindset of continuous improvement and lifelong learning.
- Celebrating Successes and Learning from Challenges: Recognising and celebrating the successes of educators who implement new technologies or strategies can motivate others to take risks. Conversely, sharing challenges and failures fosters a sense of community and resilience, emphasising that growth often comes from overcoming obstacles.
- Leadership Support: Strong support from school leadership is essential for fostering a culture of continuous improvement. Leaders should model lifelong learning, allocate resources for professional development, and promote an environment where innovation is valued.

Case Studies and Examples

Successful technology integration in educational settings can significantly enhance student engagement, improve learning outcomes, and foster innovative teaching practices (Mishra & Koehler, 2006; Tamim et al., 2011). This section presents three case studies that exemplify effective technology integration strategies in schools (Garrison & Vaughan, 2008; Ching et al., 2018).

Successful Technology Integration in Schools

Case Study 1: Flipped Classroom in High School Science

Background:

At Oakridge High School, a dedicated science teacher implemented the flipped classroom model to transform her traditional teaching methods. This approach involved assigning video lectures for homework and using class time for interactive experiments and discussions.

Implementation:

- **Homework:** Students watched pre-recorded video lectures at home, which covered key concepts and introduced new topics.
- **Class Time:** In class, the teacher facilitated hands-on experiments and collaborative discussions, allowing students to apply their knowledge and engage deeply with the material.

Outcomes:

- **Increased Engagement:** Students reported higher levels of engagement and interest in the subject matter. The interactive nature of class time allowed them to ask questions and collaborate with peers.
- **Improved Learning Outcomes:** Assessment data showed a significant improvement in test scores and understanding of complex scientific concepts. The teacher noted that students were better prepared for class discussions and activities, leading to a more productive learning environment.

Key Takeaways:

- The flipped classroom model can effectively enhance student engagement and understanding, particularly in subjects requiring hands-on exploration.
- Providing students with access to instructional content outside the classroom allows for deeper in-class interactions and a more personalised learning experience.

Case Study 2: Blended Learning in Middle School Math

Background:

At Maplewood Middle School, a math teacher adopted a blended learning approach to provide students with personalised learning experiences. This model combined online instructional modules with in-person collaborative activities.

Implementation:

- **Online Modules:** Students completed self-paced online modules that provided instruction, practice, and assessments tailored to their individual needs.
- **In-Person Support:** During class time, the teacher facilitated small group discussions and problem-solving sessions, allowing students to work collaboratively on challenging concepts while receiving targeted support.

Outcomes:

- **Personalised Learning:** Students appreciated the ability to learn at their own pace, leading to increased motivation and ownership of their learning process.

- **Enhanced Collaboration:** The collaborative activities during class time fostered peer-to-peer learning and support, enhancing students' problem-solving skills and confidence in math.

Key Takeaways:

- Blended learning allows for flexibility in instruction, accommodating diverse learning preferences and paces.
- The combination of online and in-person learning experiences can create a supportive environment that promotes collaboration and personalised learning.

Case Study 3: Virtual Reality in History Education

Background:

At Westfield High School, a history teacher explored innovative teaching methods by incorporating virtual reality (VR) into the curriculum. The goal was to create immersive learning experiences that would enhance students' understanding of historical events and concepts.

Implementation:

- VR Experiences: The teacher utilised VR technology to simulate historical settings and events, such as ancient civilisations and significant battles. Students were able to "visit" historical sites and witness pivotal moments in history.
- Interactive Assignments: Following the VR experiences, students engaged in discussions, reflections, and creative projects that connected their immersive experiences to their broader learning objectives.

Outcomes:

- Increased Retention: Students demonstrated a deeper understanding and retention of historical concepts. The immersive nature of VR made learning more memorable and impactful.
- Engaged Learning Environment: The novelty of VR technology sparked curiosity and excitement among students, fostering a more dynamic and interactive classroom environment.

Key Takeaways:

- Virtual reality can enhance learning by providing immersive experiences that allow students to engage with content in meaningful ways.

- Incorporating innovative technologies, such as VR, can foster student engagement and interest, transforming traditional approaches to teaching history.

These case studies illustrate the potential of effective technology integration in educational settings. By adopting models such as flipped classrooms, blended learning, and virtual reality, educators can create engaging, personalised, and impactful learning experiences for their students.

In each case, the key to success lies in aligning technology with instructional goals, actively involving students in their learning process, and providing opportunities for collaboration and exploration. These strategies can serve as valuable frameworks for educators seeking to enhance their teaching practices and improve student outcomes through technology integration.

Table 4.2: Challenges and Solutions in Integrating Technology in Teacher Education (with Global Case Studies).

Challenge	Description	Solution	Case Study Example & Outcome
Resistance to Change	Teachers' reluctance to adopt new technologies and methods.	Professional development and ongoing support.	**Case Study Example:** Shanghai Jiao Tong University (China) **Outcome:** Increased adoption through continuous training programs.
Lack of Technical Skills	Insufficient knowledge and skills to effectively use digital tools.	Training programs and user-friendly tools.	**Case Study Example:** European Schoolnet Academy (EU) **Outcome:** Enhanced teacher competency in technology integration.
Limited Access to Technology	Inadequate access to devices and internet connectivity.	Equitable resource distribution and infrastructure improvement.	**Case Study Example:** The Open University (UK) **Outcome:** Broader access through distributed resources.

Time Constraints	Teachers' limited time to learn and integrate new technologies.	Time management strategies and integration of technology in daily routines.	**Case Study Example:** AltSchool (USA) **Outcome:** Efficient integration with streamlined processes.
Privacy and Security Concerns	Risks related to data privacy and cybersecurity.	Implementing robust security policies and educating users.	**Case Study Example:** Stanford University (USA) **Outcome:** Improved data protection and user awareness.
Student Distraction	Potential for digital tools to distract rather than engage students.	Establishing clear guidelines and monitoring usage.	**Case Study Example:** Keio University (Japan) **Outcome:** Reduced distractions with effective guidelines.
Inconsistent Technology Adoption	Varying levels of technology integration across different classrooms or institutions.	Standardizing technology policies and practices.	**Case Study Example:** Monterrey Institute of Technology (Mexico) **Outcome:** Consistent adoption through unified policies.
Cost of Technology	High costs are associated with purchasing and maintaining digital tools and infrastructure.	Seeking funding opportunities and cost-effective solutions.	**Case Study Example:** University of Cape Town (South Africa) **Outcome:** Secured funding and reduced costs with strategic partnerships.

Opportunities and Challenges

Opportunities

Enhanced Engagement and Interaction

Digital tools create interactive learning environments that engage students and foster active participation. Features like multimedia content, interactive quizzes, and collaborative platforms enhance the learning experience by transforming passive learning into active exploration. Students can engage in discussions, problem-solving tasks, and interactive activities that cater to diverse learning styles, making lessons more memorable and effective.

Personalized Learning

Technology enables personalised learning experiences tailored to individual students' needs and preferences. Adaptive learning platforms adjust the pace and difficulty of lessons based on student performance, ensuring that each learner receives the appropriate level of challenge and support. This individualised approach promotes a deeper understanding of material, as students can progress at their own pace and revisit concepts as needed.

Access to a Wealth of Resources

The internet provides access to a vast array of educational resources, including online courses, e-books, research articles, videos, and simulations. This wealth of information supports self-directed learning and empowers students to explore topics beyond the classroom curriculum. With a variety of resources at their fingertips, students can engage in independent research, broaden their knowledge, and develop critical thinking skills.

Collaboration and Communication

Digital tools facilitate collaboration and communication among students and educators. Platforms like Google Classroom and Zoom enable real-time collaboration, virtual discussions, and remote learning, supporting global learning communities. Students can work together on projects, share ideas, and communicate across geographical boundaries, fostering a sense of connection and shared learning experiences.

Data-Driven Instruction

Technology allows for the collection and analysis of student performance data. Educators can use this data to inform instruction, tailor interventions, and track student progress, supporting evidence-based teaching practices. By analysing assessment results and learning analytics, educators can identify patterns, strengths, and areas for improvement, allowing for timely and targeted support.

Innovative Pedagogical Approaches

Digital tools support innovative pedagogical approaches, such as flipped classrooms, blended learning, and gamification. These approaches create flexible, student-centred learning experiences that promote critical thinking and creativity. By utilising technology to enhance traditional teaching methods, educators can create dynamic learning environments that encourage exploration and collaboration.

Professional Development and Continuous Learning

Technology provides opportunities for ongoing professional development. Online courses, webinars, and virtual conferences offer flexible professional development options that fit into educators' busy schedules. Digital communities of practice enable educators to share experiences, resources, and best practices, fostering a collaborative culture of continuous learning and improvement.

Challenges

While the integration of technology in education presents numerous opportunities, it also comes with challenges that educators must navigate to ensure effective implementation:

Digital Divide

One significant challenge is the digital divide, which refers to the gap between individuals who have access to technology and those who do not. Students from low-income backgrounds may lack reliable internet access or devices, hindering their ability to participate fully in technology-enhanced learning. Educators must address these disparities to ensure equitable learning opportunities for all students.

Technology Overload

The abundance of digital tools and resources can lead to technology overload, where educators feel overwhelmed by the number of options available. Selecting appropriate tools that align with pedagogical goals can be challenging, and without proper guidance, educators may struggle to integrate technology effectively into their teaching practices.

Resistance to Change

Some educators may resist adopting new technologies due to familiarity with traditional teaching methods or concerns about their effectiveness. Overcoming this resistance requires ongoing support, training, and professional development to build confidence and competence in using technology in the classroom.

Need for Training and Support

Successful technology integration requires educators to be well-trained in both the tools they are using and the pedagogical strategies that accompany them. Many educators may lack the necessary training or ongoing support to utilise digital tools effectively, leading to underutilization or ineffective implementation.

Balancing Screen Time

With increased reliance on digital tools, there is a growing concern about excessive screen time for students. Educators must find a balance between leveraging technology for learning and ensuring students engage in other important activities, such as physical education, social interactions, and hands-on learning experiences.

Maintaining Engagement

While technology can enhance engagement, it can also become a distraction if not used purposefully. Educators must be mindful of how technology is integrated into lessons and ensure that it serves a clear educational purpose rather than detracting from the learning experience.

Privacy and Security Concerns

As educational institutions increasingly utilise digital tools, concerns about student privacy and data security arise. Educators and administrators must be vigilant in protecting student information and ensuring that the platforms they use comply with data protection regulations.

Figure 4.3: Pie Chart of Challenges Faced by Teachers in Technology Integration.

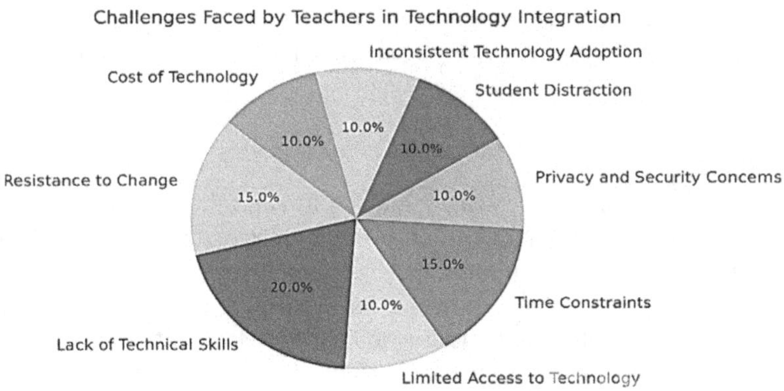

Strategies for Addressing Challenges

Investing in Infrastructure

Investing in reliable infrastructure is crucial for successful technology integration. This includes providing students with access to modern devices, ensuring strong and stable internet connectivity, and offering technical support for both educators and students. Adequate funding and resources enable educational institutions to create an environment where all students can access the necessary tools and support for their learning, regardless of their socio-economic background. Investments should also extend to maintaining and upgrading existing technology to keep pace with educational advancements.

Comprehensive Professional Development

Professional development programs must be comprehensive, practical, and ongoing to effectively address the challenges of technology integration. Training should cover both technical skills related to the use of digital tools and pedagogical strategies for incorporating technology into teaching practices. This dual focus ensures that educators can use digital tools effectively and confidently, ultimately improving student learning outcomes. Continuous support, such as coaching and mentorship, can further enhance educators' abilities to implement technology successfully in the classroom.

Fostering a Culture of Innovation

Creating a culture of innovation within educational institutions involves encouraging educators to experiment with new technologies and reflect on their teaching practices. This culture promotes a growth mindset, where educators are willing to take risks, try new methods, and share their successes and challenges with their peers. Leadership and support from school administrators are essential in fostering this environment, as they can provide the necessary resources, recognition, and encouragement for educators to innovate and improve their teaching through technology.

Balancing Screen Time

As technology becomes increasingly integral to education, it is essential for educators to balance technology use with offline activities to promote students' health and well-being. Incorporating physical activities, hands-on projects, and face-to-face interactions ensures a balanced approach to learning that addresses both cognitive and social-emotional development. Educators can establish guidelines for screen time and include strategies for engaging students in a variety of learning modalities, fostering a holistic educational experience.

Ensuring Data Privacy and Security

Implementing robust data privacy and security measures is critical for protecting student information in a technology-driven educational landscape. Educators and institutions must comply with relevant regulations, such as FERPA and GDPR, and educate students about online safety, responsible digital citizenship, and the importance of protecting their personal information. Building trust with parents and the community through transparent communication about data practices is also vital for fostering a safe learning environment.

Aligning Technology with Pedagogy

Aligning technology with pedagogical goals requires careful planning and reflection on the part of educators. When selecting digital tools, educators should ensure that they enhance instructional objectives, provide meaningful learning experiences, and support student outcomes. This alignment can be achieved through curriculum mapping, where educators identify specific learning goals and select technology that complements and reinforces those objectives. Engaging in collaborative planning with colleagues can also promote shared understanding and effective integration of technology across subjects.

Providing Technical Support

Reliable technical support is essential for addressing technical issues promptly and minimising disruptions in the learning process. Schools should have dedicated IT staff who are well-trained and available to assist both educators and students. Clear protocols for troubleshooting and resolving technical problems should be established, ensuring that educators can quickly regain focus on teaching and learning without prolonged interruptions. Providing ongoing training for IT staff can further enhance their ability to address emerging technologies and challenges.

Future Directions

Emerging Technologies in Education

Emerging technologies such as artificial intelligence (AI), augmented reality (AR), and blockchain hold immense potential to revolutionise the educational landscape. AI can facilitate personalised learning experiences by analysing individual student data and adapting instructional methods to suit each learner's needs, thus enhancing engagement and comprehension. AR can create immersive, interactive learning environments that allow students to explore complex concepts in a hands-on manner, such as visualising anatomical structures in biology or historical sites in geography. Meanwhile, blockchain technology can ensure secure, transparent record-keeping for student credentials, enabling the development of a verifiable digital identity for learners and streamlining processes like credential verification and transfer of academic records. As these technologies become more accessible, they can create opportunities for more engaging, efficient, and personalised learning experiences.

Trends in Educational Technology Research

Educational technology research is continuously exploring new frontiers, with a focus on understanding the multifaceted impacts of technology on learning outcomes. Researchers are investigating the effectiveness of various digital tools and pedagogical approaches, such as gamification, adaptive learning platforms, and blended learning environments, to determine best practices for technology integration. Additionally, studies are being conducted on how to measure student engagement, motivation, and achievement in tech-enhanced classrooms. Staying informed about these trends enables educators and policymakers to make evidence-based decisions, ensuring that the integration of technology into education is guided by solid research findings that support optimal learning outcomes.

Preparing for the Future of Digital Learning

Preparing for the future of digital learning requires educators and institutions to anticipate changes in technology and pedagogy, embrace innovation, and foster a culture of continuous improvement. This involves developing a proactive approach to professional development, where educators are not only trained in current technologies but are also encouraged to explore and experiment with emerging tools. A growth mindset among educators is crucial; being adaptable and open to change will allow them to effectively respond to new challenges and opportunities in the educational landscape. Furthermore, collaboration among educators, technology developers, and researchers is essential for shaping a future that prioritises high-quality, technology-enhanced education for all students, equipping them with the skills needed to thrive in a digital world.

Figure 4.4: Flowchart of Professional Development for Technology Integration.

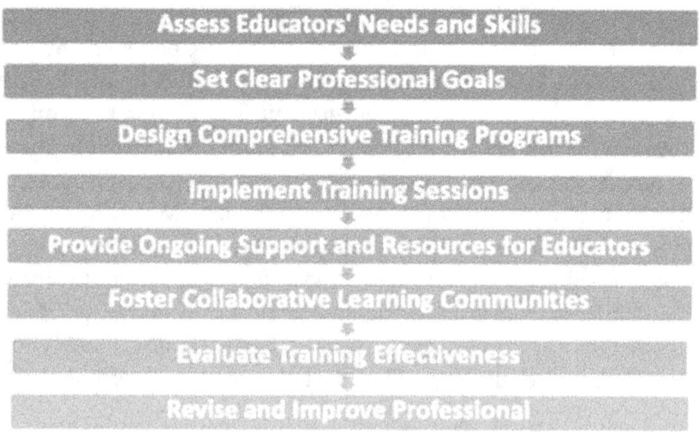

Conclusion

Summary of Key Points

Integrating digital tools into pedagogy significantly enhances teaching and learning, offering opportunities for personalised, engaging, and data-driven instruction. This effective technology integration involves carefully aligning digital tools with pedagogical goals to ensure that they enrich the learning experience rather than distract from it. Additionally, investing in robust infrastructure is essential to support the effective use of technology in educational settings. Ongoing professional development for educators is crucial, equipping them with the skills and confidence necessary to utilise these digital tools effectively.

The Impact of Digital Tools on Education

Digital tools have fundamentally transformed the educational landscape, creating interactive, flexible, and inclusive learning environments. They cater to diverse learning styles and preferences, fostering collaboration and communication among students and educators. These tools not only enhance engagement but also prepare students for success in an increasingly digital world. By integrating technology into the classroom, educators can provide students with the skills and knowledge necessary to navigate the complexities of the twenty-first century.

Moving Forward: Embracing Technology for Enhanced Learning

Embracing technology for enhanced learning requires a commitment to continuous improvement, innovation, and collaboration among educators, administrators, and stakeholders. As we move forward, leveraging the potential of digital tools will enable educators to create meaningful, effective, and future-ready learning experiences for all students. This commitment to technological integration will empower students to thrive in a dynamic, digital landscape, equipping them with the skills and competencies needed to excel in their future endeavours.

References

Anderson, R. E. (2016). Integrating ICT into teaching and learning: Reflections on pedagogical insights and best practices. *Educational Technology & Society*, 19(3), 1-14.

Anderson, T., & Shattuck, J. (2012). Design-based research: A decade of progress in education research? *Educational Researcher*, 41(1), 16-25. https://doi.org/10.3102/0013189X11428813

Bates, A. W. (2015). *Teaching in a digital age: Guidelines for designing teaching and learning.* Tony Bates Associates Ltd.

Bergmann, J., & Sams, A. (2012). *Flip Your Classroom: Reach Every Student in Every Class Every Day.* International Society for Technology in Education.

Bishop, J. L., & Verleger, M. A. (2013). The flipped classroom: A survey of the research. *ASEE National Conference Proceedings.* https://doi.org/10.18260/1-2--22585

Blaschke, L. M. (2012). Heutagogy and lifelong learning: A review of heutagogical practice and self-determined learning. *The International Review of Research in Open and Distributed Learning,* 13(1), 56-71. https://doi.org/10.19173/irrodl.v13i1.1076

Bolliger, D. U., & Halupa, C. (2012). Student perceptions of satisfaction and anxiety in an online doctoral program. *Distance Education, 33*(1), 81–98. https://doi.org/10.1080/01587919.2012.667961

Ching, Y. H., Hsu, Y. C., & Baldwin, S. (2018). Developing a virtual reality experience for pre-service teachers: Learning about classroom management. *Journal of Educational Technology Systems,* 47(2), 241-258.

Darling-Hammond, L., Zielezinski, M. B., & Goldman, S. (2014). *Using technology to support at-risk students' learning.* Alliance for Excellent Education.

Davies, R. S., & West, R. E. (2014). Technology integration in schools. In J. M. Spector, M. D. Merrill, J. Elen, & M. J. Bishop (Eds.), *Handbook of research on educational communications and technology* (pp. 841-853). Springer. https://doi.org/10.1007/978-1-4614-3185-5_68

Ertmer, P. A., & Ottenbreit-Leftwich, A. T. (2010). Teacher technology change: How knowledge, confidence, beliefs, and culture intersect. *Journal of Research on Technology in Education, 42*(3), 255–284. https://doi.org/10.1080/15391523.2010.10782551

Ertmer, P. A., Ottenbreit-Leftwich, A. T., Sadik, O., Sendurur, E., & Sendurur, P. (2012). Teacher beliefs and technology integration practices: A critical relationship. *Computers & Education,* 59(2), 423-435. https://doi.org/10.1016/j.compedu.2012.02.001

Fullan, M. (2013). *The new meaning of educational change* (4th ed.). Teachers College Press.

Garrison, D. R., Anderson, T., & Archer, W. (2000). Critical inquiry in a text-based environment: Computer conferencing in higher education. *The Internet and Higher Education,* 2(2-3), 87-105. https://doi.org/10.1016/S1096-7516(00)00016-6

Garrison, D. R., & Kanuka, H. (2004). Blended learning: Uncovering its transformative potential in higher education. *The Internet and Higher Education,* 7(2), 95-105. https://doi.org/10.1016/j.iheduc.2004.02.001

Garrison, D. R., & Vaughan, N. D. (2008). *Blended Learning in Higher Education: Framework, Principles, and Guidelines.* Jossey-Bass. https://doi.org/10.1002/9781118269558

Greenhow, C., Robelia, B., & Hughes, J. E. (2009). Learning, teaching, and scholarship in a digital age: Web 2.0 and classroom research: What path should we take now? *Educational Researcher,* 38(4), 246-259. https://doi.org/10.3102/0013189X09336671

Harris, J., & Hofer, M. (2009). Grounded tech integration: An effective approach based on content, pedagogy, and teacher planning. *Learning & Leading with Technology*, 37(2), 22-25.

Harris, J., Mishra, P., & Koehler, M. (2009). Teachers' technological pedagogical content knowledge and learning activity types. *Journal of Research on Technology in Education*, 41(4), 393–416. https://doi.org/10.1080/15391523.2009.10782536

Hobbs, R. (2010). *Digital and media literacy: A plan of action*. The Aspen Institute.

Jensen, J. L., Kummer, T. A., & Godoy, P. D. (2014). Improvements from a flipped classroom may simply be the fruits of active learning. *CBE—Life Sciences Education*, 13(2), 223-231.

Khalil, M., & Elkhider, I. (2020). The role of Nearpod in improving student engagement and academic performance in education. *Education and Information Technologies*, 25(6), 5301-5318.

Kirkwood, A., & Price, L. (2014). Technology-enhanced learning and teaching in higher education: What is 'enhanced' and how do we know? A critical literature review. *Learning, Media and Technology*, 39(1), 6–36. https://doi.org/10.1080/17439884.2013.770404

Koehler, M. J., & Mishra, P. (2009). What is technological pedagogical content knowledge (TPACK)? *Contemporary Issues in Technology and Teacher Education*, 9(1), 60-70.

Lai, K. W. (2011). Digital technology and the culture of teaching and learning in higher education. *Australasian Journal of Educational Technology, 27*(8), 1263-1275. https://doi.org/10.14742/ajet.892

Larson, L. C., & Miller, T. N. (2011). 21st-century skills: Prepare students for the future. *Kappa Delta Pi Record*, 47(3), 121–123. https://doi.org/10.1080/00228958.2011.10516575

Lawless, K. A., & Pellegrino, J. W. (2007). Professional development in integrating technology into teaching and learning: Known, unknowns, and ways to pursue better questions and answers. *Review of Educational Research*, 77(4), 575-614. https://doi.org/10.3102/0034654307309921

Mishra, P., & Koehler, M. J. (2006). Technological pedagogical content knowledge: A framework for teacher knowledge. *Teachers College Record*, 108(6), 1017-1054. https://doi.org/10.1111/j.1467-9620.2006.00684.x

Mouza, C., & Lavigne, N. C. (2013). *Introduction to emerging technologies for the classroom: A learning sciences perspective*. Springer. https://doi.org/10.1007/978-1-4614-4696-5_1

Ng, W. (2012). Can we teach digital natives digital literacy? *Computers & Education*, 59(3), 1065–1078. https://doi.org/10.1016/j.compedu.2012.04.016

Puentedura, R. R. (2014). *SAMR: A contextualized introduction*. Retrieved from http://www.hippasus.com/rrpweblog/archives/2014/06/29/SAMR_ContextualizedIntroduction.pdf

Prensky, M. (2001). Digital natives, digital immigrants. *On the Horizon*, 9(5), 1-6. https://doi.org/10.1108/10748120110424816

Ribble, M. (2012). *Digital citizenship in schools: Nine elements all students should know*. International Society for Technology in Education (ISTE).

Roblyer, M. D., & Doering, A. H. (2013). *Integrating educational technology into teaching* (6th ed.). Pearson.

Shulman, L. S. (1986). Those who understand: Knowledge growth in teaching. *Educational Researcher,* 15(2), 4-14. https://doi.org/10.3102/0013189X015002004

Spires, H. A., Hervey, L. G., Morris, G., & Stelpflug, C. (2012). Energizing project-based inquiry: Middle grade students read, write, and create videos. *Journal of Adolescent & Adult Literacy,* 55(6), 483-493. https://doi.org/10.1002/JAAL.00058

Tamim, R. M., Bernard, R. M., Borokhovski, E., Abrami, P. C., & Schmid, R. F. (2011). What forty years of research says about the impact of technology on learning: A second-order meta-analysis and validation study. *Review of Educational Research,* 81(1), 4-28. https://doi.org/10.3102/0034654310393361

Twining, P., Heller, R. S., Nussbaum, M., & Tsai, C. C. (2017). Some guidance on conducting and reporting qualitative studies. *Computers & Education,* 106, A1-A9. https://doi.org/10.1016/j.compedu.2016.12.002

Voogt, J., & Roblin, N. P. (2012). A comparative analysis of international frameworks for 21st century competences: Implications for national curriculum policies. *Journal of Curriculum Studies,* 44(3), 299-321. https://doi.org/10.1080/00220272.2012.668938

Zhao, Y., & Frank, K. A. (2003). Factors affecting technology uses in schools: An ecological perspective. *American Educational Research Journal, 40*(4), 807-840. https://doi.org/10.3102/00028312040004807

Chapter Five

Transformative Technology Integration: Practical Approaches for Engaging Classrooms

Grace Okubo

Association of Professional Futurists, Nigeria

Abstract

Grace Okubo examines how advancements in technology are reshaping education. It offers educators practical strategies for using technology to enhance student engagement, collaboration, and learning outcomes across various subjects. The manuscript emphasises cross-disciplinary integration, where digital tools connect different areas of study, creating a holistic understanding of complex themes. It also highlights the importance of collaborative platforms, peer learning, and global classrooms in developing critical twenty-first-century skills. Case studies like Worldreader, uLesson, BYJU'S, and Virtual School Victoria showcase successful technology integration in diverse educational settings.

Keywords: Educational technology, student engagement, cross-disciplinary integration, collaborative learning, flipped classroom, project-based learning, twenty-first-century skills.

***

Introduction

In recent years, education has been profoundly transformed by rapid technological advancements. The rise of Artificial Intelligence (AI) has introduced AI-driven tools like intelligent tutoring systems and automated grading, while Virtual and Augmented Reality (VR/AR) are creating immersive learning environments. The COVID-19 pandemic accelerated the global shift to online learning, with schools quickly adopting digital technologies.

Today's students, as digital natives, expect their education to integrate technology. To keep them engaged, teachers must incorporate tools like interactive whiteboards, VR, and adaptive learning software, which make learning dynamic and inclusive. The vast access to information online requires teachers to embed digital literacy in their curricula, guiding students in navigating and evaluating content.

As the job market increasingly values technological skills, education must prepare students by teaching essential competencies like coding, data analysis, and digital communication. Early exposure to technology facilitates the development of these skills. Technology also expands learning beyond the classroom, fostering global collaboration through online platforms and allowing personalised education tailored to individual needs and paces. Interactive and gamified experiences further enhance motivation and inspire a lifelong love of learning. This chapter provides educators with strategies to enhance student engagement, collaboration, and learning outcomes, preparing students for success in both academic and digital spheres.

Field-Tested Approaches for Integrating Technology

Teachers' responses to classroom technology vary considerably. While some readily embrace digital tools, others demonstrate resistance, often driven by insufficient training, limited exposure, and low self-efficacy. Effective technology integration requires alignment with teachers' underlying pedagogical philosophies (Hartman et al., 2019). Even in resource-constrained environments, educators can facilitate integration by leveraging creativity, building collaborative networks, and advocating for institutional support. Demonstrating incremental successes often strengthens the case for broader stakeholder investment in educational technology (Education and Career News, 2021).

Cross-Disciplinary Technology Integration

Cross-disciplinary integration, also known as interdisciplinary learning, involves the strategic use of digital tools and resources across multiple subjects to promote an interconnected understanding of various topics. This approach allows teachers to create cohesive, immersive, and interactive learning experiences that transcend subject boundaries, enabling students to apply skills and knowledge from one area to another seamlessly (Chiaro, 2021).

For instance, take the theme of "sustainability." In a cross-disciplinary model of instruction, teachers design lessons that enable students to read and analyse articles and stories on environmental issues in language arts, apply mathematical concepts to solve real-world problems related to resource allocation, investigate the impact of human activities on ecosystems in science, and analyse the evolution of environmental policies and their outcomes in social studies. Integrating technology enables students to engage with the theme through collaborative discussions, hands-on projects, and critical analysis. Students develop a deep, multifaceted understanding of sustainability from both local and global perspectives (Borda, Haskell, & Boudreaux, 2020).

Several digital platforms have been successfully used to facilitate cross-disciplinary learning, and the following sections will explore some of these examples in detail.

Language Arts: Utilising E-books and Interactive Storytelling Platforms

Language Arts educators can significantly enhance their curriculum by incorporating e-books and interactive storytelling platforms, as these resources provide engaging, multimodal content that improves reading comprehension and supports literary analysis. Here are some practical steps for implementation:

- **Select age-appropriate e-books:** Choose titles that align with specific learning objectives. Platforms such as Amazon Kindle, Google Play Books, and Apple Books offer diverse selections with features like text-to-speech, translation tools, and read-aloud capabilities, which encourage interactive engagement.
- **Utilise multimedia elements:** Leverage audio, video, and interactive graphics to make reading more engaging. Platforms like Epic!, Oxford Owl, and Storyline Online offer interactive books that support learning through these elements. Teachers can:
    - Organise reading sessions.
    - Integrate quizzes.
    - Incorporate culturally diverse books to broaden students' perspectives.
- **Incorporate annotation tools:** Use e-book readers with annotation features to help students deepen their understanding of the material.
- **Enhance creativity through interactive storytelling:** Platforms like Twine and Inklewriter allow students to create and illustrate their own stories, fostering creativity and narrative skills.
- **Implement AI tools in storytelling:**
    - Establish clear learning objectives.
    - Provide structured guidance.
    - Encourage critical thinking and analysis.
    - Emphasise ethical use.

- Allow students to brainstorm and draft ideas independently before using AI for assistance.

- **Facilitate collaboration:** Use digital platforms like Padlet and Google Classroom to enable students to share thoughts, ask questions, and engage in discussions, creating a dynamic learning environment.

Table 5.1. Recommended List of Platforms for Selecting E-Books for Language Studies Across Grades.

Category	Platform	Description	Key Features
General Platforms for All Grades	Amazon Kindle	Offers a wide range of e-books for all grade levels	Text-to-speech, dictionary, translation
	Google Play Books	An extensive collection of e-books for various educational levels	Highlights, notes, read-aloud
	Apple Books	A large library of e-books suitable for different age groups and educational needs	Read-aloud, interactive books
Specific Platforms for K-12	Epic!	Free for educators and offers over 40,000 high-quality children's books	Read-to-me, audiobooks
	Raz-Kids	Interactive e-books for kids with quizzes to assess comprehension	Quizzes, student progress tracking
	TumbleBookLibrary	Animated talking picture books for elementary students	E-books, read-alongs, graphic novels, videos, audiobooks
Middle School and High School	OverDrive/Sora	Provides access to e-books and audiobooks through school libraries	Customizable reading experiences, offline access
	Project Gutenberg	Free access to a vast collection of classic literature suitable for secondary education	No registration required
	Kobo	Extensive e-book library, including academic resources and popular titles	Integrates with various devices, personalized recommendations

Language Arts and Literature	Hoopla	Digital media service with a wide selection of e-books, audiobooks, comics, and more	Available through local libraries, no waitlists
	Libby by OverDrive	User-friendly app for borrowing e-books and audiobooks from local libraries	Easy access to library collections
	Scribd	Subscription-based service offering a broad range of e-books, audiobooks, and articles	Personalized recommendations, offline access
Specialized Language Learning	Book Creator	Allows students and teachers to create interactive e-books	Multimedia integration, ideal for language projects and storytelling
	Learning Ally	Audiobooks specifically designed for students with learning disabilities	Focuses on improving comprehension and literacy skills
Academic and Research	JSTOR	Digital library for academic journals, books, and primary sources	Ideal for advanced language and literature studies
	Questia	Offers access to books and journal articles for high school and college students	Research tools like notes, highlights, citations
Younger Students	Storyline Online	Streams videos of actors reading children's books aloud	Enhances comprehension and literacy through visual and auditory storytelling
	Oxford Owl	Free e-book library for primary school students	Range of reading levels and genres, interactive activities
Diverse Learning Needs	Books That Grow	Provides levelled e-books that adjust to the reader's abilities	Ideal for differentiated instruction and inclusive classrooms
	Bookshare	Accessible e-books for students with dyslexia, blindness, and other reading barriers	Free for qualified U.S. students and schools

Mathematics: Implementing Dynamic Geometry Software and Virtual Manipulatives

Integrating dynamic geometry software and virtual manipulatives can significantly enhance mathematics instruction by enabling students to visualise and grasp complex concepts more effectively. Here are practical steps for teachers:

- **Select appropriate tools:** Consider platforms like GeoGebra, Desmos, or Geometer's Sketchpad. When choosing, account for:
    - Versatility and coverage: GeoGebra covers a wide range of topics and offers community resources, saving preparation time.
    - Real-time feedback: Desmos excels in engaging students with immediate feedback.
    - Accessibility and cost: GeoGebra is free, making it accessible to all students, while Geometer's Sketchpad requires a purchase, potentially limiting access.
- **Incorporate virtual manipulatives:** Use resources like the National Library of Virtual Manipulatives to visually represent mathematical concepts. These tools:
    - Simulate physical manipulations.
    - Offer animations and interactive feedback.
    - Provide the ability to save work, adaptable for both classroom and remote learning.
- **Design interactive lessons:**
    - Integrate tools like GeoGebra in geometry lessons to explore properties of triangles or relationships between angles and sides.
    - Encourage independent or group experimentation with the software to foster exploration and discovery.
    - Guide students with specific tasks or questions to ensure that learning objectives are met.
- **Assess students' understanding:**
    - Utilise built-in assessment features or create custom quizzes within these tools.
    - For example, ask students to demonstrate how altering a triangle's side length affects its angles to reinforce learning and provide feedback on progress.

Science: Leveraging Simulation Software and Virtual Labs

Simulation software and virtual labs have transformed science education by enabling students to conduct experiments and explore phenomena within a controlled environment. Here are actionable steps for teachers:

- **Select reputable simulation software:** Consider tools like PhET Interactive Simulations, Gizmos, or Labster, which offer a wide range of simulations across various scientific topics. These platforms:
  - Eliminate the need for expensive lab equipment, making science education more accessible, especially in developing regions.
  - Support personalised learning by accommodating diverse paces and styles.
- **Integrate virtual labs into the curriculum:** Use platforms like Labster to complement hands-on experiments with immersive 3D simulations, enriching the learning experience.
- **Design inquiry-based activities:**
  - Utilise simulation software to help students investigate scientific questions. For example, PhET Interactive Simulations allows students to explore the laws of motion by manipulating variables, observing outcomes, and analysing data.
  - Facilitate experimentation to develop students' critical thinking and problem-solving skills through structured activities, guided by questions or worksheets.
- **Assess and reflect on student progress:**
  - Use the built-in assessment tools provided by the software or create custom criteria.
  - Encourage students to reflect on their experiments and engage in discussions through platforms like Google Classroom to enhance understanding and promote deeper engagement with the material.

Table 5.2. List of Simulation Software and Virtual Labs for Science Education.*

Category	Software	Description
General Science	PhET Interactive Simulations	Free interactive math and science simulations covering physics, chemistry, biology, earth science, and math.
	Labster	Virtual labs for biology, chemistry, physics, and engineering with gamified learning experiences.

	ChemCollective	Virtual labs and scenarios for teaching and learning chemistry in a simulated environment.
	Beyond Labz	Virtual labs in chemistry, physics, and biology emulating real lab experiences with various experiments.
Biology	BioDigital Human	Interactive 3D platform for exploring human anatomy and disease.
	Visible Body	Comprehensive 3D anatomy and physiology visualisations with interactive models and simulations.
	Froguts	Virtual dissection software focusing on frog dissection and other organisms.
Chemistry	Molecular Workbench	Interactive simulations for teaching molecular science concepts and visualising molecular dynamics.
	ChemReaX	Chemical reaction simulator for studying kinetics, equilibrium, and thermodynamics.
	Virtual Chemistry Laboratory	An online platform for performing virtual chemistry experiments and exploring various concepts and techniques.
Physics	Algodoo	2D physics simulation software for creating interactive simulations and exploring physics principles.
	Interactive Physics	Physics simulation software for visualising and exploring physical phenomena with custom experiments.
	Physics Classroom	Collection of interactive simulations and animations covering topics from kinematics to electromagnetism.
Earth Science	Earth and Moon Viewer	Interactive simulations for visualising the Earth and moon with real-time data and various viewing options.
	Virtual Earthquake	Simulation for understanding earthquakes and seismology, including locating epicentres and measuring magnitudes.
	Climate Interactive	Simulates the impact of climate change and policies with interactive models for exploring climate science.
Environmental Science	EcoMUVE	Virtual ecosystem simulations for exploring ecological concepts through immersive, problem-based learning.
	Waterbear	Interactive environmental science simulations focusing on topics like water cycles and ecosystems.

	GLOBE Observer	Allows students to participate in citizen science projects with tools for observing and recording environmental data.
Astronomy	Stellarium	Planetarium software for simulating the night sky, exploring stars, constellations, and planets.
	Celestia	Real-time 3D visualisation of space, including a comprehensive catalogue of stars, planets, and spacecraft.
	Universe Sandbox	Interactive space simulator for modelling and exploring the cosmos, including gravity, collisions, and climate changes.

Social Studies: Using Digital Archives and Virtual Reality

Enhance social studies education by incorporating digital archives and virtual reality (VR) technologies to bring history to life and deepen students' cultural understanding. Here's how teachers can implement these tools:

- **Leverage digital archives:** Use platforms like the Library of Congress, Smithsonian Digital Collections, and Europeana to access rich primary source materials, such as documents, photographs, and artefacts. These resources can significantly enrich classroom lessons by providing authentic historical content.
- **Incorporate virtual reality for immersive learning:** Engage students with history through VR platforms like Google Expeditions, VR Edu, or TimeLooper. These tools allow teachers to:
  - Transport students to historical sites and significant events, making history more vivid and tangible.
  - For example, pair a virtual tour of ancient Egypt's pyramids with an analysis of primary documents to create an engaging and comprehensive learning experience.
- **Encourage critical analysis:** Use digital tools for annotating and discussing primary sources and artefacts. This practice helps students develop essential historical inquiry skills.
- **Promote collaborative projects:**
  - Facilitate group work where students use digital archives to research historical events or cultural practices.
  - Guide students in creating multimedia presentations to culminate their projects, fostering teamwork, communication skills, and deeper subject understanding.

Table 5.3. Software for Transforming Social Studies Education.*

Category	Platform	Description
General Social Studies	Google Earth	Satellite imagery, maps, and 3D terrain for exploring geographical and historical landmarks.
	National Geographic Kids	Educational games, videos, and articles related to geography, animals, and cultures.
	Kahoot!	An interactive quiz platform for creating social studies quizzes and interactive learning.
History and Cultural Studies	TimeMaps	Interactive historical maps showing the rise and fall of civilisations and historical events.
	Historypin	Global archive of historical photos, videos, and audio recordings.
	World History Atlas	Interactive atlas offering detailed historical maps and timelines.
	Digital Public Library of America (DPLA)	Access to millions of photographs, manuscripts, books, and more from U.S. libraries, archives, and museums.
Geography	ArcGIS Online	Comprehensive mapping and analysis platform for creating and sharing GIS data.
	Gapminder	Visualises world statistics to show global development trends.
	Scribble Maps	Create custom maps and visualise geographic data.
Civics and Government	iCivics	Educational games and lesson plans about civics and government.
	OER Commons Civics Collection	Open educational resources for teaching civics and government.
Economics	EconEdLink	Interactive tools and resources for teaching economics and personal finance.
	Khan Academy Economics and Finance	Comprehensive lessons and interactive exercises on economics and personal finance.
Additional Tools and Resources	PBS LearningMedia	Free, standards-aligned videos, interactives, and lesson plans for social studies.

	Smithsonian Learning Lab	Access to digital resources from the Smithsonian's museums, research centres, and libraries.
	NewseumED	Educational resources on media literacy, journalism, and First Amendment studies.

Cross-disciplinary technology integration offers a powerful way to enhance education by making learning more interactive, engaging, and relevant. By utilising e-books and interactive storytelling platforms in language arts, dynamic geometry software, and virtual manipulatives in mathematics, simulation software, and virtual labs in science, and digital archives and virtual reality in social studies, teachers can create enriched learning experiences that prepare students for the complexities of the twenty-first century.

Strategies for Engaging Students

Integrating technology into the classroom significantly enhances student engagement by creating more interactive and personalised learning experiences. This allows students to explore concepts through dynamic tools and receive real-time feedback tailored to their individual learning styles. In this context, I explore three field-tested strategies for incorporating technology: interactive learning environments, flipped classroom models, and project-based learning. Each of these approaches includes practical steps and examples designed to help teachers effectively implement these strategies in their classrooms.

Interactive Learning Environments

Interactive learning environments leverage technology to create engaging and personalised educational experiences, with gamification and adaptive learning platforms as key components. Gamification involves applying game-like elements—such as points, badges, and leaderboards—to educational activities, motivating students through rewards and challenges. This approach enhances participation, retention, and a sense of achievement, leading to more immersive learning. For instance, a history teacher might use Classcraft to transform classroom activities into a role-playing game, where students earn points for assignments, discussions, and mastering historical concepts.

To implement gamification effectively, align game elements with educational goals, create a motivating reward system, and integrate educational games like Kahoot! for quizzes and reviews.

Adaptive learning platforms, like DreamBox and i-Ready, personalise education by adjusting content based on each student's needs and performance. These platforms use data analytics to tailor lessons and assessments, ensuring students receive the appropriate level of challenge and support. In practice, a math teacher might use DreamBox to provide customised math practice, with the platform adjusting problem difficulty to match each student's progress. Select a platform that aligns with your curriculum, monitor student progress, and differentiate instruction based on the platform's recommendations.

Flipped Classroom Models

Flipped classroom models reverse traditional learning structures by having students first engage with new material outside of class—typically through videos, readings, or interactive lessons—allowing classroom time to focus on hands-on activities, discussions, and problem-solving. This approach promotes active learning, deeper understanding, and greater engagement.

To implement a flipped classroom, teachers can create or curate content using tools like Screencast-O-Matic or Edpuzzle to produce or annotate video lectures. Assign these videos as homework, ensuring students understand they must watch them before class. In-class activities should then be designed to apply the concepts from the videos, such as group discussions, problem-solving tasks, or projects.

For instance, a science teacher might assign a video lecture on the phases of the moon as homework. In class, students could collaborate to create models of the lunar phases and discuss their observations. Similarly, teachers can provide access to articles, simulations, and interactive modules to support self-paced, independent study. Platforms like Google Classroom can be used to facilitate online discussions, allowing students to explore supplemental materials and engage in deeper analysis before class discussions.

Project-Based Learning

Project-based learning (PBL) is an instructional approach where students engage in real-world projects, fostering collaboration, critical thinking, and creativity. By working on complex, open-ended questions, students apply knowledge from various subjects to develop solutions or create meaningful products. PBL emphasises student-driven inquiry, hands-on learning, and practical application, making education more relevant and engaging.

PBL begins with challenging questions that require deep research and critical thinking. Students explore these questions and develop final projects—such as presentations, proposals, or chapters—that connect classroom knowledge to real-world contexts. This method encourages autonomy, allowing students to pursue projects that interest them, thereby enhancing engagement.

To implement PBL, teachers should instruct students in digital literacy, teaching them how to evaluate sources and conduct effective online research.

Providing research tools like Google Scholar and digital libraries supports this process. Collaboration is facilitated through platforms like Google Workspace or Microsoft Teams, with regular progress checks to ensure teamwork. For presentations, students can use tools like PowerPoint or Prezi, incorporating multimedia elements to enhance creativity. For instance, in social studies, students might research historical events using online archives for a group project, or in environmental science, collaboratively write a report on local issues using shared documents.

By leveraging interactive learning environments, flipped classroom models, and project-based learning, educators can significantly enhance student engagement and cater to diverse learning needs. These strategies not only make learning more interactive and personalised but also equip students with critical skills necessary for their future academic and professional success.

However, while these approaches are crucial for individual learning, fostering collaboration among students is equally important in today's interconnected world. Effective collaboration nurtures essential skills such as communication, teamwork, and problem-solving. The following section will delve into field-tested approaches for integrating technology to foster collaboration, focusing on the use of collaborative platforms, peer learning, and global classrooms. These methods will help educators create a more interactive, cooperative, and inclusive learning environment.

Fostering Collaboration

In today's educational landscape, fostering collaboration among students is critical for developing essential skills such as communication, teamwork, and problem-solving. Technology has transformed traditional methods of teaching, providing innovative tools that enhance collaborative learning experiences. I shall now explore field-tested approaches for integrating technology into the classroom to foster collaboration, focusing on collaborative platforms, peer learning, and global classrooms.

Collaborative Platforms

Collaborative platforms like Google Classroom, Microsoft Teams, and Learning Management Systems (LMS) such as Canvas or Blackboard are crucial for enhancing communication and teamwork. These tools allow students to collaborate effectively, even remotely.

Google Classroom simplifies class management by enabling teachers to create classes, distribute assignments, and foster peer interaction through discussion tools. Microsoft Teams offers video conferencing, file sharing, and real-time document editing, with channels for structured communication.

LMS platforms like Canvas and Blackboard support group work, resource sharing, and peer feedback, helping students develop critical skills.

These tools create a dynamic learning environment that promotes active participation and enhances the learning experience.

Peer Learning

Peer learning leverages students' knowledge to support one another's learning, with digital platforms enhancing real-time collaboration and feedback.

Peer Tutoring pairs students with complementary strengths, using platforms like Edmodo and Padlet for virtual spaces to post questions, share resources, and offer explanations under teacher supervision. Group Projects benefit from tools like Google Docs and Microsoft OneNote, facilitating co-authorship and assigning roles to ensure fair participation. Real-Time Collaboration through Zoom or Skype enables virtual study groups and breakout sessions for small group tasks.

These strategies enhance academic understanding and develop essential social and collaborative skills.

Global Classrooms

Connecting with global classrooms through video conferencing and collaborative projects broadens students' perspectives and enhances cultural awareness. Video Conferencing tools like Skype in the Classroom and Google Meet enable students to engage with peers from different countries, discussing global issues and exchanging cultural insights. For example, an English teacher might organise an international book discussion to explore diverse interpretations.

Collaborative Projects on platforms like ePals and PenPal Schools connect students worldwide on topics such as global citizenship or environmental sustainability, fostering a sense of global community. Programs like iEARN offer structured projects aligned with curricula, engaging students in real-world problem-solving on global issues. These strategies enrich learning and prepare students for a multicultural world.

Integrating technology into the classroom to foster collaboration is essential for developing twenty-first-century skills in students. Collaborative platforms, peer learning, and global classrooms provide innovative ways to enhance communication, teamwork, and critical thinking. By implementing these field-tested approaches, teachers can create a dynamic and engaging learning environment that prepares students for future success. With practical steps and thoughtful planning, educators can harness the power of technology to transform the collaborative learning experience.

Table 5.4. Digital Forums for Classroom Engagement.*

Platform	Description	Features
Google Classroom	A free web service by Google for schools, streamlining file sharing between teachers and students	Assignments, quizzes, grading, file sharing, announcements, and discussions
Edmodo	A global education network connecting learners with people and resources, functioning as a social network	Class groups, assignments, quizzes, grade tracking, and parent communications
Schoology	An LMS providing tools for managing classroom activities, supporting online learning and engagement	Course management, attendance, gradebook, discussion forums, and collaboration tools
Flipgrid	A video discussion platform where educators create "grids" for video discussions	Video responses, feedback tools, video moderation, and integration with other educational apps
Piazza	An online platform facilitating interaction among students and instructors, useful for large classes	Anonymous posting, class Q&A, wiki-style collaboration, and instructor endorsements
Canvas	An LMS with robust discussion forum capabilities for academic conversations	Discussion forums, assignment submissions, grading, quizzes, and integration with educational tools
Moodle	An open-source LMS with strong support for discussion forums	Customizable discussion forums, assignment submissions, quizzes, and course management tools
Microsoft Teams for Education	Integrates with Office 365, providing a collaborative environment for discussions, file sharing, and projects	Channels for class discussions, file sharing, assignment submissions, video conferencing, and integration with Microsoft Office tools
Seesaw	A student-driven digital portfolio platform allowing documentation of learning and sharing with teachers/ parents	Digital portfolios, assignments, parent communication, and multimedia responses
Kialo Edu	A debate platform designed for educational purposes, fostering critical thinking and discussion skills	Argument mapping, collaborative debates, teacher moderation, and structured discussion formats

Ensuring Technology Supports Effective Learning Outcomes

Integrating technology into education can significantly enhance learning outcomes when implemented effectively. This chapter explores three key strategies: data-driven instruction, formative assessment tools, and blended learning models. By employing these approaches, teachers can create a dynamic and responsive learning environment that meets the diverse needs of their students.

Data-Driven Instruction

Data-driven instruction uses analytics from educational software to guide teaching and implement targeted interventions, enabling teachers to make evidence-based decisions that enhance student learning. Educational platforms like Khan Academy, DreamBox, and Edmodo provide built-in analytics tools that monitor student performance and engagement. These tools generate reports that help teachers identify learning gaps and adjust instruction accordingly. For example, if data reveals that students struggle with fractions, a teacher can focus on that topic and provide additional resources.

This method supports personalised learning by creating individualised plans based on student data. Tools like i-Ready and Renaissance Learning offer adaptive learning pathways, adjusting tasks to each student's performance, ensuring they receive the appropriate level of challenge. Regularly reviewing data, identifying patterns, and adjusting instruction based on insights helps teachers tailor their strategies to meet diverse student needs. Communicating progress with students and parents keeps everyone informed and engaged in the learning process.

Formative Assessment Tools

Formative assessment tools provide immediate feedback and track student progress, allowing teachers to make real-time adjustments to instruction. Tools like Quizizz, Kahoot!, and Google Forms enable teachers to create quizzes and polls that give students instant feedback, helping them quickly learn from mistakes. These tools also help teachers gauge comprehension and correct misconceptions early.

Tracking progress with platforms like Seesaw and ClassDojo allows teachers to maintain digital portfolios, monitor growth, and identify performance trends. Regular quizzes and polls assess understanding and guide instruction, while maintaining portfolios ensures timely interventions. Providing constructive, actionable feedback helps students improve and understand their learning journey.

Blended Learning Models

Blended learning combines traditional classroom methods with online resources to create a flexible and effective learning environment. This approach enhances student engagement by merging in-person and digital instruction, catering to diverse learning styles.

Blended learning's flexibility allows students to learn at their own pace, accessing resources anytime through platforms like Canvas, Blackboard, and Moodle. The use of multimedia resources, interactive simulations, and virtual labs on platforms like Nearpod and Edpuzzle keeps students engaged and motivated. This dynamic approach fosters a responsive educational experience, balancing digital and traditional activities, offering supplemental online materials, encouraging online discussions, and providing continuous support to students.

Integrating technology into the classroom to ensure effective learning outcomes requires thoughtful planning and execution. Data-driven instruction, formative assessment tools, and blended learning models offer powerful strategies for enhancing student learning and engagement. By leveraging analytics to inform instruction, providing immediate feedback through digital assessments, and combining traditional teaching methods with online resources, teachers can create a dynamic and responsive learning environment. Implementing these field-tested approaches will enable educators to meet the diverse needs of their students, fostering a culture of continuous improvement and academic success.

Case Studies

WorldReader

Worldreader is an international nonprofit organisation that promotes literacy by encouraging children to read at least 25 books a year. Its digital platform, BookSmart, integrates into classrooms to support cross-disciplinary technology use, engage students, and enhance learning outcomes for children ages 3-12. BookSmart provides access to thousands of books in various languages, available anytime on any device.

Leveraging mobile technology, which reaches over 90% of the global population, BookSmart offers interactive and multimedia-rich content to improve understanding and retention. The platform's mobile and e-reader access helps boost reading motivation and literacy rates with diverse, culturally relevant content.

Worldreader also fosters collaboration through tools that enable shared reading and group discussions, helping students build critical thinking and communication skills. The platform includes analytics for teachers to track student progress and tailor instruction, resulting in improved reading proficiency and academic performance.

In Ghana, Worldreader's program was integrated into the curriculum to provide additional resources that supplement traditional textbooks, enriching the learning experience and bridging gaps in educational content (Abingya, 2023). Similarly, in Nigeria, Worldreader partnered with local schools and organisations to implement reading programs encouraging students to explore new genres and topics, significantly increasing their engagement and interest in reading (Worldreader, 2018). Teachers in several West African schools reported that students who engage in group reading activities using Worldreader's digital library show improved teamwork and collaborative learning behaviours (Worldreader, 2016). The successful implementation of Worldreader's program in Ghana also led to measurable improvements in literacy rates, indicating that technology incorporation effectively supports and enhances learning outcomes (Lowe, 2020).

uLesson

uLesson was founded in Jos, Nigeria, to provide accessible and high-quality educational content to secondary school students. The app aims to bridge the educational gap in Africa by leveraging technology to offer personalised learning experiences. This mission is particularly significant in regions where access to educational resources is limited due to infrastructural challenges such as inconsistent power supply and lack of internet connectivity (The uLesson Group, 2021).

uLesson integrates into classrooms by providing comprehensive content for mathematics, science, and English through video lessons, quizzes, and interactive exercises. This cross-disciplinary approach allows for seamless transitions between subjects, helping students draw connections and apply their knowledge across different areas of study. uLesson employs various strategies to effectively engage students. The app engages students with relatable video content, regular quizzes, and gamification elements such as points and badges, which help maintain interest and motivation. It adapts to individual learning paces and styles, enhancing the learning experience.

Figure 5.1. Empowering Education: uLesson's Commitment to Accessible, High-Quality Learning for Secondary School Students.

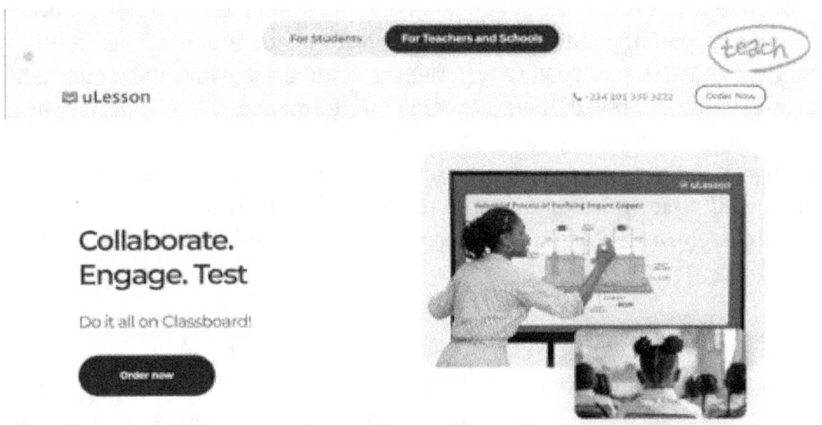

Note: uLesson provides accessible and high-quality educational content to secondary school students. Screenshot from the uLesson platform, accessed July 2024. Used under Fair Use.

While uLesson primarily offers individual learning experiences, it also fosters collaboration among students. The platform encourages group study sessions where students can watch lessons together and discuss concepts, promoting collaborative learning. uLesson provides forums where students can ask questions, share knowledge, and collaborate on solving problems, enhancing peer-to-peer interaction and learning. Educators can use the app to facilitate group activities and discussions, ensuring that students collaborate effectively and learn from each other's insights.

uLesson ensures that technology supports effective learning outcomes through several key methods. The app tracks students' progress and performance, allowing teachers to identify areas where students struggle and need additional support. This data-driven approach helps in tailoring instruction to meet individual needs. Digital assessments and quizzes provide immediate feedback to students, helping them understand their mistakes and learn from them promptly. uLesson combines traditional teaching methods with its digital resources, creating a blended learning environment that leverages the best of both worlds.

This model ensures flexibility and adaptability in teaching and learning, accommodating various learning styles and needs. uLesson's popularity soared during the COVID-19 pandemic when traditional schooling was disrupted. Launched just weeks before the lockdown in March 2020, the app quickly garnered over half a million downloads, demonstrating its critical role in ensuring continuity in education.

Despite the resumption of regular schooling schedules, uLesson continued to grow as parents and schools integrated it into their educational routines. Students have shared positive experiences, emphasising how the platform has significantly improved their academic performance. They reported using uLesson to achieve top grades and transform their academic results. The ability to revisit lessons and practice exams was particularly appreciated, as it enhanced their understanding and exam readiness. These stories underscore the app's role in fostering student engagement and academic success (uLesson, 2022).

BYJU's

BYJU'S, an innovative global edtech leader at the intersection of content, media, and technology, has revolutionised classroom learning. Originating in India, BYJU'S now has a presence in over 100 countries, serving 150+ million students who rely on it as a trusted learning partner for technology-enabled, personalised, and engaging educational journeys (BYJU's, n.d).

BYJU'S cutting-edge technology is directly integrated into the classroom environment, making learning more dynamic and effective. Teachers can capitalise on its comprehensive digital learning platform covering a wide range of subjects, and utilising advanced AI models to create personalised learning experiences tailored to individual student needs. Models like BADRI, MathGPT, and TeacherGPT help students understand and connect concepts across different subjects by customising content delivery based on each student's unique learning requirements (Dayal, 2023).

Teachers use BYJU'S in various ways to keep students engaged and actively participating in their learning process. Interactive video lessons, quizzes, and gamified learning experiences make education enjoyable and interactive, significantly boosting student engagement. The introduction of BYJU'S WIZ further enhances classroom learning by providing hyper-personalised experiences that adapt to each student's learning pace and style.

Collaborative learning is a cornerstone of BYJU'S approach. In partnership with Intel, BYJU'S has developed programs to upskill teachers, equipping them with diverse teaching techniques that promote student collaboration. The platform supports virtual group activities and discussions, allowing students to work together, share insights, and learn collectively (Elets Digital Learning, 2023).

BYJU'S technology also improves learning outcomes by providing real-time feedback and performance tracking. This data-driven approach enables teachers to identify areas where students need additional support and ensure instruction is tailored to meet individual needs. AI-driven models like TeacherGPT offer guidance and support specific to each student's learning requirements, resulting in more effective and targeted education.

BYJU'S Education for All initiative profoundly impacted underprivileged children in India. Within a year of its launch, the initiative reached 3.4 million children across 26 states, providing free access to high-quality digital learning resources. The program aims to educate 10 million children by 2025. Initial studies indicate that 75% of users engage with the app for an hour daily, and 57% attribute their improved academic performance to BYJU'S (Admin, 2022).

Virtual School Victoria

The World Bank reports that over 53% of school-age children in low- and middle-income countries are "learning poor," unable to read by age 10 (Azevedo, 2019). Addressing this issue requires flexible schooling models that adapt to students' diverse circumstances.

Virtual School Victoria (VSV) exemplifies how virtual schooling can increase enrolment and provide alternative education for those unable to attend traditional schools. VSV administers technology to offer interdisciplinary, tech-driven curricula that foster a holistic learning experience. VSV's interactive virtual platform offers personalised activities and promotes collaboration, preparing students for success in future careers. Data-driven methods monitor progress, allowing tailored instruction that improves academic outcomes.

Conclusion

Educational technology has reshaped the learning landscape, offering innovative strategies to enhance student engagement, collaboration, and learning outcomes. This chapter explored various approaches educators use, such as e-books, dynamic geometry software, simulation software, and virtual reality, to create immersive and interactive learning experiences across disciplines. These technologies, combined with pedagogical strategies like interactive learning environments, flipped classrooms, and project-based learning, make education more delightful and personalised, helping students develop critical thinking, creativity, and a lifelong love of learning. The chapter also highlighted the role of collaborative platforms, peer learning, and global classrooms in championing essential twenty-first-century skills, preparing students for success in an interconnected world. Case studies like Worldreader, uLesson, BYJU'S, and Virtual School Victoria demonstrated how technology can bridge educational gaps and empower students, especially in underserved regions.

The future of educational technology promises even greater advancements that will further transform the classroom experience. As technology redefines the boundaries of the classroom, models of educational delivery will continue to evolve and extend beyond traditional physical spaces. Educators will access new ways to tailor learning experiences to individual student needs, enhance global collaboration, and prepare students for the rapidly changing demands of the modern world. Educators continue to create dynamic, inclusive, and future-ready learning environments by staying adaptable and embracing innovation.

References

Abingya, A. (2023, September 14). Worldreader brings Ghanaian families and educators together to get children reading. *Worldreader.* https://www.worldreader.org/now/accra-digireads-experience-2022/

Admin. (2022, March 22). BYJU'S 'Education for All' initiative has impacted 3.4 million children in one year of its launch; aims. *BYJU'S.* https://byjus.com/press/byjus-education-for-all-initiative-has-impacted-3-point-4-million-children-in-one-year-of-its-launch-aims-to-educate-10-million-children-by-2025/

Baig, M. I., & Yadegaridehkordi, E. (2023). Flipped classroom in higher education: A systematic literature review and research challenges. *International Journal of Educational Technology in Higher Education, 20*(1), Article 5. https://doi.org/10.1186/s41239-023-00430-5

Borda, E., Haskell, T., & Boudreaux, A. (2020, December). *Cross-disciplinary learning: A framework for assessing application of concepts across STEM disciplines.*

BYJU'S. (n.d.). *Online learning programs for K3, K10, K12, NEET, JEE, UPSC & Bank Exams.* https://byjus.com/

BYJU'S. (2023, June 8). BYJU's launches AI suite 'BYJU's WIZ' for personalised learning. *Elets Digital Learning.* https://digitallearning.eletsonline.com/2023/06/byjus-launches-ai-suite-byjus-wiz-for-personalised-learning/

Chiaro, C. (2021, February 25). Approaches to successful cross-curricular integration. *TeachHub.* https://www.teachhub.com/teaching-strategies/2019/11/approaches-to-successful-cross-curricular-integration/

Dayal, S. (2023, June 8). Byju's WIZ GPT: Advance learning with Badri, MathGPT, and TeacherGPT. *TICE News.* https://www.tice.news/tice-trending/byjus-wiz-ai-models-badri-mathgpt-teachergpt

Deterding, S., Dixon, D., Khaled, R., & Nacke, L. (2011). From game design elements to gamefulness. *Proceedings of the 15th International Academic MindTrek Conference,* 9–15. https://doi.org/10.1145/2181037.2181040

Education and Career News. (2021, June 7). *5 ways to advocate for better technology for teachers and students.* https://www.educationandcareernews.com/empowering-our-educators/5-ways-to-advocate-for-better-technology-for-teachers-and-students/

Hartman, R. J., Townsend, M. B., & Jackson, M. (2019). Educators' perceptions of technology integration into the classroom: A descriptive case study. *Journal of Research in Innovative Teaching & Learning, 12*(3), 236–249. https://doi.org/10.1108/jrit-03-2019-0044

João Pedro Azevedo. (2019). Learning poverty measures and simulations (p. 6). *The World Bank.* https://doi.org/10.1596/1813-9450-9446

Lowe, S. J. Z. (2020, September 22). Study shows e-readers help children learn to read. *Worldreader.* https://www.worldreader.org/now/new-study-shows-e-readers-help-children-learn-to-read/

Prodigy Education. (2023, July 28). Project-based learning: Benefits, examples & classroom tips. *Prodigy Education Blog.* https://www.prodigygame.com/main-en/blog/project-based-learning/

Ramakrishnan, S. (2024, June 12). How adaptive learning platforms revolutionise L&D in 2024. *Disprz.* https://disprz.ai/blog/adaptive-learning-platform-overview

uLesson. (2022, November 23). Inside Africa: Nigeria's uLesson makes learning easier [Video]. *YouTube.* https://www.youtube.com/watch?v=_Bhep5YsYyU

Tierney, G., & Urban, R. (2023). Designing for equity: Moving project-based learning from equity adjacent to equity infused. *ICLS 2023 Proceedings.* https://doi.org/10.22318/icls2023.325936

uLesson. (2021, May 28). *uLesson learners' stories.* https://ulesson.com/blog/ulesson-learners-stories/

Worldreader. (2016, September 30). *Digital reading in Ghana schools.* https://www.worldreader.org/where-we-are/ghana/

Worldreader. (2018, February 27). *Digital reading in Nigerian schools.* https://www.worldreader.org/where-we-are/nigeria/

Chapter Six

Engaging Students in History Class through Digital Humanities Methods and Tools: Good Practices and Strategies for Integrating Technology and Producing Public History

Marina Bantiou

University of Thessaly, Greece

Abstract

This chapter examines how integrating digital humanities methods and tools into history courses enhances student engagement and promotes active learning. Digital humanities, an interdisciplinary field that combines new technologies with the study and production of history, offers innovative approaches for students to interact with historical content. By undertaking public history projects, students actively create and share historical narratives, fostering public participation and critical thinking. This approach develops historical consciousness and a deeper understanding of historical contexts. The chapter highlights practical, field-tested strategies for incorporating digital humanities tools such as data visualisation, interactive mapping, podcasting, and digital archives into history education. A review of literature and case studies from diverse educational settings provides a theoretical framework and identifies effective practices. The findings suggest that students, by acting as historians and content creators, achieve a richer connection to history, improving their learning outcomes. This study offers valuable insights for educators aiming to enhance history teaching through digital methods.

Keywords: History education, digital humanities, public History, historical consciousness, digital History.

***

Introduction

For over a century, most history teachers have predominantly relied on traditional methods of lecturing and rote memorisation, with only a few making efforts to revise their pedagogy to foster greater student engagement and learning (Wiersma, 2008). The integration of Digital Humanities methods and tools in history education represents a dynamic shift from traditional pedagogies to more interactive and engaging approaches. It has become a crucial area of exploration in contemporary pedagogy, as it offers innovative approaches for students to interact and engage with historical content. Digital humanities is defined as an interdisciplinary field that leverages digital technologies and quantitative methods to advance research in the humanities (Battershill & Ross, 2022). Digital humanities involve the integration of digital tools and methods into humanities research, allowing for the exploration and representation of cultural objects in innovative ways while addressing the challenges posed by the abundance of digital information and the necessity of encoding knowledge for computational processing (Berry, 2012). Digital humanities closely intersects with the field of Public History, which focuses on engaging the wider public in the interpretation and presentation of historical narratives by providing the digital tools and platforms necessary to create interactive, widely accessible historical content. Public history, which emerged in the United States and Canada and expanded internationally, is a practice-oriented discipline that aims to engage diverse audiences in interpreting and presenting historical narratives, promoting inclusivity and openness (Demantowsky, 2018). Defined as a complex identity discourse, public history involves collectives and individuals negotiating narratives within institutional contexts. Battershill and Ross (2022) underscore the significance of digital tools in advancing history education by providing novel methods for presenting and analysing historical content. They emphasise the effectiveness of these tools in fostering active learning by engaging students in the production of digital content that functions simultaneously as a learning mechanism and as a means of generating public history outcomes.

Public history encourages interest, public participation, and emotion, as its presentation to the general public prompts the generation of questions and the development of critical thinking, contributing to the formation of historical consciousness. By producing public history projects, students become active participants, creating and sharing historical narratives with a wider audience. Facilitated by digital tools, these projects, which are examined as case studies, encourage collaboration, critical thinking, and a deeper understanding of historical contexts. Additionally, the integration of tools and methods from the innovative field of Digital Humanities in the classroom can enhance historical learning, enabling students to actively engage not only in research (Pei-Yu et al.,

2023; Harman, 2018) but also in the production of public history. By incorporating digital methods and tools, such as recording and transcribing oral testimonies, visualizing data, creating audio-visual presentations via Tableau and Google Data Studio, interactive mapping through StoryMapJS, podcasting, creating audio historical tours of local monuments, and creating publicly accessible digital archives, students are actively involved in the process of producing historical sources and digital archives. Borgman (2009) argues that the humanities should adopt the sciences' approach by integrating initiatives that enable students to engage with and utilise data from an early age, ideally beginning in primary education. Early exposure to exploring cultural records and constructing narratives can make the study of the humanities more engaging, while also equipping students with collaborative skills and familiarity with distributed tools and sources by the time they reach higher education. By fostering skills in data analysis and collaborative work from an early age, the humanities could bridge the gap between traditional pedagogies and the demands of research, thereby ensuring that students are capable of both critical thinking and digital fluency.

This chapter is grounded in extensive research and practical experience in the intersecting fields of history education, public history, and digital humanities, alongside an analysis of diverse case studies drawn from different educational settings worldwide. It aims to explore innovative strategies for enhancing student engagement in history classes by integrating digital humanities methods and tools. By examining both theoretical frameworks and practical applications from diverse case studies, the chapter seeks to offer new approaches to teaching history that move beyond traditional pedagogies, focusing on methods that foster active participation, critical thinking, and a deeper understanding of historical contexts. Case studies were selected from a diverse range of educational contexts, including primary and secondary education institutions from various countries. The selection criteria were based on their demonstrated success in enhancing student engagement, fostering critical thinking, and promoting active learning through the use of digital tools and methods in history class. Each case study was analysed to identify the specific strategies employed, the digital tools utilised, the pedagogical approaches adopted, and the outcomes achieved. By drawing on diverse examples from different countries and contexts, this chapter highlights successful strategies that educators have used to incorporate digital tools, such as data visualisation, digital mapping, podcasting, and the creation of online archives, into the history curriculum.

These methods have proven effective in transforming traditional history teaching practices by encouraging students to become active participants in the creation and dissemination of historical knowledge. To further validate the findings, the study proceeds to a comparative analysis of these case studies, examining the commonalities and differences in approaches across different educational settings and cultural contexts, which enables the identification of good practices that can be adapted to suit various educational levels and history curricula worldwide. By addressing the challenges and opportunities presented by digital humanities, the chapter contributes to the ongoing discourse on effective history education in the digital age, aiming to empower educators with the resources and strategies needed to adapt to evolving pedagogical demands.

Practical Applications of Digital Humanities Methods and Tools in History Education

Digital humanities methods allow students to explore history in more interactive and participatory ways, being co-creators of historical knowledge, encouraging them to think critically about the historical sources they use, the narratives they construct, and the perspectives they present. The increasing availability of open-access digital archives and resources and free, user-friendly online platforms to create historical content facilitates the process for educators to integrate digital humanities methods into the history curriculum. To further illustrate the impact of digital humanities in history education, this section examines several case studies where these tools have been successfully integrated.

Digital mapping, through the creation and use of maps on digital platforms for visualising, analysing, and interacting with geographic data, is an interesting method of engaging students in the process of learning history. Digital mapping platforms, being highly user-friendly and versatile for embedding diverse data types such as images, videos, links, and public commentary, are popular tools in digital humanities pedagogy, making them ideal for introductory student assignments that focus solely on creating interactive, multidimensional maps before incorporating more extensive research or written components, allowing educators to first assess the quality of student work and refine instructional materials accordingly (Battershill & Ross, 2022). For example, in a project conducted by early young adolescents at an independent school in South Seattle, students were engaged in digital mapping, where they researched and represented local urban sites of historical significance on an interactive mapping web platform, as part of a broader study of Seattle's cultural history (Mitchell & Elwood, 2012).

The platform enabled students to create and enhance interactive maps using common GIS elements (visual layers, points, lines, symbols, data tables, etc.), while incorporating multimedia content such as photos, text, videos, interviews etc. and engaging in collaborative learning through comments and discussions, thereby building a continuously evolving record of their learning.

Through this project, students became more engaged with local history, fostering a deeper understanding of their community's historical actors and a greater interest in civic participation. Researchers in a U.S. city used GIS technology in a high school unit to help students explore local history and cultural identity by creating a digital cultural heritage map (Curby et. al., 2023). The study found that by using geospatial tools to visualise geographic concepts and collect field data, students were able to construct narratives around their cultural identities, suggesting that geospatial technologies are flexible tools for teaching content across various levels.

Digital storytelling is defined as "the practice of using digital tools to construct and share creative writing pieces (whether fiction or nonfiction) through text, image, sound, video, or (better yet) a combination thereof", offering a multimedia alternative to traditional essays that allows students to express their ideas in more dynamic ways by producing formats as podcasts, blogs, videos, etc. (Battershill & Ross, 2022). In history education, this can take the form of creating a narrative around a specific historical event, person, or period. Tools like StoryMapJS, TimelineJS, WeVideo, Audacity, Windows Movie Maker, or Canva allow students to combine text, images, audio, and video to create compelling digital stories that illustrate historical narratives in an engaging way. For example, a study in Hungary explored the use of digital storytelling in history classes with sixth- and tenth-grade students to assess its impact on learning and engagement (Lanszki, 2022). Over four weeks, students used digital tools to create narratives centred on historical figures or events from a personal perspective, integrating text, images, audio, and video, which helped enhance their reading, writing, and critical thinking skills while fostering a personal connection to history.

Oral history projects involve collecting and preserving firsthand accounts of historical events from individuals who experienced them. Digital tools for transcription, audio and text editing in order to create digital outcomes provide students with the means to record, transcribe, and publish these oral histories. Integrating oral history projects and digital humanities methods into educational settings provides students with immersive learning experiences that bridge historical research and public engagement. For example, a Year 4 class at Tardebigge VA First School in Worcestershire conducted an oral history project by interviewing 12 former students from the 1920s to the 1960s, and these interviews were then transformed into a community play and a radio program, featuring extracts from the students' oral histories, which was based on both the oral histories and research from the school's log books and registers; this approach illustrates how student-generated content can create a profound connection to local history while engaging the broader community (Letts & Lloyd, n.d.).

Similarly, Year 5 pupils collaborated with a local museum to interview carpet workers, with excerpts from their interviews featured in an interactive audio installation that enriched the museum's exhibits with authentic, student-produced narratives, and another notable example in secondary education involved students who, after interviewing relatives about life in wagons, scripted a digital audio tour for a local museum housing gypsy wagons, effectively integrating oral history with museum curation and interpretation. These projects collectively underscore the potential of oral history and digital humanities methods in educational contexts. They demonstrate how students can develop critical research skills, engage with historical content creatively, and contribute to public history through various digital mediums. The "Telling Their Stories: Oral History Archives Project (OHAP)" at the Urban School of San Francisco exemplifies an innovative pedagogical model in which high school students actively engage in the process of documenting and publishing oral histories from elders who experienced significant twentieth century events, such as the Holocaust, Japanese American internment, and the Civil Rights Movement, through training in oral history methods such as video recording, transcription, editing, and digital archiving on a public website (Levin, 2011). This approach not only integrates traditional historical inquiry with digital humanities tools but also fosters deep student engagement with primary sources, cultivates critical thinking and technical skills, and strengthens the connection between students and lived historical experiences, suggesting a replicable framework for enhancing history education across diverse settings. Additionally, engaging students in activities that involve exploring digital archives and repositories critically and reflectively helps them understand the evolving forms and norms of textual production and dissemination (Battershill & Ross, 2022).

The creation of digital archives by students is an emerging practice in history education that offers significant benefits in developing both historical understanding and digital literacy. By engaging in the process of building digital archives, students move beyond passive learning and become active participants in the preservation and interpretation of historical knowledge. They collect, organise, and curate primary source materials, such as oral histories, photographs, letters, and documents, thereby gaining firsthand experience in historical research methods. This hands-on approach encourages deeper engagement with the content as students are tasked with critical evaluation, contextualization, and the ethical considerations of working with diverse historical sources. Engaging students in the creation of metadata using platforms such as Omeka enables them to understand the connections between digital objects within archives and promotes consideration of how metadata affects the interpretation of data, its context and wider systems, power, and technology they represent.

The evolving role of museums as agents of public history in the digital age, highlighting their significant influence on education and historical understanding, needs to be discussed, exploring how they incorporate digital technologies to enhance public engagement and extend their educational reach. Of interest is the case of a three-month project at a Washington, DC school with eighth-grade students exploring how identity formation is understood and represented through public institutions such as museums (Trofanenko, 2014). Students examined the role of museums in shaping personal and collective identities, incorporating social media and web technologies into their learning along with museum visits and created a five-minute video mash-up, combining digital archival material, music, altered photographs, and exhibition objects, allowing them to articulate their own meta-narratives of how the museum's exhibits represent US collective identity and its connection to war and military involvement. Additionally, they used social media and web-based technologies, such as Google, YouTube, and Flickr, to gather supplementary information and enhance their understanding of history, and accessed online collections from institutions as the Smithsonian, the Library of Congress, and the National Archives to support their research. Students noted that while the objects are important, they only offer partial information and can lead to passive learning, as opposed to more engaging and insightful exploration of history through digital humanities methods and tools, making history in the classroom more interactive.

Good Practices and Challenges

Given that many young people engage with history through public platforms such as museums, historic sites, and digital resources, it is crucial to teach them how to critically interpret public history (Kelly, 2018). However, there is a significant gap in curricula globally, where public history skills are often neglected, leaving students unprepared to discern the intersection of history and politics in these public spaces. To address this educational deficit, incorporating digital humanities methods and tools into history education offers innovative ways to both enhance engagement and promote critical interpretation skills. After examining indicative specific case studies and pedagogical strategies, the following discussion delves into good practices for integrating digital humanities methods and tools into history education, exploring how these approaches can enhance student engagement and understanding, but also into challenges that educators may encounter when implementing these methods.

Across the above case studies, several clear commonalities in the application of digital humanities tools to history education are observed. First, digital tools offer interactive platforms that enhance students' active participation in the learning process. In each case, the use of technology enables students to move beyond passive learning by researching, creating, and disseminating historical narratives. In addition to promoting engagement, the digital humanities projects in these case studies emphasise critical thinking. Whether through mapping, digital storytelling or oral history projects, students are meant to evaluate and interpret historical sources and data, encouraging them to reflect on the meaning of the information they encounter, fostering a critical engagement with the past. Finally, another common feature is the use of collaboration in these projects. Many of the case studies highlight how digital tools encourage teamwork, allowing students to collaborate on research and data visualisation.

While commonalities provide a basis for understanding the effectiveness of digital humanities in history education, significant differences in application and context also emerged in these case studies. A notable difference is the level of complexity and technological access involved in these projects. For example, some oral history projects were technically more complex than simple digital storytelling tasks, requiring students to develop advanced technical skills such as digital archiving, transcription, and public dissemination. These projects often involved specialised software and tools, requiring a higher level of digital literacy and access to robust technology infrastructure. In contrast, digital storytelling and digital mapping projects were observed to involve fewer technical requirements, focusing primarily on the creative expression of historical narratives through media elements such as video, images, and text. This difference highlights the fact that some projects aimed to build, along with history knowledge, technical proficiency, while others prioritise creativity and historical narrative construction, making them more adaptable to different technological and school environments.

Finally, there are significant differences in how public participation is incorporated into these projects. Some projects, particularly those involving oral history, directly involve the community by inviting local participants or publishing the results in a public space, thereby creating a tangible connection between students and the wider public. These projects are often designed to contribute to public history efforts by making historical narratives accessible to a wider audience. In contrast, other projects were more internally focused, with the classroom or school community as the primary audience. This variation reflects different educational goals: some projects aim to cultivate public participation and civic engagement, while others focus more on individual learning and creative expression within the school environment.

Specifically, interactive maps provide modest learning benefits over traditional paper maps, particularly for conceptual learning tasks such as finding, comparing, and explaining geographic information, but are less effective for memorising facts or evaluating simple patterns (Taylor & Plewe, 2006). Digital mapping tools such as StoryMapJS, ArcGIS, and Google Earth have proven effective in helping students visualise historical events, for example, migration patterns and understand the geographic contexts of history. Educators can start by conducting a comprehensive tutorial on the chosen mapping tool to familiarise students with its functionalities and then assign projects that involve mapping historical events, and ensure that students integrate primary sources from online digital collections into their work. In addition, using these tools in collaborative projects enhances peer discussions and critical analysis, and promotes team and communication skills. Encouraging students to undertake local history projects, educators can help them map community landmarks or historical sites, integrating multimedia elements such as photos, videos, and oral histories to enrich their understanding.

Another effective approach is digital storytelling, which encourages students to use multimedia elements like images and sounds in their work, which can make history more engaging and enjoyable (Guiliano, 2022). However, students must also be intentional about their choices, considering how each element contributes to their historical narrative and its impact on the audience and the story being told. Also, it is highlighted that digital storytelling can improve students' core competencies, though challenges such as limited digital resources and time constraints can affect its implementation (Lanszki, 2022). Digital storytelling, by integrating multimedia elements, serves as a powerful educational tool that enhances a wide range of skills and literacies in students: not only supports traditional educational objectives but also fosters collaborative and student-centred learning, improves various language and critical thinking skills, and bridges cultural gaps (Shahid & Khan, 2022).

Additionally, podcasting tools enable students to produce history-themed podcasts, where they can conduct interviews or narrate historical events, developing at the same time their research and storytelling skills. However, Barsch's study argues that, despite the potential of digital storytelling to engage students and enhance their media skills, it does not necessarily lead to a more critical evaluation of historical content in educational videos, since students focused more on the video's appearance rather than its historical reliability, suggesting that digital storytelling training alone does not improve critical media literacy in history education (Barsch, 2020). Apart from these interactive timelines, tools such as TimelineJS could be used effectively in order to help students develop chronological thinking and contribute to their historical understanding by adding primary sources and even creating their own historical multimedia-rich timelines.

Oral history projects effectively bridge historical research and public history by engaging students in collecting, transcribing, and disseminating firsthand accounts. It is very important to encourage students to select topics of personal or local significance to increase their investment in the project. Using digital tools for transcription, audio editing and archiving is considered a good practice because it allows students to develop technical skills alongside historical research. Online archives and digital collections play a significant role in modern history education. Also, encouraging students to create and manage digital archives provides hands-on experience in historical research and digital literacy in order to understand the importance of metadata and the impact of digital representation on historical interpretation. Students can collect, curate, and share on digital platforms, such as Omeka, primary source documents like photographs, oral histories, and artefacts related to the historical topic they work on.

Extending on this, students can be encouraged to create their virtual museum or exhibition, enabling them to showcase their project to a wider audience. However, the implementation of oral history projects in schools and the creation of digital archives by the students themselves is a time-consuming process, which requires careful planning and constant guidance from educators. Regarding the digital archives Mazé (2014) argues that the transition to digital accessibility for oral histories at Baylor's Institute for Oral History has revolutionized the preservation, understanding, fostering deeper empathy and understanding across cultures, and global reach of oral histories, while presenting ongoing challenges related to technological adaptation, increased funding, and the need for strategic planning to balance ethical, legal, and historiographical considerations while ensuring sustainability in a resource-constrained environment.

Despite the potential for enhancing student engagement in history education, digital literacy may be a significant challenge in general, since students may lack the necessary skills to effectively utilise digital tools, which can impede their ability to engage deeply with historical content and limit their overall learning experience. Effective use of digital humanities tools requires a certain level of digital literacy among both students and educators. Educators may need additional training to effectively integrate these tools into their curricula and support students in their use. Similarly, students may need training in digital literacy skills to navigate and use these tools effectively. Also important is the question of the quality and utilisation of digital content, which must be considered. Students may face issues regarding the accuracy and reliability of the information they gather and present, making it important for educators to ensure that students practice the skill of critically evaluating digital sources. Also, integrating digital humanities methods with traditional pedagogical approaches and incorporating them into the history curriculum can be challenging, since it requires careful planning and consideration of both pedagogical goals and available resources. Finally, a critical issue in the implementation of these projects is the age and educational level of the participating students.

Projects can and should be adapted to different age groups, ensuring that the complexity of the tasks and the digital tools used are age-appropriate. For younger students in primary education, simpler tools and methods such as basic digital storytelling, interactive maps and timelines or simple data visualisation can be effective. These projects emphasise creativity and engagement while introducing students to historical thinking and digital literacy in an accessible way. Teachers working with younger students can focus on creating multimedia historical narratives that allow them to explore local history, using images, short text, and audio, encouraging an early interest in historical thinking without burdening them with complex technical tasks. For older students in secondary education, more complex digital tools can be introduced. Projects such as oral history archives or interactive mapping that require research, data collection and technical skills such as transcription, editing or archiving are more appropriate for this age group. These projects encourage deeper critical thinking, allowing students to engage more directly with historical research methodologies and public history practices. Ultimately, educators can effectively adjust digital humanities projects to suit the developmental stage and skill level of their students, ensuring that all age groups can benefit from interactive and engaging approaches to history education.

Some other limitations to integrating digital humanities tools and methods into the history course are resource limitations and technical issues. Schools often encounter significant financial barriers when attempting to acquire the requisite digital tools, software, and infrastructure. These financial constraints can impede the ability of educational institutions to fully implement digital humanities projects, as they may lack the necessary funds to purchase advanced technology or maintain up-to-date digital resources. In addition to financial limitations, technical issues pose another substantial challenge. Software glitches, compatibility problems, and other technical disruptions can interfere with the seamless execution of digital humanities projects. Such issues can hinder the learning process by causing interruptions, diminishing the effectiveness of the digital tools, and creating frustration among both students and educators. Therefore, reliable and responsive technical support becomes essential to address and mitigate these issues, ensuring that technical difficulties do not detract from the educational experience.

Conclusions

Integrating digital humanities methods and tools into history education in primary and secondary education represents a more interactive and participatory learning model. Drawing on the various indicative case studies and practical applications discussed above, digital humanities offer numerous benefits for engaging students in digital research and public history and enhancing their historical understanding. Through digital methods and tools, students are provided with opportunities to interact with historical content through innovative approaches, allowing them to become active participants in the creation and dissemination of historical knowledge. Furthermore, the expanding corpus of academic literature on digital humanities pedagogy offers a comprehensive array of examples, case studies, and best practices that can serve as a valuable resource for educators, leading to the development of effective school history projects and equipping educators with strategies to anticipate and address potential challenges. Despite the advantages of incorporating digital humanities methods into history education, significant challenges remain. Digital literacy is crucial, as both students and teachers need adequate training to use these tools effectively. Educators require training to integrate digital humanities into their curricula, while students need training to navigate and critically evaluate digital resources. In addition, the quality and reliability of digital content raise concerns, which require careful evaluation and critical analysis. Additionally, balancing the use of digital tools with traditional teaching methods requires careful planning and adaptation in order to provide a coherent and comprehensive history curriculum. In conclusion, integrating the digital humanities into history education offers promising opportunities for enhancing student engagement, promoting active learning, and deepening historical understanding.

References

Barsch, S. (2020). Does experience with digital storytelling help students to critically evaluate educational videos about history? *History Education Research Journal, 17*(1), 67–80. https://doi.org/10.18546/HERJ.17.1.06

Battershill, C., & Ross, S. (2022). *Using digital humanities in the classroom: A practical introduction for teachers, lecturers, and students.* Bloomsbury Publishing. https://doi.org/10.5040/9781350180932

Berry, D. M. (2012). *Introduction: Understanding the digital humanities. In D. M. Berry (Ed.), Understanding digital humanities* (pp. 1–20). Palgrave Macmillan. https://doi.org/10.1057/9780230371934_1

Borgman, C. L. (2009). The digital future is now: A call to action for the humanities. *Digital Humanities Quarterly, 3*(4). https://www.digitalhumanities.org/dhq/vol/3/4/000077/000077.html

Curby, A., Brown, K., Weinburgh, M., & Valverde, E. (2023). Whose story is being told? Local history, cultural heritage, and digital maps. In E. Langran, P. Christensen, & J. Sanson (Eds.), *Proceedings of Society for Information Technology & Teacher Education International Conference* (pp. 1501–1508). Association for the Advancement of Computing in Education (AACE). https://www.learntechlib.org/primary/p/222023/

Demantowsky, M. (2018). What is public history. In M. Demantowsky (Ed.), *Public history and school: International perspectives* (pp. 3–37). De Gruyter Oldenbourg. https://doi.org/10.1515/9783110466133

Guiliano, J. (2022). *A primer for teaching digital history: Ten design principles.* Duke University Press. https://doi.org/10.1515/9781478022299

Harman, E. K. (2018). The transformative power of digital humanities in teaching family history online. *Journal of University Teaching & Learning Practice, 15*(3), 1–18. https://doi.org/10.53761/1.15.3.7

Kelly, T. M. (2018). The politics of public history education. In M. Demantowsky (Ed.), *Public history and school: International perspectives* (pp. 205–211). De Gruyter Oldenbourg. https://doi.org/10.1515/9783110466133-015

Kelly, T. M. (2016). *Teaching history in the digital age.* University of Michigan Press.

Lanszki, A. (2022). *Using of digital storytelling in the teaching of history.* Historia Scholastica, 8(1), 169–200. https://doi.org/10.15240/tul/006/2022-1-008

Letts, J., & Lloyd, H. (n.d.). *Schools and young people inspiring ideas.* Oral History Society. https://www.ohs.org.uk/inspiring-ideas/

Levin, H. (2011). Authentic doing: Student-produced web-based digital video oral histories. *The Oral History Review, 38*(1), 6–33. https://doi.org/10.1093/ohr/ohr046

Mazé, E. (2014). Deconstruction without destruction: Creating metadata for oral history in a digital world. In D. A. Boyd & M. A. Larson (Eds.), *Oral history and digital humanities: Voice, access, and engagement* (pp. 145–156). Palgrave Macmillan. https://doi.org/10.1057/9781137322029_9

Mitchell, K., & Elwood, S. (2012). Engaging students through mapping local history. *Journal of Geography, 111*(4), 148–157. https://doi.org/10.1080/00221341.2011.624189

Guo, P.-Y., Li, M.-C., Chen, C.-M., & Chen, C.-Y. (2023). Inquiry-based learning with a digital humanities research platform's support to facilitate students' historical learning performance and satisfaction. In *14th IIAI International Congress on Advanced Applied Informatics* (pp. 760–761). https://doi.org/10.1109/IIAI-AAI59060.2023.00164

Shahid, M., & Khan, R. M. (2022). Use of digital storytelling in classrooms and beyond. *Journal of Educational Technology Systems, 51*(1), 603–617. https://doi.org/10.1177/00472395221112599

Taylor, W., & Plewe, B. (2006). The effectiveness of interactive maps in secondary historical geography education. *Cartographic Perspectives, 55*, 16–33. https://doi.org/10.14714/CP55.325

Trofanenko, B. (2014). Playing into the past: Reconsidering the educational promise of public history exhibits. In K. Lee (Ed.), *Pastplay: Teaching and learning history with technology* (pp. 257–269). University of Michigan Press. https://doi.org/10.2307/j.ctv65swr0.16

Wiersma, A. (2008). A study of the teaching methods of high school history teachers. *The Social Studies, 99*(3), 111–116. https://doi.org/10.3200/TSSS.99.3.111-116

Chapter Seven

Seamless Classroom Integration Strategies for Effective Teaching and Learning

Bonashri Gogoi

Tezpur University, Assam, India

Abstract

This chapter explores the concept of Seamless Classroom Integration strategies aimed at enhancing teaching and learning experiences in educational settings. The chapter delves into the importance of integrating various tools, technologies, and methodologies to create a cohesive and interactive learning environment for students. By seamlessly blending traditional teaching methods with modern technological advancements, educators can cater to diverse learning styles and engage students more effectively. The chapter begins by discussing the significance of incorporating digital tools and resources into the classroom to supplement traditional teaching approaches. It highlights the benefits of using multimedia presentations, interactive whiteboards, educational apps, and online resources to create dynamic and engaging lessons. By leveraging technology, educators can personalise learning experiences, provide real-time feedback, and foster collaboration among students. Furthermore, the chapter explores the role of active learning strategies in promoting student engagement and participation. It emphasises the importance of incorporating hands-on activities, group projects, and discussions to encourage critical thinking and problem-solving skills. By creating a student-centred learning environment, educators can empower students to take ownership of their learning journey and develop essential twenty-first-century skills. The chapter discusses the significance of differentiated instruction in meeting the diverse needs of students in the classroom. It explores various strategies such as tiered assignments, flexible grouping, and personalised learning paths to accommodate different learning styles, abilities, and interests. By tailoring instruction to individual students, educators can ensure that all learners have the opportunity to succeed and reach their full potential.

Keywords: educational technologies, integration strategies, inclusive education, student-centred learning environment.

***

Introduction

The rapidly evolving educational landscape, driven by technological advancements and a growing recognition of diverse learning needs, necessitates innovative approaches to teaching and learning. In response to these shifts, the concept of "Seamless classroom integration" has emerged as a pedagogical framework for creating dynamic, effective, and student-centred learning environments. This integrated approach involves the thoughtful fusion of traditional teaching methods, digital tools, and active learning strategies to foster a cohesive and engaging learning experience.

Grounded in the principles of student-centred learning, personalised instruction, and fostering the development of critical thinking, collaboration, and communication skills essential for success in the twenty-first century. Empirical evidence from diverse educational settings highlights the potential of seamless integration to enhance student learning outcomes and teacher pedagogical practices. It emphasises the use of technology as a tool to enhance, rather than replace, the teacher's role in facilitating learning. For instance, the Los Angeles Unified School District has implemented a comprehensive technology strategy, utilising devices and platforms to create a more engaging and responsive classroom environment (Sheninger, 2014). A middle school in New York implements a blended learning program, where students rotate between online and face-to-face instruction. The program uses adaptive software to provide personalised learning pathways for each student (Rosenberg, 2001). In addition, there were several studies that stated the importance of integrating technology into the classroom. The use of Virtual Reality (VR) not only increases student engagement but also supports experiential learning that can improve understanding and retention of material (Chang et al., 2020). One of the main strengths of Virtual Reality in education is its ability to create immersive learning experiences (McFaul & FitzGerald, 2020). For example, in history lessons, students can experience historical events through historical reconstructions of place and time, such as visiting Ancient Rome or witnessing the signing of the Declaration of Independence. This approach allows students to experience and understand the historical context in a richer and more detailed way than traditional methods. An elementary school in Canada adopted a project-based learning approach, where students work on interdisciplinary projects that integrate science, technology, engineering, arts, and mathematics (STEAM) concepts. The teacher uses digital tools to facilitate collaboration and provide feedback (Thomas, 2000). This integrated approach benefits both teachers and students. Educators gain access to powerful tools for differentiation, assessment, and communication, while students benefit from increased engagement, personalised learning pathways, and the development of essential skills for future success.

The Significance of Incorporating Digital Tools and Resources into the Classroom

Digital tools and resources are not intended to replace the vital role of teachers but rather to serve as powerful allies in creating richer, enhanced student engagement, motivation, and academic performance (Rafiq et al., 2024). Technology, when thoughtfully integrated, has the potential to transform traditional teaching methods and personalise learning in unprecedented ways.

It's crucial to understand that technology serves as a powerful ally to educators, not a replacement for the essential role they play in the classroom. The integration of digital tools should always aim to enhance, not erase, the human element of teaching. Technology can transform traditional methods by:

Amplifying Teacher Expertise

Rather than replacing teachers, technology frees them from tasks like rote instruction or repetitive grading, allowing them to focus on their strengths. A study examines the impact of a technology-enhanced professional development program on teachers' expertise and confidence in using technology, and finds that the program was effective in improving teachers' ability to design and deliver high-quality instructional materials (Blanchard et al., 2016). In essence, technology empowers teachers to work smarter, not harder, creating more engaging and effective learning experiences for all students.

Creating Blended Learning Environments

Blended learning refers to an educational approach that combines traditional face-to-face learning with online learning experiences (Hrastinski, 2019; Kumar et al., 2021). A teacher might use a dynamic online simulation to introduce a complex concept (engaging visual learning) and then lead a thought-provoking class discussion to solidify understanding (encouraging critical thinking and collaboration). For example, Virtual Reality (VR) has emerged as a technology that is changing the educational paradigm, providing new opportunities for more interactive and engaging learning (Huttar & Brintzenhofe-Szoc, 2020). With its ability to simulate realistic environments and place users in the middle of a controlled learning experience, Virtual Reality opens the door to a revolutionary approach to education. The use of

Virtual Reality not only increases student engagement but also supports experiential learning that can improve understanding and retention of material (Chang et al., 2020). One of the main strengths of Virtual Reality in education is its ability to create immersive learning experiences (McFaul & FitzGerald, 2020).

Providing Data-Driven Insights

Technology equips teachers with real-time data on student progress, allowing for more informed instructional decisions. This data can help identify struggling students early on, personalise interventions, and tailor lessons to meet diverse learning needs. The integration of Artificial Intelligence (AI) into the classroom has the potential to revolutionise the way students learn and teachers teach. AI algorithms can provide students with personalised feedback and recommendations, allowing for a more engaging and effective learning experience (Melo, 2023).

Engaging Content Delivery

Gone are the days of static lectures and textbook-driven lessons. Multimedia presentations, interactive whiteboards, and educational apps breathe life into learning, making it more dynamic and engaging. For instance, a study by Almousa et al. (2019) demonstrated the positive impact of using interactive simulations in a science classroom. The researchers found that students who engaged with the simulations showed a deeper understanding of complex scientific concepts and reported higher levels of motivation and engagement compared to traditional instruction methods.

Personalized Learning

Technology empowers educators to tailor learning experiences to individual student needs and paces. Adaptive learning platforms, such as Khan Academy, provide personalised learning pathways based on student performance, offering targeted support and practice. The global impact of such platforms is significant. In India, for example, Khan Academy has partnered with schools and NGOs to provide free, high-quality education to millions of students, bridging gaps in access and tailoring learning to individual needs.

Real-Time Feedback and Assessment

Digital tools provide immediate feedback to students, fostering a continuous learning loop. Online quizzes, polling tools, and interactive exercises allow teachers to gauge student understanding in real-time and adjust instruction accordingly. For example, a teacher using an online polling tool during a lesson can quickly identify areas of confusion and address them immediately, ensuring that all students are on track.

Fostering Collaboration

Technology transcends geographical boundaries, connecting students and educators across the globe. Online platforms, shared documents, and video conferencing tools facilitate collaborative projects, fostering communication, teamwork, and cross-cultural understanding. Imagine a group of students from different continents working together on a research project, sharing ideas, conducting virtual interviews, and co-creating presentations – all made possible through the power of technology.

By embracing the potential of digital tools and resources, educators can create learning environments that are engaging, personalised, and collaborative, preparing students for success in an increasingly interconnected world.

Active Learning Strategies for Enhanced Engagement

Active learning is an instructional approach that emphasises student-centred learning, where students are actively engaged in the learning process. It fosters critical thinking skills, collaboration, and application of knowledge, which are essential for success in the twenty-first century. There are some active learning strategies that help to enhance classroom engagement:

Student-Centeredness

Learning is focused on students' needs, interests, and experiences. By shifting the focus from teacher-centred to student-centred approaches, active learning strategies enhance engagement, motivation, and deeper understanding of subject matter. These strategies promote student engagement, foster critical thinking, and develop essential skills for the twenty-first century.

Collaborative Learning

Students engage in activities that promote critical thinking, problem-solving, and collaboration, for example, discussions and debates, group projects and collaborations, case studies and problem-based learning, simulations, and role-plays. Collaborative learning is a powerful active learning strategy that fosters engagement, communication, and problem-solving skills. Benefits include

- Group work: Encourages diverse perspectives, promotes teamwork, and develops communication skills.
- Peer teaching: Enhances understanding, builds confidence, and develops leadership skills.

- Collaborative projects: Foster problem-solving, critical thinking, and time management skills.

Examples of collaborative learning activities:

- Jigsaw method: Students work in small groups to complete a task, each contributing a unique piece of information.
- Group debates: Students engage in structured debates to develop critical thinking and public speaking skills.
- Collaborative research projects: Students work in teams to conduct research, analyse data, and present findings.

Collaborative learning activities promote active engagement, deeper understanding, and the development of essential skills, leading to enhanced academic achievement and preparation for the real world.

Hands-on Experiences

Students learn through direct experiences and applications. Some examples of active learning strategies include flipped classrooms, interactive lectures, and Virtual Reality (VR), etc. For example, Virtual Reality (VR) has emerged as a technology that is changing the educational paradigm, providing new opportunities for more interactive and engaging learning (Huttar & Brintzenhofe-Szoc, 2020). With its ability to simulate realistic environments and place users in the middle of a controlled learning experience, Virtual Reality opens the door to a revolutionary approach to education. The use of Virtual Reality not only increases student engagement but also supports experiential learning that can improve understanding and retention of material (Chang et al., 2020). One of the main strengths of Virtual Reality in education is its ability to create immersive learning experiences (McFaul & FitzGerald, 2020). Hands-on activities and projects are effective active learning strategies that foster deeper understanding and engagement. Examples include

- Experiments: Conducting science experiments, such as building bridges or creating volcanoes, to illustrate key concepts.
- Simulations: Using software or real-world scenarios to mimic real-life situations, like stock market simulations or mock trials.
- Project-based learning: Assigning students to work on real-world projects, such as designing a sustainable community or developing a marketing campaign.
- Maker education: Encouraging students to design, create, and innovate using technologies like 3D printing or coding.

- Case studies: Analysing real-world scenarios or business cases to develop problem-solving and critical thinking skills.

These hands-on activities and projects promote active learning, critical thinking, and collaboration, leading to enhanced engagement and deeper understanding of subject matter.

Critical Thinking and Problem-Solving

Active learning strategies are instructional methods that engage students in the learning process, promoting critical thinking, analytical skills, and creative problem-solving. These strategies encourage students to participate, collaborate, and think deeply about the subject matter.

Differentiated Instruction: Meeting Diverse Needs

Differentiated instruction is not a one-size-fits-all approach, but instead should be shaped to meet the needs of all students. It is a comprehensive teaching approach that intends to maximise the learning outcomes of all students in the classroom and decrease the achievement gap (Gheyssens et al., 2020; Griful-Freixenet et al., 2020). The differentiated instruction strategies in meeting the diverse needs of students in the classroom are discussed below:

Tiered assignments: Tiered instruction is grouping students for instruction based on their prior background knowledge in a given subject area (Richards & Omdal, 2007). This approach allows educators to provide a more personalised and effective learning experience for students at various skill levels.

Designing Tiered Assignments

1. Identify learning objectives: Clearly define what students should know and be able to do.
2. Assess student readiness: Determine students' current skill levels and understanding.
3. Create tiers: Develop assignments with varying levels of complexity, depth, and abstractness.
4. Tier 1: Foundational: For students who need additional support, provide scaffolding and explicit instruction.
5. Tier 2: Developing: For students who are progressing, offer moderate complexity and guided practice.
6. Tier 3: Advanced: For students who are advanced, provide challenging, open-ended tasks with minimal guidance.

Examples of Tiered Assignments

1. Mathematics: For a mathematics lesson on fractions, Tier 1 students might complete a worksheet with simple fraction problems, Tier 2 students might complete a word problem involving fractions, and Tier 3 students might design and create a real-world project involving fractions.
2. Science: For a science lesson on cells, Tier 1 students might complete a diagram labelling activity, Tier 2 students might complete a short research project on cell structure, and Tier 3 students might design and conduct an experiment to investigate cell function.
3. English Language Arts: For an English language arts lesson on literature analysis, Tier 1 students might complete a reading comprehension worksheet, Tier 2 students might complete a literary analysis essay, and Tier 3 students might design and create a multimedia presentation analysing the literature.

Flexible Grouping

Flexible grouping is a teaching strategy that involves grouping students flexibly based on their needs, interests, or skills. This approach allows for dynamic and temporary grouping arrangements that can change as students' needs and abilities evolve. Research has shown the positive effects of differentiated instruction and flexible grouping strategies (Guay et al., 2017).

Benefits of Flexible Grouping

1. Improved student engagement: Flexible grouping can increase student motivation and participation.
2. Enhanced student outcomes: Differentiated instruction and flexible grouping strategies have been shown to improve student achievement (Guay et al., 2017).
3. Increased teacher efficiency: Flexible grouping can help teachers manage instructional time and resources more effectively.

Examples of Flexible Grouping Strategies

1. Learning centres: Students rotate through learning centres, each with a different activity or task.
2. Jigsaw method: Students work in small groups, each member responsible for a different aspect of the topic.

3. Literature circles: Students work in small groups, each member responsible for a different role in the discussion.

Technology-enabled personalised learning involves using digital tools and platforms to tailor instruction to individual students' needs, abilities, and learning styles. This approach can help improve student outcomes, increase engagement, and enhance teacher efficiency.

Technology-Enabled Personalised Learning

1. Improved student outcomes: Personalised learning can lead to increased student achievement and better preparation for post-secondary education and careers.
2. Increased student engagement: Interactive and immersive learning experiences can increase student motivation and participation.
3. Enhanced teacher efficiency: Automated grading, real-time feedback, and data analytics can help teachers manage instructional time and resources more effectively.

Technology-Enabled Personalised Learning Tools

1. Learning Management Systems (LMS): Platforms like Canvas, Blackboard, and Moodle facilitate personalised learning.
2. Adaptive assessments: Tools like Kahoot, Quizlet, and Dream Box adjust difficulty levels based on student performance.
3. Intelligent Tutoring Systems (ITS): AI-powered systems like Carnegie Learning and Knewton provide real-time feedback and guidance.
4. Learning analytics: Analyse student data to inform instruction and adjust learning paths.
5. Gamification and Simulation: Engage students with interactive experiences, promoting deeper learning.

Case Studies Related to Technology-Enabled Personalised Learning

1. Arizona State University's Online Degree Program: Implemented a personalised learning platform using adaptive assessments and learning analytics, resulting in a 25% increase in student retention (ASU, 2020).
2. Chicago Public Schools' Personalised Learning Initiative: Utilised learning management systems and adaptive assessments to provide personalised learning experiences, resulting in a 10% increase in student achievement (CPS, 2019).

3. KIPP Charter Schools' Blended Learning Model: Implemented a blended learning model using intelligent tutoring systems and learning analytics, resulting in a 15% increase in student achievement (KIPP, 2018).

Examples of Successful Implementations

Khan Academy: A free online learning platform that uses video lectures, practice exercises, and personalised learning dashboards to provide tailored instruction.

Role of Assistive Technology in Supporting Students with Disabilities

Assistive technology plays a vital role in supporting students with disabilities by providing them with equal access to education and opportunities for academic success. It helps to bridge the gap between their abilities and the demands of the learning environment.

Example: Text-to-Speech Software

For example, a student with dyslexia or visual impairment can benefit from text-to-speech software like Claro Read or Natural Reader. This tool converts written text into spoken words, allowing the student to:

- Listen to texts, articles, and documents, reducing reading difficulties
- Focus on comprehension and understanding, rather than struggling with decoding words
- Complete assignments and readings with increased independence and confidence

By using assistive technology, students with disabilities can overcome barriers, build confidence, and reach their full potential.

Challenges of differentiated instruction in large classrooms

Differentiated instruction is an approach to teaching that recognises and responds to the diverse learning needs of students. However, implementing differentiated instruction in large classrooms can be challenging. Here are some common challenges, and how to address them:

1. Managing multiple learning groups: In large classrooms, managing multiple learning groups can be overwhelming.
2. Providing individualised attention: With many students, providing individualised attention to each student can be difficult.

3. Assessing student learning: Assessing student learning in large classrooms can be time-consuming and challenging.
4. Managing classroom behaviour: Managing classroom behaviour in large classrooms can be challenging, especially when implementing differentiated instruction.

Strategies to overcome

1. Use technology to support differentiated instruction: Utilise digital tools, such as learning management systems, online assessments, and educational software, to support differentiated instruction and manage multiple learning groups.
2. Implement a station-rotation model: Divide students into small groups and have them rotate through different learning stations, each with a different activity or task.
3. Use formative assessments: Regularly use formative assessments to monitor student learning and adjust instruction accordingly.
4. Provide choices and autonomy: Offer students choices and autonomy in their learning, such as choosing topics or methods of presentation.
5. Use peer-to-peer learning: Encourage peer-to-peer learning by having students work in pairs or small groups to complete tasks or projects.
6. Utilise parent and community volunteers: Recruit parent and community volunteers to provide additional support and individualised attention to students.
7. Use data to inform instruction: Use data from formative and summative assessments to inform instruction and make data-driven decisions.

Fostering a Collaborative Learning Community

One of the most powerful methodologies gaining prominence is collaborative learning. Collaborative learning involves students working together in small groups to achieve common learning goals, fostering a rich environment for knowledge exchange, critical thinking, and social development.

Here are some specific examples of successful collaborative projects that enhance global relevance:

Collaborative Projects Across Disciplines

1. Interdisciplinary STEM project: Students from engineering, computer science, and biology departments collaborated to design and develop a low-cost prosthetic limb for individuals in developing countries.

2. Business and arts collaboration: Students from business and arts schools worked together to develop a sustainable fashion brand, creating clothing lines from recycled materials and selling them online.
3. Environmental science and policy project: Students from environmental science and policy departments collaborated to develop a comprehensive plan to reduce carbon emissions in their local community.

Collaborative Projects Across International Borders

1. Global health project: Students from universities in the United States, Africa, and Asia collaborated to develop a mobile health application to track and prevent the spread of infectious diseases.
2. International business plan competition: Students from business schools in Europe, North America, and Asia collaborated to develop and pitch business plans for sustainable energy solutions.
3. Cultural exchange and documentary project: Students from universities in the United States and Japan collaborated to produce a documentary film exploring the cultural differences and similarities between the two countries.

Online Platforms for Collaborative Projects

1. GitHub: A web-based platform for collaborative software development and version control.
2. Google Workspace: A suite of productivity tools for collaborative document editing, spreadsheet analysis, and presentation design.
3. Trello: A visual project management tool for organising and tracking collaborative tasks and projects.

Benefits of Collaborative Projects

1. Develops teamwork and communication skills: Collaborative projects help students develop essential skills for working effectively in teams.
2. Fosters creativity and innovation: Collaborative projects encourage students to think creatively and develop innovative solutions to complex problems.
3. Enhances global relevance and cultural awareness: Collaborative projects across disciplines and international borders help students develop a deeper understanding of global issues and cultural differences.

Challenges and Considerations for Seamless Integration

Seamless integration of technology in schools can be challenging, especially in low-income areas. Here are some challenges and considerations:

Funding for Tech Resources

Securing funding for tech resources can be a significant challenge for schools, especially in low-income areas. Here are some strategies:

1. Grants and funding opportunities: Research and apply for grants and funding opportunities from government agencies, foundations, and private organisations.
2. Public-private partnerships: Collaborate with local businesses, organisations, and community groups to secure funding and resources.
3. Crowdfunding: Use crowdfunding platforms to raise funds for specific tech projects or initiatives.
4. Budget reallocation: Reallocate existing budget funds to prioritise tech resources and infrastructure.

Strategies for Schools with Limited Access to Technology

Schools with limited access to technology can still provide high-quality education using alternative strategies:

1. Shared devices: Share devices among students, such as laptops or tablets, to maximise access.
2. Offline resources: Utilise offline resources, such as educational games, videos, and software, to provide learning opportunities without internet access.
3. Mobile learning: Leverage mobile devices, such as smartphones, to provide access to educational resources and learning opportunities.
4. Community partnerships: Partner with local libraries, community centres, or businesses to provide access to technology and the internet outside of school hours.

Strategies for Schools in Regions Lacking Tech Infrastructure

Schools in regions with a lack of tech infrastructure can still provide high-quality education using alternative strategies:

1. Satellite internet: Utilise satellite internet to provide internet access in areas with limited or no connectivity.

2. Mobile hotspots: Use mobile hotspots to provide internet access for students and teachers.
3. Offline learning platforms: Utilise offline learning platforms, such as Calibri or RACHEL, to provide access to educational resources without internet access.
4. Solar-powered learning: Utilise solar-powered learning solutions, such as solar-powered laptops or tablets, to provide access to technology in areas with limited electricity.

By understanding these challenges and considerations, schools can develop effective strategies to seamlessly integrate technology into their curriculum, despite limited resources or infrastructure.

Conclusion

In conclusion, seamless classroom integration offers a transformative approach to education, yielding numerous benefits for student engagement, personalised learning, and the development of essential skills. By harmoniously blending technology and traditional teaching methods, educators can create a dynamic learning environment that fosters active participation, tailors instruction to individual needs, and cultivates critical thinking, collaboration, and communication skills. However, a thoughtful and balanced approach to integration is crucial to avoid over-reliance on technology and ensure alignment with learning objectives. As we look ahead to the future of education, it is clear that integrated learning environments will continue to evolve, prioritising student-centred learning, flexibility, and autonomy. By embracing this vision, educators can empower students to thrive in an ever-changing world, equipped with the skills and adaptability necessary for lifelong learning and success.

References

Almousa, O., Prates, J., Yeslam, N., Mac Gregor, D., Zhang, J., Phan, V. and Qayumi, K. (2019). Virtual reality simulation technology for cardiopulmonary resuscitation training: An innovative hybrid system with haptic feedback. *Simulation & Gaming*, 50(1): 6-22. https://doi.org/10.1177/1046878118820905

Blanchard, M. R., LePrevost, C. E., Tolin, A. D. and Gutierrez, K. S. (2016). Investigating Technology-Enhanced Teacher Professional Development in Rural, High-Poverty Middle Schools. *Educational Researcher*, 45(3): 207-220. https://doi.org/10.3102/0013189X16644602.

Chang, S.-C., Hsu, T.-C. and Jong, M. S.-Y. (2020). Integration of the peer assessment approach with a virtual reality design system for learning earth science. *Computers & Education*, 146: 103758. https://doi.org/10.1016/j.compedu.2019.103758

Gheyssens, E., Consuegra, E., Engels, N. and Struyven, K. (2020). Good things come to those who wait: The importance of professional development for the implementation of differentiated instruction. *Frontiers in Education* (Vol. 5, p. 96). https://doi.org/10.3389/feduc.2020.00096

Griful-Freixenet, J., Vantieghem, W., Gheyssens, E. and Struyven, K. (2020). Connecting beliefs, noticing and differentiated teaching practices: a study among pre-service teachers and teachers. *International journal of inclusive education,* 1-18. https://doi.org/10.1080/13603116.2020.1862404

Guay, F., Roy, A. and Valois, P. (2017). Teacher structure as a predictor of students' perceived competence and autonomous motivation: The moderating role of differentiated instruction. *British Journal of Educational Psychology,* 87(2): 224-240. https://doi.org/10.1111/bjep.12146

Hamid, A., Setyosari, P., Ulfa, S. and Kuswandi, D. (2019). The implementation of mobile seamless learning strategy in mastering students' concepts for elementary school. *Journal for the Education of Gifted Young Scientists,* 7(4): 967-982. https://doi.org/10.17478/jegys.622416

Hrastinski, S. (2019). What do we mean by blended learning?. *TechTrends,* 63(5): 564-569. https://doi.org/10.1007/s11528-019-00375-5

Huttar, C. M. and BrintzenhofeSzoc, K. (2020). Virtual Reality and Computer Simulation in Social Work Education: A Systematic Review. *Journal of Social Work Education,* 56(1): 131–141. https://doi.org/10.1080/10437797.2019.1648221

Kumar, A., Krishnamurthi, R., Bhatia, S., Kaushik, K., Ahuja, N. J., Nayyar, A., and Masud, M. (2021). Blended learning tools and practices: A comprehensive analysis. *Ieee Access,* 9: 85151-85197. https://doi.org/10.1109/ACCESS.2021.3085844

McFaul, H. and FitzGerald, E. (2020). A realist evaluation of student use of a virtual reality smartphone application in undergraduate legal education. *British Journal of Educational Technology,* 51(2): 572-589. https://doi.org/10.1111/bjet.12850

Melo, N. (2023). Incorporating artificial intelligence into the classroom: An examination of benefits, challenges, and best practices.

Rafiq, S., Iqbal, S. and Afzal, A. (2024). The Impact of Digital Tools and Online Learning Platforms on Higher Education Learning Outcomes. *Al-Mahdi Research Journal,* 5(4): 359-369.

Richards, M. R. E. and Omdal, S. N. (2007). Effects of tiered instruction on academic performance in a secondary science course. *Journal of advanced academics,* 18(3): 424-453. https://doi.org/10.4219/jaa-2007-499

Rosenberg, M. J. (2001). E-learning: Strategies for delivering knowledge in the digital age. New York: Mcgraw-Hall.

Sheninger, E. (2019). Digital leadership: Changing paradigms for changing times. Corwin Press. https://doi.org/10.21125/inted.2019.2528

Thomas, J. W. (2000). A review of research on project-based learning. San Rafael, CA: Autodesk Foundation.

Chapter Eight
Integrating AI in Teacher Education

Ivy Shen

Southeast Missouri State University, Missouri, USA

Jana Gerard

Southeast Missouri State University, Missouri, USA

Abstract

Artificial intelligence (AI) technology has altered the educational landscape. This chapter summarises the history of AI technology and chatbots and explores the potential benefits of AI in supporting teacher education. The authors argue that incorporating AI in teacher preparation programs can improve the technology literacy of pre-service educators and help foster critical and interdisciplinary thinking. The authors provide practical recommendations and best practices on how AI technology should be integrated into teacher education programs, highlight the challenges associated with using AI, and offer solutions. The challenges pertain to academic integrity, bias, data privacy, over-reliance on technology, and the digital divide. In conclusion, the authors suggest that addressing these concerns requires clear guidelines, policies, and an ongoing dialogue among educators, policymakers, and educational institutions to ensure that the benefits of AI in education are realised without compromising fundamental ethical standards.

Keywords: artificial intelligence, teacher education, ethics.

***

Introduction

Artificial intelligence (AI) is a subfield of computer science concerned with building programs that simulate human intelligence processes, such as speech recognition, decision-making, and language translation. Powered by AI technology, chatbots are software applications designed to simulate human conversation through text or voice interactions. The first generation of chatbots, developed before the twenty-first century, follows predefined rules and pairs input words with a list of possible scripted responses, leading to clunky and awkward exchanges. Chatbots introduced in the new century incorporate machine learning, a subset of AI that involves the development of algorithms that allow computers to learn from and make predictions or decisions based on data. Chatbots that use natural language processing (NLP) can understand, interpret, and generate human language. These chatbots are called intelligent assistants or virtual assistants, which are machine learning-powered software that allows human users to quickly find relevant information or complete tasks (Azvine et al., 2000). Many companies soon began to implement chatbots in customer service and messaging platforms to reduce labour and save costs. To effectively disseminate public health information during the COVID-19 pandemic, government agencies across the globe employed online chatbots to answer high volumes of questions and provide pandemic-related practices, accelerating the technology.

Industries with customers demanding around-the-clock services, such as banking, healthcare, finance, and customer service, are particularly eager to develop virtual assistants to provide customer care. For example, Bank of America launched a chatbot named Erica as a virtual financial assistant for their clients in 2018, and as of April 2024, Erica has responded to 800 million inquiries from over 42 million clients (Bank of America, 2024). For another example, Progressive Insurance uses chatbot Flo to improve the efficiency of their customer service operations. Flo can handle multiple inquiries simultaneously and provide relevant responses in real time. Interacting with Flo online is like conversing with a human since Flo can chatbot about almost anything with the customer (Progressive Insurance, 2024).

Newer chatbots developed in recent years can create sophisticated text based on user input, write coherent essays, solve math problems, write computer code, and generate images and videos. Many in-service teachers today use chatbots to form lesson plans, design in-class activities, grade student work, and deliver instruction. Chatbots offer diverse activities to assist student learning, such as interactive games, quizzes, flashcards, and discussions. Incorporating AI in teacher education programs has a few benefits. First, it helps improve the technology literacy of these future teachers and equips them with the ability to instill technology literacy in their future students. As AI becomes increasingly prevalent in the workplace, the percentage of future jobs requiring AI skills is on the rise. It is critical to expose pre-service educators to the latest technology so they can be well-prepared for their future endeavours. Second, chatbots operate on a vast database. They can generate almost limitless amounts of information, making them a library of teaching materials. The impressive amount of data AI contains offers great opportunities for students to analyse and evaluate information, practising their critical thinking and problem-solving skills. Third, chatbots facilitate personalised learning that meets various individual needs. Introverts or learners with social skill deficits can engage in non-judgmental interactions with a chatbot, reducing anxiety sometimes associated with communications with other humans. Physically challenged learners can access chatbots around the clock without having to travel to campus. Chatbots offer text display or audio delivery, making learning easy for learners with visual or hearing disabilities. Further, empirical evidence suggests that chatbots can assist social skill training, which is beneficial for future teachers. Teachers interact with students, parents, and administrators regularly. It is imperative for them to be able to handle difficult social situations. Chatbots offer a risk-free environment for interaction, which helps the user practice and develop social skills. This chapter aims to give an overview of chatbots' utilities in teacher education and highlight their challenges. The authors argue that teacher education programs should harness the power of chatbots regardless of their downfalls as AI slowly becomes the norm in our everyday lives.

AI Technology and Chatbots

Lately, AI technology has attracted much attention from educators due to its promising potential. AI-supported chatbots offer a wealth of information from a wide range of disciplines. Chatbots combine AI and natural language processing or other technology that enables them to interact to a certain level of conversation with a human user through text or voice (Pérez et al., 2020). Powered by AI programs, chatbots can simulate human behaviour by conversing and interacting with users using natural language. The first chatbot, ELIZA, was created by MIT professor Joseph Weizenbaum in 1966. ELIZA is a computer program that interacts with users in a typed conversation by looking for keywords in the initially entered statement (New Jersey Institute of Technology, 2023). Specifically, it processes the words, pairs them with a list of possible scripted responses from a pre-entered database and finds the most relevant response. Chatbots developed in the twenty-first century showcased machine learning and advanced algorithms. Machine learning focuses on the use of data and algorithms to automate an analytical model using input and output data, find intrinsic structures in input data, and imitate the way that humans learn, making the response more accurate (IBM, 2023). Chatbots developed in the 2010s are more sophisticated. These chatbots, often called virtual assistants or intelligent assistants (e.g., Alexa and Siri), are able to perform tasks that previously required human touch, such as placing phone calls, setting reminders, and looking up the internet.

During the COVID-19 pandemic, communication and chatbot technologies accelerated to address public needs. In the face of lockdowns and social distancing, the government and public authorities used social media platforms extensively to disseminate pandemic-related mandates. Citizens also used social media to engage with others for relationship maintenance. During the pandemic, 90% of Americans considered the internet essential or important in their daily life (Pew Research Center, 2021). The extensive use of the internet and social media prompted the rapid development of chatbot technology. Institutions like the Centers for Disease Control and Prevention (CDC) and the World Health Organization (WHO) employed chatbots to monitor public health and answer high volumes of citizen questions related to the coronavirus around the clock (Miner et al., 2020). The COVID-19 public health emergency has ended, but the prevalent use of chatbots remains.

In recent years, AI technology has been widely used in higher education. Universities and colleges develop chatbots to interact with students and faculty members and provide information ranging from course enrolment and campus resources to academic advising and parking. Some universities have developed emotional support chatbots to help freshmen transition to college life. Not only can chatbots be used as administrative tools, but they have academic value as well. Chatbots can generate texts, answer written prompts, and interpret questions, making them popular among students (Shen, 2024).

Chatbots have been villainised due to some students' overreliance on them, which leads to compromised academic integrity. One of the most commonly used chatbots in higher education, ChatGPT, can write emails, generate texts, write computer code, solve math problems, interpret questions, and provide accurate and thorough answers. Soon after OpenAI, an American AI research laboratory, launched ChatGPT in November 2022, the chatbot quickly gained traction. Impressively, the chatbot reached one million users in the first five days and surpassed 100 million after two months, making it the fastest-growing application in history (University of Rochester, 2023). In addition to generating information and solving problems, ChatGPT can achieve high scores on a range of tests. In a recent study (Kung et al., 2022), researchers put ChatGPT through the United States Medical Licensing Exam, a three-step examination for medical licensure, and learned that ChatGPT performed at or near the passing threshold for all three exams, leaving many in awe that a chatbot makes a good medical test-taker without any medical training. The newer language model of ChatGPT, ChatGPT-4, outperforms most human test-takers on difficult academic exams (OpenAI, 2023). For example, ChatGPT-4 excelled on the SAT Reading and Writing section with a score of 710 out of 800, which put it in the ninety-third percentile, and earned a 700 out of 800 on the math test. On the bar exam, ChatGPT-4 scored in the ninetieth percentile with a score of 298 out of 400 (OpenAI, 2023).

College students discovered ChatGPT with great excitement and began taking advantage of its remarkable information-generating features. According to a recent survey of 1,000 U.S. college students (Government Technology, 2023), nearly a third of them have used ChatGPT to complete homework assignments, and close to 60% use it on more than half of their work. In addition to copying and pasting chatbot-generated content, many students use ChatGPT to think for them. A student at Columbia University revealed to the *Chronicle of Higher Education* (2023) that the user could use ChatGPT to generate ideas based on a writing prompt, any one of which the user can choose to include in their assignment. The student also explained that the user could tell the algorithm what the topic is and ask for a central claim, then have ChatGPT provide an outline to argue this claim. It is nearly impossible for educators to detect the use of ChatGPT since ChatGPT creates original work each time, and its responses are not in the databases of plagiarism detectors.

Colleges and universities began taking preventive measures to combat academic dishonesty. One of the most common strategies initially implemented was banning chatbots on their networks and computers to eliminate plagiarized work and discourage cheating. However, students are able to circumvent the strategy because they are still able to access chatbots through their own devices, such as laptops and smartphones or complete assignments off-campus. Unable to ban access to chatbots, educators alter their learning outcome assessments by replacing writing assignments with presentations, oral exams, podcasts, and drawings. Further, some educators resort to the old-school paper and pencil method to test students' knowledge or have students type in the classroom using browsers that restrict and track computer activity. Despite these challenges, AI technology can be a beneficial pedagogical tool in teacher education.

Technology Literacy

Incorporating AI in teacher education programs not only helps improve the technology literacy of pre-service educators but also equips them with the ability to instill technology literacy in their future students. Technology literacy, often referred to as digital literacy, is the ability to effectively use technology tools, understand digital systems, and critically evaluate information from various digital sources (Gilster, 1997). It encompasses a wide range of skills and knowledge that enable individuals to navigate the digital world. The concept of literacy has broadened over time. As the application of AI becomes prevalent, AI literacy has also entered the public vocabulary. Long and Magerko (2020, p. 2) define it as "a set of competencies that enables individuals to critically evaluate AI technologies; communicate and collaborate effectively with AI; and use AI as a tool online, at home, and in the workplace." In other words, AI literacy involves the knowledge and skills required to understand, use, and interact with AI systems effectively. AI has become increasingly integrated into various aspects of society. Recent surveys by the Organization for Economic

Cooperation and Development (OECD) (2024) indicate that employers today use AI technology to sift through resumes, interact with customers, evaluate work, and provide training. AI is being woven into the workplace at an unexpected scale, with 75% of knowledge workers using AI at work today (Microsoft, 2024). The overwhelming majority of users praised AI for helping them save time, focus on their most important work, be more creative, and enjoy their work more (Microsoft, 2024). AI can empower pre-service educators to easily analyse student data, allowing them to identify learning needs, personalise learning, and make data-driven decisions that improve student outcomes.

Digital literacy is a component of digital competence. Digital competence is a combination of knowledge, skills, and attitudes toward the use of technology as well as the ability to use technology to perform tasks, solve problems, communicate, manage information, and collaborate (Palacios Hidalgo et al, 2020). Digital competence also incorporates the ability to create and share content effectively, appropriately, securely, critically, creatively, independently, and ethically (Palacios Hidalgo et al, 2020). Pre-service educators who are digitally competent are prepared to use AI to enhance student learning and help students be more successful (Instefjord & Munthe, 2016). Digital portfolios are an effective way to assess teacher and student digital competence, including assessing the effective integration of AI into teaching and learning (Gerard & Giasi, 2024). Digital portfolios allow teachers and students to collate digital work samples that demonstrate various aspects of digital competence, including examples of digital literacy, digital fluency, and digital citizenship (Sillat et al., 2021). Compiling a digital portfolio allows teachers and students an opportunity to both critically evaluate and reflect on their learning (Prieto et al., 2020).

Smart devices and the internet have evidently become the norm in modern everyday life. Some mistake device ownership with digital literacy and believe that college students today are already digitally literate since they grew up in the internet age. In reality, knowing how to operate a device does not always equip one with the ability to use it for academic purposes. Further, as smart devices and the internet become ubiquitous, the disparity in access exacerbates the digital gap. Many low-income and nontraditional students lack adequate technical skills due to their lack of access. Nontraditional students are defined as older students (mostly over 24 years old) who work full-time, have children, and/or have other life circumstances that can interfere with the successful completion of educational objectives (National Center for Education Statistics, 2020). Approximately 38% of all the students attending colleges and universities nationwide in the Fall 2020 semester were nontraditional (National Center for Education Statistics, 2021). Many nontraditional students also have a low-income background and work minimum-wage jobs. These students may only have access to technology when they are on campus.

Further, on a global level, digital divide refers to the gap between individuals, communities, and countries that have access to modern information and communication technologies and those that do not (Fang et al., 2019). High-speed internet access varies drastically between developed and developing countries. In many rural areas of low-income countries, internet infrastructure is inadequate or non-existent, while urban and wealthier regions in developed countries enjoy reliable, high-speed broadband (Chen & Wellman, 2004). In many parts of the world, the cost of computers, smartphones, or tablets remains prohibitive, especially in low-income countries. Even when connectivity is available, many people cannot afford the necessary equipment to access the digital world. Digital literacy is often built on a foundation of basic education. In regions with high illiteracy rates, people may not have the reading and writing skills needed to use digital devices effectively. International students who study in the U.S. to become future educators may not be equipped with the rudimentary skills to navigate the technology-saturated world. If higher education is to prepare students to adequately meet the demands of an increasingly technological world, it must integrate technology into the academic curriculum (Daggett, 2010). Incorporating AI technology in teacher education programs not only helps to hone students' technical skills and prepare them for their teaching endeavours but also enables them to instill technology literacy in their future students.

Critical Thinking

Critical thinking is defined as "reasonable, reflective thinking that is focused on deciding what to believe and do" (Ennis, 1987, p. 81). Cultivating students' critical thinking ability is a major goal of higher education because such an ability empowers students to analyse and evaluate information, which enables them to engage in purposeful judgment (Behar-Horenstein & Niu, 2011). Critical thinking allows students to become better problem solvers, improves their ability to make informed decisions, and facilitates lifelong learning. Universities and colleges highly value students' critical thinking abilities. According to a study conducted by the Association of American Colleges and Universities (2011), 95% of the chief academic officers from surveyed higher education institutions and 81% of the surveyed employers view critical thinking as one of the most important intellectual skills for college students.

Critical thinking encourages pre-service educators to link teaching and learning and consider how they will encourage P-12 students to think critically (Lorencova et al., 2019). To support this goal, teacher education programs integrate critical thinking and problem-solving learning strategies into teacher education curricula (Lithoxoidou & Georgiadou, 2023). Cooperative learning, which emphasises communication, collaboration, and group problem-solving, enhances critical thinking when integrated with analysis and reflection. This combined approach requires students to synthesise information, brainstorm, and evaluate potential solutions, which fosters critical thinking skills (Veenman et al., 2002). Authentic learning engages students in real-life problems, prompting them to analyse issues, reflect on these issues, and then apply personal experiences and subject matter knowledge to solve the problems (Lombardi & Oblinger, 2007). Pre-service educators then transfer learning from their authentic learning experiences to the field and clinical practice, which allows them to self-regulate and apply critical thinking in unexpected situations such as student conflict (Theelen et al., 2022). Experiential learning promotes critical thinking and lifelong learning by providing opportunities for pre-service educators to experiment and refine their teaching practices (Williams & Sembiante, 2022). By simulating real-life situations with elements of ambiguity, realism, and complexity, pre-service educators develop complex decision-making skills that require the integration of critical thinking and problem-solving with subject matter knowledge (Lithoxoidou & Georgiadou, 2023). Problem-based learning combines elements of authentic learning and experiential learning and encourages students to use prior experience, subject matter knowledge, critical thinking skills, and reflection to address real-world problems and devise practical solutions (Akl, 2014). Participatory learning actively encourages students to think for themselves. Rather than passively absorbing information, the participatory approach motivates students to be a part of the learning process and take responsibility for planning their own learning sessions (Joseph et al., 2017). By incorporating these learning strategies, teacher education programs ensure pre-service educators understand how to think critically and solve problems in their future classrooms.

Cooperative Learning

In a teacher education course focused on cooperative learning, pre-service educators are placed in small groups. These groups focus on the collaborative creation of learning materials and lesson plans while collaborating, thinking critically, and practically applying pedagogical concepts (Darling-Hammond, 1997). One challenge often faced in group work is ensuring all group members are equally engaged and contributing meaningfully to a project (Guskey, 2000). AI can play a pivotal role in group work accountability in many ways. First is intelligent task distribution (Slavin, 1995), where the AI can analyse the strengths, weaknesses, and learning preferences of group members based on coursework, assessments, and self-reported data (Hattie, 2009). The AI can then suggest task distribution that leverages individuals' strengths while ensuring group members have the opportunity to develop new skills. For example, if a group is tasked with creating a lesson plan, the AI could suggest that a pre-service educator who excels at content creation but struggles with designing assessments takes the lead on content while also closely collaborating on the assessment portion.

Another way that AI can support cooperative learning is as the group works on their lesson plan, the AI monitors their progress and group interactions. The AI can provide real-time feedback on each group member's contributions, including suggesting resources, highlighting inconsistencies, and offering tips for more effective collaboration. For instance, if the AI detects that the group discussion is unproductive or off-topic, the AI can gently steer the group back to a more constructive discussion by focusing attention on points that need to be addressed.

Further, the AI can collect and analyse data on each group's collaborative process, including participation rates, communication patterns, and task completion times. This data can be used to generate personalised reports for students and instructors (Mayer, 2014). Students then receive insight into their collaboration skills and recommendations for improvement, and instructors can use these reports to identify groups that need additional support. This data can also help instructors facilitate reflective discussions on effective teamwork. By integrating AI into cooperative learning, pre-service educators are exposed to cutting-edge technology while being supported in their development of essential collaborative skills. The AI acts as a mediator, ensuring equitable participation, fostering constructive feedback, and providing personalised guidance (Fullan, 2007). Using AI in cooperative learning enhances learning experiences and prepares pre-service educators to integrate this type of technology in their own classrooms, leading to more effective and inclusive teaching practices (Chan et al., 2021).

Authentic Learning

Authentic learning is vital in teacher education courses designed to prepare pre-service educators for real-world classroom challenges (Lombardi, 2007). Authentic learning includes engaging with real-world problems, reflecting on practice, and applying theory in practical settings (Jonassen, 2011). Creating and managing authentic learning experiences can be challenging, especially when trying to provide personalised feedback and ensuring that every pre-service educator encounters a diverse range of scenarios. Using AI can be transformative in supporting these learning experiences through AI-powered simulations. An AI-driven simulation platform can feature AI-driven avatars that simulate students with diverse learning needs (Dewey, 1938), behaviours, and cultural identities. These avatars can react dynamically to pre-service educators' instructional and classroom management strategies and interpersonal interactions. AI simulations can also present pre-service educators with varying classroom scenarios, such as teaching a mixed-ability group (Herrington et al., 2010), managing classroom disruptions, or adapting lessons for English language learners. The AI can tailor these scenarios based on the pre-service educator's progress, ensuring that each learning experience is both relevant and challenging. As the scenarios are designed to mimic complex and unpredictable real-world classroom situations, these scenarios require pre-service educators to think critically, make quick decisions, and then reflect on their choices.

As pre-service educators navigate through the simulations, the AI can provide immediate and context-sensitive feedback (Moore & Kearsley, 2012) that is both formative and summative, which helps pre-service educators understand the impact of their decisions. The AI can also suggest alternative strategies, such as differentiating instruction or increasing positive reinforcement. After each scenario, the AI can guide pre-service educators through a reflective process, asking questions that encourage deeper thinking about the simulation and how the simulation relates to educational theories and practices. As the AI system continuously gathers data on pre-service educators' interactions, decision-making processes, and areas of struggle or success, the data can be used to generate detailed reports that help both pre-service educators and teacher educators understand the development of skills over time. Such insights can further inform personalised learning experiences that ensure each pre-service educator receives targeted support in areas where such support is most needed (Bichler et al., 2024). By incorporating AI into authentic learning experiences, teacher preparation programs can provide pre-service educators with a safe yet challenging environment to practice and refine their skills. AI-driven simulations offer realism and variability (McAleese, 1986) that is difficult to achieve with traditional methods alone. The adaptive feedback and data-driven insights provide a highly personalised learning experience that ensures each pre-service educator can engage deeply with course materials and develop the competencies needed to be effective educators. Using AI-based simulations enhances the quality of educator preparation while equipping pre-service educators with the confidence to handle the complexities of real-world classrooms.

Experiential Learning

Experiential learning allows pre-service educators to apply theoretical knowledge in real-world settings (Kolb, 2015), such as internships, field experiences, and student teaching. One of the challenges of experiential learning, however, is ensuring pre-service educators receive consistent quality feedback and opportunities for reflection during these hands-on experiences (Dewey, 1938). AI can be a powerful tool that enhances experiential learning through personalised support, facilitated reflection, and data-driven insights. An AI-driven mentorship platform designed to support pre-service educators during their field placements can integrate AI-driven analytics, real-time feedback, and personalised learning pathways to maximise the impact of experiential learning (Guskey, 2000). For example, a pre-service educator can wear a device connected to an AI platform to record their lessons. The AI then analyses the recordings in real time, focusing on aspects of teaching such as instructional strategies, student engagement, classroom management, and questioning techniques. The AI provides immediate and detailed feedback on specific areas for growth, like lesson pacing, distributing attention among students, or the effectiveness of formative assessments. For example, if the AI detects that the pre-service educator consistently calls on the same students, the AI may suggest strategies to involve more students in the discussion, which promotes inclusivity and engagement.

The AI can also create personalised learning pathways for each pre-service educator that recommend specific resources, additional practice scenarios, or targeted professional development activities. If a pre-service educator struggles with classroom management, the AI may suggest reading relevant articles, watching pertinent videos, or engaging in simulated classroom management scenarios to build classroom management skills. As reflection is a critical component of experiential learning (Radovic et al., 2021), the AI can assist pre-service educators in maintaining a reflective journal where the pre-service educators can document experiences, challenges, and insights gained during their field experiences (Lombardi, 2007). The AI can prompt reflective questions, analyse the content of reflections (Boud et al., 1985), and suggest connections between the pre-service educators' experiences and educational theories. Further, AI can also highlight patterns in reflections, helping pre-service educators identify recurring themes or challenges.

Lastly, an AI system can aggregate data from various sources, such as classroom observations, reflective journals, mentor feedback, and self-assessment, to provide a comprehensive view of a pre-service educator's development over time. The pre-service educators and teacher educators can access data through a dashboard that visualises progress in key areas like instructional effectiveness, classroom management, and student engagement. This data-driven approach enables teacher educators to enact more targeted interventions. For example, if data shows that a pre-service educator has made progress with instructional strategies but is struggling with differentiation, the teacher educator can focus on coaching sessions on that specific skill. By integrating AI into experiential learning, teacher preparation programs can offer pre-service educators more personalised, reflective, and data-informed experiences (Kolb, 2015). AI-enhanced mentorship ensures that pre-service educators receive consistent and meaningful feedback even in diverse or remote field experiences. AI-enhanced mentorship also supports continuous professional growth by providing custom learning opportunities and facilitating deep reflection on practice. This approach enriches experiential learning and prepares pre-service educators to be reflective practitioners adept at using technology to enhance their teaching and learning (Boud et al., 1985).

Problem-Based Learning

Problem-based learning (PBL) is a pedagogical approach that presents pre-service educators with complex problems that require critical thinking, collaboration, creativity, and the application of knowledge across various domains (Barrows et al., 1980). In teacher preparation, PBL assists pre-service educators in developing skills needed to solve the kinds of multifaceted challenges they will encounter as in-service educators (Jonassen, 2011). However, facilitating effective PBL is resource-intensive and requires constant ongoing guidance, personalised feedback, and continual scaffolding (Savery & Duffy, 1995). AI can support PBL through real-time assistance, adaptive resources, and data-driven insights (Trust, 2020). An AI-powered support system designed to assist pre-service educators as they navigate PBL activities can integrate intelligent tutoring, dynamic resource recommendations, and collaborative tools to enhance PBL experiences. The AI can analyse a group's approach to a problem, identifying when they are on the right track or veering off course. For example, if a group is struggling to identify the cause of a classroom management issue in a PBL scenario, the AI can prompt the group with questions to reconsider assumptions or explore alternative explanations. AI can also provide real-time suggestions if a group encounters roadblocks, helping the group to progress without giving a solution. This approach ensures that learning remains student-driven while still providing the necessary scaffolding.

Further, the AI can continuously assess the group's needs and provide resource recommendations that might include academic articles, instructional videos, case studies, or expert advice from virtual mentors (Herrington, 2010). The AI adapts recommendations based on a group's progress and challenges, ensuring that the resources are relevant and immediately applicable to the problem. The AI platform can include advanced collaboration tools that facilitate effective teamwork; the AI can monitor group communication patterns and provide feedback on collaborative dynamics, such as suggesting more equitable participation or helping resolve group conflicts. For example, if the AI detects that a group member is dominating discussions, the AI may suggest strategies to promote contributions, such as assigning roles or using structured discussion protocols. The AI can also help a group document the problem-solving process, capturing key decisions, rationales, and reflections.

This documentation supports the group's current project but also serves as a valuable resource for future learning. At the conclusion of the PBL learning activity, the AI can generate a detailed report of the group's performance, including data on how the group approached the problem, the effectiveness of group collaboration, resources the group utilised, and the quality of the solution (Hattie, 2009). The AI can then provide reflective prompts based on this data, which encourages pre-service educators to critically analyse the group's problem-solving process and consider how lessons learned can be applied to future challenges (Amerstorfer & von Münster-Kistner, 2021). By integrating AI into PBL, teacher preparation programs can provide pre-service educators with more supported and enriched learning experiences. AI-powered PBL support systems not only enhance the effectiveness of the PBL process but also help pre-service educators develop critical problem-solving skills, foster deeper engagement, and promote reflective practice. This ensures that pre-service educators are well-prepared to address complex real-world problems that will be encountered as in-service educators, as well as integrate similar AI-driven tools into teaching practices (Benedicto & Andrade, 2022).

Participatory Learning

AI-assisted technology enhances learners' critical thinking skills by providing a wealth of information and facilitating participatory learning. First, chatbots are effective tools in training critical thinking abilities because they provide a wide range of information with high accuracy and speed, which helps students reflect on the topics they are already familiar with. However, it is important to note that, as sophisticated as chatbots are, they occasionally generate incorrect information. Chatbots draw information from pre-programmed databases, which were created by humans, meaning human bias and prejudice can appear in the created answers (Ferrer et al., 2021). For example, OpenAI (2023) admits that despite its incredible capacities, ChatGPT maintains a tendency to perform tasks incorrectly and make up facts when there is not enough information available on a subject from the database. According to a recent survey that interviewed students who have experimented with ChatGPT (The New York Times, 2023), although the program is useful for assignments, the accuracy of the information is questionable. This, however, may serve as an opportunity for students to see the mistakes and biases in the text generated and achieve a more comprehensive understanding of the subject matter.

Second, scholars have identified that active and participatory learning stimulates the development of critical thinking, analytical, and problem-solving skills (Bransford et al., 1999; Ziegler, 2001; Acker, 2003). Participatory learning is vital in teacher education programs because it encompasses multiple engaging learning experiences. First, it requires active learning, which is essential for developing a deep understanding of content and retaining knowledge (Metzger & Langley, 2020). Additionally, participatory learning provides pre-service educators with practical experience in classroom management, lesson planning, and instructional strategies (Allen & Wright, 2014) as well as fostering communication and collaboration with peers and mentors, helping them develop critical teamwork skills (Murray-Harvey et al., 2013). Moreover, participatory learning builds pre-service educators' confidence, enabling them to adopt more student-centred learning approaches in their teaching as well as promoting adaptability and innovation, which encourages pre-service educators to explore new teaching methods and address evolving educational needs of students (Kickbusch et al., 2020). Finally, participatory learning encourages pre-service educators to reflect on their practice, critically analyse their teaching, identify their strengths and weaknesses, and continually improve their skills as a teacher (Matsumoto-Royo et al., 2022).

In participatory learning, learners are actively involved in the learning process. AI-powered chatbots have a long history of being used as participatory learning agents. In the 1970s, Intelligent Tutoring Systems (ITSs) were developed to assist human tutors and provide students with access to course resources outside of regular class hours. ITSs are AI-powered computer programs that provide intelligent tutors who know what they teach, whom they teach, and how to teach (Alkhatlan & Kalita, 2019). ITSs simulate human tutors and offer interactive educational experiences around the clock. At a fraction of the cost of hiring human teachers, conversational ITSs deliver content in a one-on-one setting by asking and answering questions and conducting discussions, identifying gaps in knowledge, and addressing these gaps in natural language (Ilagan & Ilagan, 2023). Instead of being fed information passively, ITSs empower and engage learners by posing questions and providing instant feedback, which can have an immediate impact on learning progress. An Arizona State University augmented intelligence researcher (VanLehn, 2011) compared computer tutors and human tutors for their impact on learning gains and confirmed that ITSs are nearly as effective as human tutoring. One recent example of ITS is AutoTutor, a computer tutor developed by the Institute for Intelligent Systems at the University of Memphis. AutoTutor teaches content through natural language conversations. It is able to track the cognition and emotions of the student through dialogue patterns, facial expressions, and body posture. Using these cues, it then responds in a manner that adapts to the student (University of Memphis, 2024).

In addition to emotion recognition and natural language conversations, chatbots facilitate inclusive learning by creating an environment that meets different students' needs. Chatbots provide on-demand support, making learning materials accessible to all learners regardless of their physical location or time zone. Physically disabled students can access chatbots through their devices without travelling around campus. Chatbots offer screen readers for visually impaired students and text-based options for hearing-impaired learners. For students with social disabilities who feel uncomfortable asking questions to a human teacher, chatbots offer non-judgmental and encouraging interactions. Further, for intellectually disabled students who require additional accommodations, chatbots provide various formats for content consumption, such as breaking down information in different ways and repetition. Chatbots engage all types of learners in interactive conversations and simulations, enhancing their learning experience. Moreover, they can offer immediate assistance and guidance, helping learners overcome obstacles and stay motivated throughout their educational journey, which is crucial to participatory learning.

AI can also assist pre-service educators in understanding when to combine learning strategies to maximise student engagement and reinforce student success. For example, if students are designing a sustainable city as part of a PBL project, pre-service educators can use AI to determine what other learning strategies could be combined with PBL to increase student engagement and enhance student learning. AI may suggest combining cooperative learning with PBL, having each group of learners be a different city department, which requires all groups to work together to design the city. AI can give pre-service educators suggestions for further enhancing learning with models such as a flipped classroom, gamification, and jigsaw or ideas for combining more traditional learning, such as writing an essay, with more interactive learning, such as augmented or virtual reality.

Interdisciplinary Learning and Thinking

The 4Cs of Future Ready Learning, also known as the 4Cs of twenty-first century learning, encompass communication, collaboration, critical thinking, and creativity (P21 Partnership for 21st Century Learning, n.d.). Interdisciplinary learning helps pre-service educators cultivate these essential skills vital for their intellectual growth and for the learning experiences they will provide to their future students (Darling-Hammond & Bransford, 2005). Integrating the 4Cs into teacher preparation is crucial, as it prepares pre-service educators to navigate and thrive in an increasingly complex and interconnected world. By mastering these skills, pre-service educators can design learning environments that encourage students to engage in meaningful dialogue, work effectively in teams, solve complex problems, and think creatively. Interdisciplinary learning plays a pivotal role in cultivating these essential skills. Interdisciplinary learning allows pre-service educators to draw connections across different fields of study, thereby broadening their intellectual horizons and enabling them to approach problems from multiple perspectives (Guskey, 2000). This kind of learning is vital for intellectual growth and for the learning experiences these pre-service educators will provide to their future students. The process of interdisciplinary learning promotes self-recognition, where pre-service educators become aware of their strengths and areas for growth (Fullan, 2007). This self-awareness is crucial for developing skills needed to become effective educators who can adapt to various teaching contexts and student needs (Santaolalla et al., 2020).

Moreover, interdisciplinary learning links activities to different areas of knowledge, encouraging students to make connections between subjects, which enhances critical thinking and fosters a comprehensive understanding of complex topics (Newell, 2001). By engaging with content that spans multiple disciplines, pre-service educators are challenged to think beyond the confines of traditional subject boundaries, leading to richer insights and more innovative approaches to teaching. This not only promotes critical thinking but also encourages creativity, as students are required to synthesise information in novel ways and apply this information to real-world scenarios. For instance, a project that integrates elements of science, technology, and social studies may require pre-service educators to explore the environmental impact of technological advancements, evaluate ethical considerations involved, and propose sustainable solutions. Through such interdisciplinary projects, pre-service educators practice the 4Cs in meaningful contexts, learning to communicate ideas effectively, collaborate with peers from differing disciplines, critically evaluate complex information, and devise creative solutions to global issues (Wagner, 2012). During this process, pre-service educators not only enhance their own professional competencies but also model integrative, future-ready learning that they will facilitate in their classrooms. The integration of the 4Cs of future-ready learning through interdisciplinary learning in teacher preparation programs ensures that pre-service educators are well-prepared to foster holistic and future-ready learning environments. This approach equips pre-service educators with the necessary tools to nurture the next generation of students capable of navigating an increasingly complex world with confidence, creativity, and critical insight.

Chatbots are particularly useful in facilitating interdisciplinary learning because they can immediately answer questions, offer explanations, and provide resources. Through conversational interfaces, chatbots can seamlessly integrate concepts from various disciplines into discussions. Chatbot research concludes that it has a strong interdisciplinary character, spanning fields such as management and marketing, media and communication science, STEM (science, technology, engineering, and math) subjects and humanities, and psychology and sociology (Følstad et al., 2021). Users can command the chatbot to provide an explanation of a specific concept, drawing theoretical frameworks from multiple fields. For example, the author typed in the prompt "provide an explanation of social services using nonprofit and business theories," and ChatGPT instantaneously generated paragraphs of text mentioning both nonprofit and business terms such as stakeholder engagement, social impact measurement, volunteerism, innovation and adaption, and revenue generation. Powered by enormous databases, chatbots can generate information from a wide range of disciplines, allowing students to see connections between different fields of study and promoting interdisciplinary learning. Students can query chatbots on how concepts from one discipline relate to another, fostering cross-disciplinary thinking (Ifelebuegu et al., 2023).

Chatbots are highly customizable to meet various learner needs and styles, as indicated above. The adaptive and personalised learning environment with a high capacity for interaction is advantageous for learners to digest and retain new materials. Further, chatbots themselves are products of interdisciplinary research integrating technology from various STEM domains. Empirical studies demonstrate the practical effectiveness of chatbots in interdisciplinary learning. For example, Iku-Silan and colleagues (2023) conducted an experiment in an environment education course in a junior high school in northern Taiwan. Learners in the control group participated in conventional technology-assisted learning, while the experimental group used a chatbot for learning. The findings indicated that the experimental group significantly outperformed the control group on a variety of metrics such as learning achievements, collective efficacy when performing interdisciplinary or cross-disciplinary tasks, cognitive engagement, and emotional engagement (Iku-Silan et al., 2023). For another example, a group of scholars from North Carolina State University designed and tested a comprehensive AI curriculum through a series of activities (e.g., finding the shortest path between cities on a map with and without guidance, instructing students how to act as a GPS on a simple map) for middle-grade students (Akram et al., 2022). The results suggest that the activities were successful in helping students understand the interdisciplinary nature of AI through the practice of integrating scientific knowledge and practices.

Interdisciplinary learning can translate to interdisciplinary thinking, which is a critical career skill in today's increasingly complex world filled with challenges and opportunities (Northwestern University, 2024). Interdisciplinary thinking involves drawing multiple viewpoints on a given topic and integrating knowledge and methods from different disciplines to create a comprehensive understanding of a complex problem. It allows one to gain new understandings and brainstorm creative solutions. Interdisciplinary learning and thinking also help develop integrative thinking skills, consisting of taking disconnected materials or ideas and synthesising them into something new (Sill, 2001). Interdisciplinary approaches are grounded in disciplines by constructing integrative knowledge that is greater than the sum of its distinctly disciplinary parts (Newell, 1998). Integrative thinking allows the decision-maker to yield a synthesised solution.

Challenges and Solutions

Work-Life Balance for Teachers

Work-life balance has become a major factor in teacher retention, with many in-service educators citing the overwhelming demands of the profession as a key reason for leaving education (Elevate K-12, 2023). The pressure of meeting educational standards, addressing diverse student needs, and managing administrative tasks often leaves teachers feeling overburdened. A significant proportion of teachers report that there is not enough time during the school day to complete essential tasks such as grading, lesson planning, and preparing instructional materials. According to a recent survey, 54% of teachers say it's "very or somewhat difficult" for them to find work-life balance, a challenge that can lead to burnout and ultimately, a decision to leave the profession (Pew Research Center, 2024). The integration of AI tools in education offers a promising solution to this pervasive problem by automating routine and time-consuming tasks, allowing educators to focus more on instruction and student interaction and less on administrative burdens. AI tools like Diffit or MagicSchool are at the forefront of this transformation, designed to significantly reduce the time teachers spend on tasks such as grading, lesson planning, and the preparation of instructional materials (Diffit, 2024). For example, Diffit uses AI to generate differentiated instructional materials tailored to varying abilities of students; MagicSchool offers automated lesson planning and resource generation, which assists educators in crafting comprehensive lesson plans (MagicSchool, 2023). These tools benefit both in-service and pre-service educators and are being increasingly integrated into educator preparation programs. Pre-service educators, often balancing coursework, field placements, and personal commitments, can use these tools to manage workloads more efficiently. By learning to leverage AI in professional practice, pre-service educators can develop strategies to maintain a healthier work-life balance from the start of their careers. This proactive approach is crucial in fostering a sustainable teaching practice that prioritises well-being alongside professional efficacy (ESL Interactive, 2023; eMediaVA, 2023). However, the use of AI must be balanced as the overuse of AI can lead to decreased critical thinking skills, reduced creativity, over-dependence on technology, and less human interaction (Offor et al., 2024). Further, overusing AI also increases the likelihood of perpetuating implicit biases (Bushweller, 2024) and increases the privacy and security risks.

The use of AI in managing routine tasks does more than save time; it can also enhance the quality of work (Hattie, 2009). For instance, AI-driven grading tools can provide consistent and objective feedback to students while offering detailed analytics to teachers to help identify trends and areas where students may need additional support. The introduction of AI tools into daily teaching practices encourages a culture of innovation and continuous improvement. Freeing up time otherwise spent on administrative tasks allows in-service and pre-service educators to engage in more meaningful professional activities such as collaborating with colleagues, engaging in professional development, and building stronger relationships with students. This shift not only contributes to improved work-life balance but also enhances overall job satisfaction and professional fulfilment, which are critical factors in teacher retention (Desimone, 2009). The integration of AI tools into daily teaching practices encourages a culture of innovation and continuous improvement. As both in-service and pre-service educators become more comfortable using AI tools, they are more likely to explore other ways that technology can support teaching, from personalised learning for students to more effective communication with families and guardians. Ongoing engagement with technology helps keep educators motivated and invested in their careers, reducing the likelihood of burnout and attrition (Ingersoll et al., 2014). AI tools like Diffit and MagicSchool are revolutionising the teaching profession by helping educators manage their workload more efficiently and effectively. By alleviating the burden of routine tasks, these tools support a better work-life balance, which is essential for both teacher well-being and retention (Guskey, 2000). Some specific uses of AI in classrooms include boosting problem-based learning, clarifying ideas for writing, creating anticipatory sets, changing the depth of knowledge (DOK) of questions, and writing student outcomes. As AI tools continue to evolve, AI's role in supporting both in-service and pre-service educators will become more significant, making AI an indispensable component of future-ready teaching practices.

AI Ethics

The integration of AI into education is a complex issue with significant ethical considerations. Looking through the lens of equity of education and educational access, the unevenness of the distribution of AI tools and access to these tools can exacerbate existing educational inequalities, often referred to as the Digital Divide (Lutz, 2019). Further, implicit biases in AI learning algorithms can perpetuate biases included in the data AI tools are using to learn, which can lead to limited opportunities for marginalised populations (Shelton & Lanier, 2023). Another equity issue with AI use in education is AI use in personalising learning. While using AI to personalise learning can benefit many students, the overuse of AI by educators to create individualised learning can lead to further marginalising students with specific learning challenges (Wang, 2021).

Privacy concerns must also be discussed when considering the integration of AI in education. AI systems collect and process copious amounts of student data; some of this data is protected by the Family Educational Rights and Privacy Act (FERPA) as well as other federal and state statutes. This means that the use of any student data with AI needs to be closely monitored to ensure that Personally Identifiable Information (PII) is being kept secure and private (Yang & Beil, 2024). Another privacy concern with AI and education is transparency, as applied to data collection and usage. Students and parents must be provided with this information, and AI tools must be clear about what data is being collected and how that data is being used. Last, there are concerns that AI can be used to surveil students instead of being used to analyse data and assist with data-driven decisions (Sharma et al., 2023).

The rapid development and wide application of AI raise important ethical questions about its impact on teaching values. AI ethics refers to the set of moral principles and guidelines that govern the development, deployment, and use of artificial intelligence (Hagendorff, 2020). As indicated earlier, AI programs such as ChatGPT can solve math problems, write emails, construct ideas, and compose essays. The recent model of ChatGPT can even answer multiple-choice questions when the user plugs in the question and options. AI programs have become popular among students as a handy tool to complete assignments. Each time the user types in a prompt, ChatGPT draws content from its massive, preprogrammed database and generates new text, making it nearly impossible for educators to determine plagiarism. Although many AI-writing detectors are available, they are not always reliable. They may sometimes falsely flag original human-written content. AI programs like ChatGPT and Google Gemini cannot only create text but also provide ideas. The robot can spit out a list of ideas to support a thesis. Instead of thinking and writing themselves, students rely on AI programs to do the academic tasks for them, which greatly threatens academic integrity. Before the advent of AI technology, a typical writing assignment may have required hours of diligent research and writing, but students today can watch their paper being written for them within a few short moments. Introducing AI into classrooms might prompt discussions about its ethics. Students will need to grapple with questions about AI's role in society, personal responsibilities, and academic honesty. The main objective of getting an education is to learn knowledge and obtain technical skills. Relying on AI to complete assignments defies this purpose and frustrates educators.

One strategy to tackle this challenge is to assign work that is more difficult to cheat on, such as oral exams and in-person presentations. Some teachers also resort to old-school pencil and paper methods to gauge students' mastery of educational content. Another way to circumvent the problem is to assign specific and narrow-scoped prompts. Chatbots can accurately produce general information, but they do not have a strong background in hyperlocal issues. AI programs cannot answer lecture-based questions and provide adequate information on subjects grounded in the local community. Nevertheless, some AI-proof assignments, such as verbal presentations, may require additional effort to design and grade. Further, educators should emphasise to students both the value of college education and the ethical use of AI technology. While AI can help with assignments, it may not provide an understanding of the material. To fully understand the course material, students must actively engage with the content and think critically. After all, the ultimate value of a college education is to expand one's knowledge, which cannot be achieved through any shortcuts.

In addition to their negative effect on academic honesty, AI systems are often trained on large datasets that may reflect existing societal biases, which can lead to unfair or discriminatory outcomes (Ferrer et al., 2021). AI models can unintentionally perpetuate racial, gender, or socioeconomic biases present in their training data, leading to unequal treatment of certain groups. Educators should help students understand how AI systems work, including how they are trained on data and how biases can be embedded in algorithms. By teaching students about AI's limitations and potential biases, they can become more discerning users of AI tools. Training students to critically evaluate AI-generated content and using studies where AI bias has had a noticeable impact can help students grasp the importance of the ethical use of the technology.

Social Skills

As AI has become integral to daily life through its widespread application and integration across various sectors, it is critical to strike a balance between the use of technology and human interactions in learning environments. Although some may argue that overreliance on AI technology reduces opportunities for in-person interaction with teachers and classmates, the literature suggests that technology can enhance one's social skills. Combs and Slaby (1977) define social skills as "positive skills that are at least minimally acceptable according to societal norms and that are not harmful to others" (p. 162). A socially skilled individual can interpret a given social setting, respond accordingly, and behave acceptably. Statistics suggest that social skills are increasingly important in the workplace. Between 1980 and 2012, the proportion of jobs in the U.S. labour market that required high levels of social interaction increased by nearly 12 percentage points (Deming, 2017). The Pew Research Center (2020) found that employment is growing fastest in roles that demand strong social skills. In 2018,

42% of all workers were employed in jobs where social skills were crucial, compared to just 19% in positions where social skills were less important. Schools need to provide students with both academic and essential skills to succeed in the future. Integrating social skills into the educational experience is vital for a student's overall success (Cartledge & Milburn, 1978). According to education researcher Donald Deshler and his colleagues (1996), social skills are as important as academic skills because deficiencies in social skills can lead to a negative self-concept and social isolation, which in turn limit learning opportunities.

Teacher preparation programs must emphasise the development of social skills among pre-service educators. Effective teaching consists of more than subject matter expertise; effective teaching stems from building strong relationships with students, colleagues, and stakeholders (Phillipson & Phillipson, 2020). Social skills, such as empathy, conflict resolution, and communication, are necessary when creating inclusive learning environments (Leach, 2022). According to Aldrup et al. (2022), these social skills enable teachers to establish rapport with students, understand diverse perspectives, and navigate interpersonal dynamics. Strong social skills also contribute to teacher resilience and teacher job satisfaction, which contribute to student success. Dicke and associates (2020) assert that "the working environment job satisfaction factor was related to student achievement" and "[teacher] job satisfaction is associated with well-being, lower attrition, and better student outcomes" (p. 1061). Teacher preparation programs must prepare pre-service educators with a solid foundation in social skills, so they are capable of meeting the demands of the twenty-first-century classroom.

As indicated earlier, AI-powered chatbots are high customizable and smart in detecting human emotions. AI-driven platforms can simulate real-life social scenarios, helping users practice and improve their social interactions. AI can assist individuals with communication challenges, such as those with autism or social anxiety, by suggesting appropriate responses or helping to interpret social cues. AI tools can analyse facial expressions, voice tone, and other cues to understand user emotions, aiding users in responding in a more empathetic and non-judgmental way. The user can practice interactions without the anxiety of speaking with a real human being. These applications of AI can make social skills training accessible and adaptive.

Conclusion

Since the first chatbot, ELIZA, was created in the 1960s, AI technology has undergone substantial transformations. Generative AI has revolutionised the way we approach education. Chatbots are computer programs that combine AI and natural language processing. They can simulate natural human conversations through text or voice. Newly developed chatbots can write emails, solve math problems, write computer code, interpret questions, and generate beautifully written essays. AI has emerged as one of the most debated subjects in education, as chatbots potentially encourage cheating and promote laziness. Some educators worry that student reliance on AI will disrupt the learning experience and compromise academic integrity. While these concerns are valid, teacher education programs should harness the power of chatbots despite the downfalls. The challenges chatbots pose to academic integrity can be overcome by replacing unsupervised writing assignments with other formats that require original human effort, such as presentations, oral exams, and in-person writing assignments.

Chatbots hold great potential as tools to support teacher education because they are incredibly innovative, customizable, interactive, and engaging. Powered by a massive amount of data, chatbots provide a wealth of information for learners and educators. Chatbots' ability to provide immediate resources and integrate content from various disciplines makes them suitable for interdisciplinary learning. Incorporating chatbots can help foster critical and interdisciplinary thinking, which are key skills in pre-service teachers. As AI continues to evolve, the role of chatbots and generative AI in teacher education will likely expand. By leveraging the capabilities of chatbots, teacher educators can create more personalised and adaptive learning experiences that will meet the needs of diverse pre-service educators. Integrating generative AI and chatbots could enhance the accessibility and inclusivity of education, but also better prepare pre-service educators to engage with emerging technologies in their future classrooms. Embracing chatbots in teacher preparation programs is an opportunity to equip the next generation of educators with tools to thrive in a rapidly changing world.

To fully realise the potential of AI in education and educator preparation programs, there are many pieces that need to fit together. The U.S. Department of Education needs to establish clear guidelines for the development and implementation of AI in education that address critical issues such as privacy, transparency, and bias, and government agencies need to invest in research to advance and investigate AI and its uses in education. These same agencies need to collectively create data privacy regulations that specifically address the unique privacy concerns connected to AI, and funding must be made available to universities and school districts to support professional development on the appropriate and effective implementation of AI in education. Universities and other educational institutions must integrate AI into curricula and provide adequate time and funding to allow instructors to receive the necessary professional development to ensure that AI is being implemented equitably and effectively in teaching. As the institutions of higher education integrate AI into their curricula, they must also collaborate with industry partners to ensure that students are prepared to effectively implement AI in their career contexts. For educator preparation programs, this means closely cooperating with service region school districts as students complete field and clinical experiences to ensure seamless transitions from pre-service to in-service educators and the use of AI in the P-12 classroom. Specific action steps to achieve these goals include the creation of AI-specific digital literacy standards, the creation of AI ethics committees at all levels of education, and monitoring and evaluating AI implementation at all levels of education. Further action steps include creating a framework to ensure teacher educators, pre-service educators, and in-service educators can stay as up-to-date as possible with AI advancements and understand how to adjust research-based educational best practices to incorporate such advancements. Last, a systematic method must be developed to allow international collaboration between educators and educational institutions that encourages global adoption of AI standards for education and allows the sharing of innovative uses of AI in education.

References

Acker, J. R. (2003). Class acts: Outstanding college teachers and the difference they make. *Criminal Justice Review, 28*(2), 215–231. https://doi.org/10.1177/073401680302800202

Akl, A. (2014). Problem based learning (PBL) in teacher education: A review of the effect of PBL on pre-service teachers' knowledge and skills. *European Journal of Educational Sciences, 1*(1), 7–18.

Akram, B., Yoder, S., Tatar, C., Boorugu, S., Aderemi, I., & Jiang, S. (2011). Towards an AI-infused interdisciplinary curriculum for middle-grade classrooms. *Proceedings of the AAAI Conference on Artificial Intelligence, 36*(11), 12681–12688. https://doi.org/10.1609/aaai.v36i11.21544

Aldrup, K., Carstensen, B., & Klusmann, U. (2022). Is empathy the key to effective teaching? A systematic review of its association with teacher-student interactions and student outcomes. *Educational Psychology Review, 34*(3), 1177–1216. https://doi.org/10.1007/s10648-021-09649-y

Alkhatlan, A., & Kalita, J. (2018). Intelligent tutoring systems: A comprehensive historical survey with recent developments. *International Journal of Computer Applications, 181*(43), 1–20. https://doi.org/10.5120/ijca2019918451

Allen, J. M., & Wright, S. E. (2014). Integrating theory and practice in the pre-service teacher education practicum. *Teachers and Teaching, 20*(2), 136–151. https://doi.org/10.1080/13540602.2013.848568

Amerstorfer, C. M., & von Münster-Kistner, C. F. (2021). Student perceptions of academic engagement and student-teacher relationships in problem-based learning. *Frontiers in Psychology, 12*, Article 713057. https://doi.org/10.3389/fpsyg.2021.713057

Amin, A. M., Adiansyah, R., & Hujjatusnaini, N. (2023). The contribution of communication and digital literacy skills to critical thinking. *Indonesian*

Journal of Science Education, 11(3), 697–712. https://doi.org/10.24815/jpsi.v11i3.30838

Azvine, B., Djian, D., Tsui, K. C., & Wobcke, W. (2000). The intelligent assistant: An overview. In B. Azvine, N. Azarmi, & D. D. Nauck (Eds.), *Intelligent systems and soft computing: Prospects, tools and applications* (pp. 215–238). Springer-Verlag. https://doi.org/10.1007/10720181_9

Bank of America. (2024, April). BofA's Erica surpasses 2 billion interactions, helping 42 million clients since launch. https://newsroom.bankofamerica.com/content/newsroom/press-releases/2024/04/bofa-s-erica-surpasses-2-billion-interactions--helping-42-millio.html

Barrows, H. S., & Tamblyn, R. M. (1980). *Problem-based learning: An approach to medical education.* Springer.

Behar-Horenstein, L. S., & Niu, L. (2011). Teaching critical thinking skills in higher education: A review of the literature. *Journal of College Teaching & Learning, 8*(2), 25–42. https://doi.org/10.19030/tlc.v8i2.3554

Benedicto, P. N., & Andrade, R. (2022). Problem-based learning strategies and critical thinking skills among pre-service teachers. *International Journal of Science, Technology, Engineering and Mathematics, 2*(2), 1–28. https://doi.org/10.53378/352885

Bichler, S., Sailer, M., Bauer, E., Kiesewetter, J., Härtl, H., Fischer, M. R., & Fischer, F. (2024). Promoting diagnostic reasoning in teacher education: The role of case format and perceived authenticity. *European Journal of Psychology of Education*, 1–26. https://doi.org/10.1007/s10212-024-00874-x

Boud, D., Keogh, R., & Walker, D. (2013). *Reflection: Turning experience into learning.* Routledge. https://doi.org/10.4324/9781315059051

Bransford, J. B., Brown, A. L., & Cocking, R. R. (Eds.). (1999). *How people learn: Brain, mind, experience, and school.* National Academy Press.

Bushweller, K. (2024). It terrifies me: Clinical psychologist on tech overuse in the age of AI. *Education Week, 43*(22), 17–18.

Cartledge, G., & Milburn, J. F. (1978). The case for teaching social skills in the classroom: A review. *Review of Educational Research, 48*(1), 133–156. https://doi.org/10.3102/00346543048001133

Chan, S., Maneewan, S., & Koul, R. (2021). Cooperative learning in teacher education: Its effects on EFL pre-service teachers' content knowledge and teaching self-efficacy. *Journal of Education for Teaching, 47*(5), 654–667. https://doi.org/10.1080/02607476.2021.1931060

Chen, W., & Wellman, B. (2004). The global digital divide–within and between countries. *IT & Society, 1*(7), 39–45.

Cheung, A. C. K., & Slavin, R. E. (2013). The effectiveness of educational technology applications for enhancing mathematics achievement in K-12 classrooms: A meta-analysis. *Educational Research Review, 9*, 88–113. https://doi.org/10.1016/j.edurev.2013.01.001

Chua, B. L., Tan, O. S., & Liu, W. C. (2016). Journey into the problem-based learning process: Student perspectives. *Innovations in Education and Teaching International, 53*(2), 191–202. https://doi.org/10.1080/14703297.2014.964196

Çiftçi, S. K., & Karadag, E. (2023). Twenty years of digital literacy in education: A bibliometric review and research agenda. *Education and Information Technologies, 28*, 11913–11941. https://doi.org/10.1007/s10639-023-11629-5

Clark, A. E., D'Ambrosio, C., & Ghislandi, S. (2021). Growing up in the recession: Beliefs and the economic crisis. *Scottish Journal of Political Economy, 68*(1), 1–23. https://doi.org/10.1111/sjpe.12254

Cochrane, T., & Narayan, V. (2021). Designing authentic professional learning experiences for educators. *Australasian Journal of Educational Technology, 37*(5), 69–82. https://doi.org/10.14742/ajet.6513

Collins, A., Brown, J. S., & Newman, S. E. (1989). Cognitive apprenticeship: Teaching the craft of reading, writing, and mathematics. In L. B. Resnick (Ed.), *Knowing, learning, and instruction: Essays in honor of Robert Glaser* (pp. 453–494). Lawrence Erlbaum Associates.

Conole, G. (2013). *Designing for learning in an open world.* Springer. https://doi.org/10.1007/978-1-4419-8517-0

Coulson, D., & Harvey, M. (2013). Scaffolding student reflection for experience-based learning: A framework. *Teaching in Higher Education, 18*(4), 401–413. https://doi.org/10.1080/13562517.2012.752726

Cress, U., & Kimmerle, J. (2008). A systemic and cognitive view on collaborative knowledge building with wikis. *International Journal of Computer-Supported Collaborative Learning, 3*, 105–122. https://doi.org/10.1007/s11412-007-9035-z

Darling-Hammond, L., Flook, L., Cook-Harvey, C., Barron, B., & Osher, D. (2020). Implications for educational practice of the science of learning and development. *Applied Developmental Science, 24*(2), 97–140. https://doi.org/10.1080/10888691.2018.1537791

Davis, N. L., Gough, M., & Taylor, L. L. (2019). Online teaching: Advantages, obstacles and tools for getting it right. *Journal of Teaching in Travel & Tourism, 19*(3), 256–263. https://doi.org/10.1080/15313220.2019.1612313

DeCoito, I., & Richardson, T. (2018). Teachers and technology: Present practice and future directions. *Contemporary Issues in Technology and Teacher Education, 18*(2), 362–378. https://citejournal.org/volume-18/issue-2-18/science/teachers-and-technology-present-practice-and-future-directions/

Dewey, J. (1938). *Experience and education.* Collier Books.

Donnelly, R., & Fitzmaurice, M. (2005). Collaborative project-based learning and problem-based learning in higher education: A consideration of tutor and student roles in learner-focused strategies. In G. O'Neill, S. Moore, & B. McMullin (Eds.), *Emerging issues in the practice of university learning and teaching* (pp. 87–98). AISHE.

Dousay, T. A. (2017). Instructional design models and learning theories. In R. D. Branch, T. A. Dousay, & M. B. J. Y. Wilson (Eds.), *Educational media and technology yearbook* (Vol. 40, pp. 77–87). Springer. https://doi.org/10.1007/978-3-319-45019-4_10

Drake, S. M., & Reid, J. L. (2020). Integrated curriculum as an effective way to teach 21st century capabilities. *Asia Pacific Journal of Educational Research, 3*(1), 31–50. https://doi.org/10.30777/APJER.2020.3.1.02

Dunlosky, J., Rawson, K. A., Marsh, E. J., Nathan, M. J., & Willingham, D. T. (2013). Improving students' learning with effective learning techniques: Promising directions from cognitive and educational psychology. *Psychological Science in the Public Interest, 14*(1), 4–58. https://doi.org/10.1177/1529100612453266

Ebner, M., & Schön, S. (2020). COVID-19 epidemic as E-learning boost? Chronological development and effects at an Austrian university against the background of the concept of "E-learning readiness." *Future Internet, 12*(6), Article 94. https://doi.org/10.3390/fi12060094

Ertmer, P. A., & Ottenbreit-Leftwich, A. T. (2010). Teacher technology change: How knowledge, confidence, beliefs, and culture intersect. *Journal of Research on Technology in Education, 42*(3), 255–284. https://doi.org/10.1080/15391523.2010.10782551

European Commission. (2020). *Digital education action plan 2021–2027: Resetting education and training for the digital age.* https://education.ec.europa.eu/focus-topics/digital/digital-education-action-plan

European Commission/EACEA/Eurydice. (2022). *Digital education at school in Europe* (Eurydice Report). Publications Office of the European Union. https://data.europa.eu/doi/10.2797/339398

Fawns, T. (2022). An ecological critical realist approach to educational technology research. *Research in Learning Technology, 30.* https://doi.org/10.25304/rlt.v30.2663

Ferguson, R., Brasher, A., Clow, D., Cooper, A., Hillaire, G., Mittelmeier, J., Rienties, B., Ullmann, T., & Vuorikari, R. (2019). Research evidence on the use of learning analytics: Implications for education policy. *Joint Research Centre.* https://doi.org/10.2760/12297

Fiok, K. M. (2021). Building a professional learning community: Online learning and professional development during the COVID-19 pandemic. *TechTrends, 65*(3), 309–315. https://doi.org/10.1007/s11528-021-00574-8

Foulger, T. S., Graziano, K. J., Schmidt-Crawford, D. A., & Slykhuis, D. A. (2020). Throw me a lifeline: A professional development resource for teacher educators. *Contemporary Issues in Technology and Teacher Education, 20*(1), 68–101. https://citejournal.org/volume-20/issue-1-20/current-practice/throw-me-a-lifeline-a-professional-development-resource-for-teacher-educators/

Freeman, A., Adams Becker, S., Cummins, M., Davis, A., & Hall Giesinger, C. (2017). *NMC/CoSN Horizon Report: 2017 K–12 Edition.* The New Media Consortium.

Fullan, M., & Langworthy, M. (2014). *A rich seam: How new pedagogies find deep learning.* Pearson. https://michaelfullan.ca/wp-content/uploads/2014/01/3897.Rich_Seam_web.pdf

Garrison, D. R., Anderson, T., & Archer, W. (2000). Critical inquiry in a text-based environment: Computer conferencing in higher education. *The Internet and Higher Education, 2*(2–3), 87–105. https://doi.org/10.1016/S1096-7516(00)00016-6

Garrison, D. R., & Vaughan, N. D. (2008). *Blended learning in higher education: Framework, principles, and guidelines.* Jossey-Bass.

Garrison, D. R., & Vaughan, N. D. (2013). *Institutional change and leadership associated with blended learning innovation: Two case studies.* The Commonwealth of Learning.

George, S., Gruson, B., Couronné, T., Chaffar, S., Laforcade, P., Cherti, M., Gilliot, J.-M., Garlatti, S., Prévôt, P., & Gillet, D. (2021). Designing dashboards for learning analytics: A methodology based on a multi-stakeholder approach. *European Journal of Open, Distance and E-learning, 24*(1), 81–104. https://doi.org/10.2478/eurodl-2021-0006

Girvan, C., & Savage, T. (2019). Virtual worlds: A new medium for teacher professional development. *TeachMeet Research Journal, 1*, 37–43. https://teachmeet.pbworks.com/w/file/fetch/137071249/TMRJ1_Girvan_Savage.pdf

Goksel, N., & Bozkurt, A. (2019). Connected learning in teacher professional development: The case of the eTwinning community. *Contemporary Educational Technology, 10*(4), 330–344. https://doi.org/10.30935/cet.634181

González-Fernández, B., & González-Sanmamed, M. (2021). Active methodologies in initial teacher training: Towards a sustainable pedagogical model. *Education Sciences, 11*(2), Article 54. https://doi.org/10.3390/educsci11020054

Goodyear, P. (2015). Teaching as design. *HERDSA Review of Higher Education, 2*, 27–50. https://www.herdsa.org.au/herdsa-review-higher-education-vol-2/27-50

Goodyear, P., & Carvalho, L. (2013). *The architecture of productive learning networks*. Routledge. https://doi.org/10.4324/9780203075291

Greller, W., & Drachsler, H. (2012). Translating learning into numbers: A generic framework for learning analytics. *Educational Technology & Society, 15*(3), 42–57. https://www.jstor.org/stable/jeductechsoci.15.3.42

Gronseth, S. L., & Hebert, D. (2019). Advancing teacher education and curriculum development through study abroad programs. In V. C. X. Wang (Ed.), *Handbook of research on education for participative citizenship and global prosperity* (pp. 218–238). IGI Global. https://doi.org/10.4018/978-1-5225-9621-9.ch010

Gros, B., García-Peñalvo, F. J., & Escofet, A. (2012). Beyond the net generation debate: A comparison of digital learners in face-to-face and virtual universities. *International Review of Research in Open and Distance Learning, 13*(4), 190–210. https://doi.org/10.19173/irrodl.v13i4.1305

Gros, B., Maina, M. F., & Barberà, E. (2016). The architecture of smart learning environments. In B. Gros, M. F. Maina, & E. Barberà (Eds.), *The architecture of smart learning environments* (pp. 1–11). Springer. https://doi.org/10.1007/978-3-319-20872-2_1

Gutiérrez, I., & Serrano, J. L. (2016). The design of scenarios as a key component of the training of teachers in the use of ICT for learning. *Comunicar, 24*(47), 31–39. https://doi.org/10.3916/C47-2016-03

Hashweh, M. Z. (2005). Teacher pedagogical constructions: A reconfiguration of pedagogical content knowledge. *Teachers and Teaching, 11*(3), 273–292. https://doi.org/10.1080/13450600500105502

Haythornthwaite, C. (2009). Participatory transformations. In A. N. Joinson, K. Y. A. McKenna, T. Postmes, & U.-D. Reips (Eds.), *The Oxford handbook of Internet psychology* (pp. 128–140). Oxford University Press.

Haythornthwaite, C. (2015). Rethinking learning networks: Collaborative possibilities in new media. *Educational Technology & Society, 18*(3), 385–396. https://www.jstor.org/stable/jeductechsoci.18.3.385

Haythornthwaite, C., & Andrews, R. (2011). *E-learning theory and practice.* SAGE Publications.

Hernández-Leo, D., Martinez-Maldonado, R., Pardo, A., & Muñoz-Cristóbal, J. A. (2019). Learning about collaborative learning design. *British Journal of Educational Technology, 50*(6), 2829–2831. https://doi.org/10.1111/bjet.12879

Hernández-Leo, D., Prieto, L. P., & Santos, P. (2018). Preface to the special section on learning design tools. *British Journal of Educational Technology, 49*(6), 1041–1045. https://doi.org/10.1111/bjet.12702

Howard, S. K., Tondeur, J., Ma, J., & Yang, J. (2021). What to teach? Strategies for developing digital competency in pre-service teacher training. *Computers & Education, 165,* Article 104149. https://doi.org/10.1016/j.compedu.2021.104149

Hughes, J. E., & Burke, A. (2014). The digital disconnect: The widening gap between Internet-savvy students and their schools. In L. Harasim (Ed.), *Learning theory and online technologies* (pp. 245–266). Routledge.

Ilomäki, L., Paavola, S., Lakkala, M., & Kantosalo, A. (2016). Digital competence: A conceptual framework for teachers. *Oulu University Press.* https://doi.org/10.1007/s11423-016-9438-1

Instituto Nacional de Tecnologías Educativas y de Formación del Profesorado (INTEF). (2017). *Marco Común de Competencia Digital Docente.* Ministerio de Educación, Cultura y Deporte. https://aprende.intef.es/marco-comun-de-competencia-digital-docente

Instituto Nacional de Tecnologías Educativas y de Formación del Profesorado (INTEF). (2022). *Marco de referencia de la competencia digital docente.* Ministerio de Educación y Formación Profesional. https://intef.es/competencia-digital-docente/marco-de-referencia-de-la-competencia-digital-docente

Instituto Nacional para la Evaluación de la Educación (INEE). (2019). *Panorama educativo de México: Indicadores del Sistema Educativo Nacional 2019. Educación básica y media superior.* https://www.inee.edu.mx/wp-content/uploads/2019/12/P1D317.pdf

Janssen, J., Berlanga, A. J., Vogten, H., & Koper, R. (2008). Towards a learning path specification. *International Journal of Continuing Engineering Education and Life Long Learning, 18*(1), 77–97. https://doi.org/10.1504/IJCEELL.2008.016181

Janssen, J., Berlanga, A. J., Vogten, H., & Koper, R. (2012). Towards flexible learning path specification. In R. Koper (Ed.), *Learning Network Services for Professional Development* (pp. 113–127). Springer. https://doi.org/10.1007/978-3-642-20178-1_8

Janssen, J., Berlanga, A. J., & Koper, R. (2011). Evaluation of integrated support for authoring unit-of-learning designs in IMS Learning Design. *Educational Technology & Society, 14*(1), 123–135. https://www.jstor.org/stable/jeductechsoci.14.1.123

Janssen, J., Berlanga, A. J., & Koper, R. (2012). Learning path specification and its role in achieving educational goals. *British Journal of Educational Technology, 43*(6), 988–1005. https://doi.org/10.1111/j.1467-8535.2011.01257.x

Jonassen, D. H. (1999). Designing constructivist learning environments. In C. M. Reigeluth (Ed.), *Instructional design theories and models: A new paradigm of instructional theory* (Vol. II, pp. 215–239). Lawrence Erlbaum Associates.

Jonassen, D. H. (2000). Toward a design theory of problem solving. *Educational Technology Research and Development, 48*(4), 63–85. https://doi.org/10.1007/BF02300500

Jonassen, D. H. (2011). *Learning to solve problems: A handbook for designing problem-solving learning environments.* Routledge.

Jordan, K. (2014). Initial trends in enrolment and completion of massive open online courses. *The International Review of Research in Open and Distributed Learning, 15*(1), 133–160. https://doi.org/10.19173/irrodl.v15i1.1651

Kalz, M. (2015). Lifelong learning and its support with new technologies. *International Encyclopedia of the Social & Behavioral Sciences* (2nd ed., Vol. 13, pp. 93–99). Elsevier. https://doi.org/10.1016/B978-0-08-097086-8.92002-4

Kalz, M., & Specht, M. (2014). If MOOCS are the answer, did we ask the right questions? Implications for the design of large-scale online courses. *International Journal of Educational Technology in Higher Education, 11*(2), 74–86. https://doi.org/10.7238/rusc.v11i2.2105

Kasch, N., & Maier, R. (2012). How to evaluate and measure the value of Web 2.0 learning services in SMEs. *Journal of Workplace Learning, 24*(5), 380–394. https://doi.org/10.1108/13665621211239818

Kay, R. H., & Leung, S. (2013). Response to Web-Based Learning Tools: An Exploratory Study of Adult Online Learners. *The Canadian Journal of Learning and Technology, 39*(1). https://doi.org/10.21432/T2CC8K

Kay, R. H., & Leung, S. (2015). Examining the benefits and challenges of using audience response systems: A review of the literature. *Computers & Education, 85*, 52–64. https://doi.org/10.1016/j.compedu.2015.02.019

Kember, D. (2001). Beliefs about knowledge and the process of teaching and learning as a factor in adjusting to study in higher education. *Studies in Higher Education, 26*(2), 205–221. https://doi.org/10.1080/03075070120052116

Kennedy, G. E., Judd, T. S., Churchward, A., Gray, K., & Krause, K. L. (2008). First year students' experiences with technology: Are they really digital natives? *Australasian Journal of Educational Technology, 24*(1), 108–122. https://doi.org/10.14742/ajet.1233

Khalil, H., & Ebner, M. (2014). MOOCs completion rates and possible methods to improve retention—A literature review. In C. Busch (Ed.), *Proceedings of World Conference on Educational Multimedia, Hypermedia and Telecommunications* (pp. 1236–1244). Association for the Advancement of Computing in Education (AACE). https://www.learntechlib.org/primary/p/147452/

Khalil, H., & Ebner, M. (2015). "How satisfied are you with your MOOC?"—A research study on interaction in huge online courses. In S. Zvacek, M. Restivo, J. Uhomoibhi, & M. Helfert (Eds.), *Computer Supported Education. CSEDU 2014. Communications in Computer and Information Science* (Vol. 510, pp. 101–115). Springer. https://doi.org/10.1007/978-3-319-29585-5_7

Khalil, H., & Ebner, M. (2015). MOOCs completion rates and possible methods to improve retention—A literature review. In *Proceedings of World Conference on Educational Multimedia, Hypermedia and Telecommunications 2014* (pp. 1236–1244). Association for the Advancement of Computing in Education (AACE). https://www.learntechlib.org/primary/p/147452/

Kizilcec, R. F., Piech, C., & Schneider, E. (2013). Deconstructing disengagement: Analyzing learner subpopulations in massive open online courses. In *Proceedings of the Third International Conference on Learning Analytics and Knowledge* (pp. 170–179). ACM. https://doi.org/10.1145/2460296.2460330

Kop, R., Fournier, H., & Mak, J. S. F. (2011). A pedagogy of abundance or a pedagogy to support human beings? Participant support on massive open online courses. *The International Review of Research in Open and Distributed Learning, 12*(7), 74–93. https://doi.org/10.19173/irrodl.v12i7.1041

Kop, R., & Hill, A. (2008). Connectivism: Learning theory of the future or vestige of the past? *The International Review of Research in Open and Distributed Learning, 9*(3), 1–13. https://doi.org/10.19173/irrodl.v9i3.523

Krause, S. D., & Lowe, C. (2014). Invasion of the MOOCs: The promises and perils of massive open online courses. Parlor Press.

Lane, L. M. (2012). Three reasons why MOOCs will not transform higher education. *e-Literate.* https://eliterate.us/three-reasons-why-moocs-will-not-transform-higher-education/

Laurillard, D. (2002). *Rethinking university teaching: A conversational framework for the effective use of learning technologies* (2nd ed.). RoutledgeFalmer.

Laurillard, D. (2007). Pedagogical forms for mobile learning. In N. Pachler (Ed.), *Mobile learning: Towards a research agenda* (pp. 153–175). WLE Centre, Institute of Education.

Laurillard, D. (2016). The educational problem that MOOCs could solve: Professional development for teachers of disadvantaged students. *Research in Learning Technology, 24,* Article 29369. https://doi.org/10.3402/rlt.v24.29369

Laurillard, D., Charlton, P., Craft, B., Dimakopoulos, D., Ljubojevic, D., Magoulas, G., & Masterman, E. (2013). A constructionist learning environment for teachers to model learning designs. *Journal of Computer Assisted Learning, 29*(1), 15–30. https://doi.org/10.1111/j.1365-2729.2011.00458.x

Law, P. (2015). Digital badging at The Open University: Recognition for informal learning. *Open Learning: The Journal of Open, Distance and e-Learning, 30*(3), 221–234. https://doi.org/10.1080/02680513.2015.1069244

Law, P., Perryman, L.-A., & Law, A. (2015). Open educational practices and attitudes to openness across India: Reporting the findings of the Open Education Research Hub Pan-India Survey. *Journal of Interactive Media in Education, 2015*(1), Article 21. https://doi.org/10.5334/jime.ax

Law, P., Perryman, L.-A., & Law, A. (2022). Supporting low-income learners to become OER producers. *Journal of Interactive Media in Education, 2022*(1), Article 2. https://doi.org/10.5334/jime.662

Littlejohn, A., & Hood, N. (2018). Reconceptualising learning in the digital age: The [un]democratising potential of MOOCs. In R. Bennett, M. Kent, & R. Baker (Eds.), *Massive Open Online Courses and higher education: What went right, what went wrong and where to next?* (pp. 20–35). Routledge.

Margaryan, A., Bianco, M., & Littlejohn, A. (2015). Instructional quality of massive open online courses (MOOCs). *Computers & Education, 80,* 77–83. https://doi.org/10.1016/j.compedu.2014.08.005

Marshall, S. (2014). Exploring the ethical implications of MOOCs. *Distance Education, 35*(2), 250–262. https://doi.org/10.1080/01587919.2014.917706

Mason, J. (2002). *Qualitative researching* (2nd ed.). Sage.

McAuley, A., Stewart, B., Siemens, G., & Cormier, D. (2010). *The MOOC model for digital practice.* SSHRC Knowledge Synthesis Grant on the Digital Economy. https://www.researchgate.net/publication/241519877_The_MOOC_Model_for_Digital_Practice

Milligan, C., & Littlejohn, A. (2016). How health professionals regulate their learning in massive open online courses. *The Internet and Higher Education, 31,* 113–121. https://doi.org/10.1016/j.iheduc.2016.07.005

Milligan, C., Littlejohn, A., & Margaryan, A. (2013). Patterns of engagement in connectivist MOOCs. *Journal of Online Learning and Teaching, 9*(2), 149–159. http://jolt.merlot.org/vol9no2/milligan_0613.pdf

Milligan, C., Littlejohn, A., & Hood, N. (2016). Learning in MOOCs: A comparison study of 18–50- and 50+ year-old learners. *British Journal of Educational Technology, 48*(5), 1193–1206. https://doi.org/10.1111/bjet.12561

Mishra, S. (2023). A teacher's professional development is incomplete without OER. *Open Praxis, 15*(1), 1–4. https://doi.org/10.5944/openpraxis.15.1.1

Mishra, S., & Singh, A. (2017). MOOCs and curriculum development in India. In S. Bolliger & D. Kumar (Eds.), *Handbook of research on digital content, mobile learning, and technology integration models in teacher education* (pp. 238–248). IGI Global. https://doi.org/10.4018/978-1-5225-1856-3.ch012

Mishra, S., & Singh, A. (2019). Higher education faculty attitude, motivation and perception towards open and online education in India. *Distance Education, 40*(1), 62–78. https://doi.org/10.1080/01587919.2018.1553567

Moore, M. G. (2013). *Handbook of distance education* (3rd ed.). Routledge.

Mukhopadhyay, R. (2019). Developing OERs for higher education in India: The collaborative practices of the National Repository of Open Educational Resources project. *Open Praxis, 11*(4), 401–412. https://doi.org/10.5944/openpraxis.11.4.1022

Mulder, F., & Jansen, D. (2015). MOOCs for opening up education and the OpenupEd initiative. In C. J. Bonk, M. M. Lee, T. C. Reeves, & T. H. Reynolds (Eds.), *MOOCs and open education around the world* (pp. 71–81). Routledge.

Naidu, S. (2021). Supporting the digital transformation of higher education in the Asia-Pacific. *Distance Education, 42*(2), 160–174. https://doi.org/10.1080/01587919.2021.1910497

Naidu, S., & Dube, B. (2022). Online learning and the future of schooling: Challenges and possibilities. *Prospects,* 1–15. https://doi.org/10.1007/s11125-022-09604-2

Nkuyubwatsi, B. (2014). Quality assurance and recognition of open learning in Rwanda: Tensions, opportunities and the way forward. *Journal of Learning for Development, 1*(1), 47–58. https://jl4d.org/index.php/ejl4d/article/view/11

OECD. (2024). *Using AI in the workplace: Opportunities, risks and policy responses.* https://www.oecd.org/publications/using-ai-in-the-workplace-73d417f9-en.htm

Offor, U. I., Nwaru, P., & Offiah, C. (2024). Challenges of the use of artificial intelligence on academic performance of undergraduate students in public universities in Anambra State. *African Journal of Educational Management, Teaching and Entrepreneurship Studies, 12*(2), 176–190.

OpenAI. (2023). *GPT-4 technical report.* https://cdn.openai.com/papers/gpt-4.pdf

Palacios Hidalgo, F. J., Gómez Parra, M. E., & Huertas Abril, C. A. (2020). Digital and media competences: Key competences for EFL teachers. *Teaching English with Technology, 20*(1), 43–59.

Pérez, J. Q., Daradoumis, T., & Marquès Puig, J. M. (2020). Rediscovering the use of chatbots in education: A systematic literature review. *Computer Applications in Engineering Education, 28*(6), 1549–1565. https://doi.org/10.1002/cae.22326

Phillipson, S. N., & Phillipson, S. (2020). *The power of expert teaching: Lessons for modern education.* Routledge. https://doi.org/10.4324/9781351139762

Prieto, L. P., Magnuson, P., Dillenbourg, P., & Saar, M. (2020). Reflection for action: Designing tools to support teacher reflection on everyday evidence. *Technology, Pedagogy and Education, 29*(3), 279–295. https://doi.org/10.1080/1475939X.2020.1762721

Progressive Insurance. (2024). Chatting with Flo FAQs. https://www.progressive.com/contact-us/chatbot-faqs/

Radović, S., Hummel, H. G. K., & Vermeulen, M. (2021). The challenge of designing 'more' experiential learning in higher education programs in the field of teacher education: A systematic review study. *International Journal of Lifelong Education, 40*(5–6), 545–560. https://doi.org/10.1080/02601370.2021.1994664

Santaolalla, E., Urosa, B., Martín, O., Verde, A., & Díaz, T. (2020). Interdisciplinarity in teacher education: Evaluation of the effectiveness of an educational innovation project. *Sustainability, 12*(17), Article 6748. https://doi.org/10.3390/su12176748

Savery, J. R., & Duffy, T. M. (1995). Problem based learning: An instructional model and its constructivist framework. *Educational Technology, 35*(5), 31–38.

Shen, I. (2024). The robot wrote my college papers: Integrating chatbots to assist higher education. In S. Hai-Jew (Ed.), *Generative AI in teaching and learning* (pp. 311–327). IGI Global. https://doi.org/10.4018/979-8-3693-0074-9.ch013

Sillat, L. H., Tammets, K., & Laanpere, M. (2021). Digital competence assessment methods in higher education: A systematic literature review. *Education Sciences, 11*(8), Article 402. https://doi.org/10.3390/educsci11080402

Sharma, A., Khan, S., & Dar, T. A. (2023). Challenges and issues in privacy of student education record in the U.S.A. *Privacy Matters,* 235.

Shelton, K., & Lanier, D. (2023). *The promises and perils of AI in education.* Lanier Learning.

Sill, D. J. (2001). Integrative thinking, synthesis, and creativity in interdisciplinary studies. *The Journal of General Education, 50*(4), 288–311. https://doi.org/10.1353/jge.2001.0032

Slavin, R. (1995). *Cooperative learning: Theory, research, and practice.* Allyn & Bacon.

The Association of American Colleges and Universities. (2011). *The LEAP vision for learning: Outcomes, practices, impact, and employers' view.* Author.

The Chronicle of Higher Education. (2023, August 16). I'm a student. You have no idea how much we're using ChatGPT. No professor or software could ever pick up on it. *The Chronicle of Higher Education.* https://www.chronicle.com/article/im-a-student-you-have-no-idea-how-much-were-using-chatgpt

The New York Times. (2023, February 2). What students are saying about ChatGPT. *The New York Times.* https://www.nytimes.com/2023/02/02/learning/students-chatgpt.html

Tinmaz, H. (2023). Assessment challenges of using ChatGPT in higher education. *Sustainability, 15*(12), Article 9393. https://doi.org/10.3390/su15129393

Trust, T. (2020). The 3 big questions of edtech leadership. *Educational Leadership, 77*(8), 58–63. https://www.ascd.org/el/articles/the-3-big-questions-of-edtech-leadership

UNESCO. (2021). *AI and education: Guidance for policy-makers.* https://unesdoc.unesco.org/ark:/48223/pf0000376709

UNESCO. (2023). *ChatGPT and artificial intelligence in higher education: Quick start guide.* https://unesdoc.unesco.org/ark:/48223/pf0000385038

van Dis, E. A. M., Bollen, J., Zuidema, W., van Rooij, R., & Bockting, C. L. H. (2023). ChatGPT: Five priorities for research. *Nature, 614*(7947), 224–226. https://doi.org/10.1038/d41586-023-00288-7

Van Gog, T., & Sweller, J. (2015). Not new, but nearly forgotten: The redundancy principle in multimedia learning. *Educational Psychology Review, 27*(2), 265–278. https://doi.org/10.1007/s10648-014-9295-9

Vardi, M. Y. (2023). Generative AI: Is this time different? *Communications of the ACM, 66*(4), 5. https://doi.org/10.1145/3583078

Veletsianos, G. (2023, June 21). Yes, students are using ChatGPT to cheat. Here's why that's a good thing. *EdSurge.* https://www.edsurge.com/news/2023-06-21-yes-students-are-using-chatgpt-to-cheat-here-s-why-that-s-a-good-thing

Wang, Y. (2023). Using AI tools in academic writing: How to maintain originality and avoid plagiarism. *Journal of Academic Ethics.* Advance online publication. https://doi.org/10.1007/s10805-023-09467-2

West, D. M., & Allen, J. R. (2023). *Turning point: Policymaking in the era of artificial intelligence.* Brookings Institution Press.

Wheeler, S. (2012). e-Learning and digital learning. In N. M. Seel (Ed.), *Encyclopedia of the sciences of learning* (pp. 1074–1076). Springer. https://doi.org/10.1007/978-1-4419-1428-6_660

Williamson, B., & Eynon, R. (2020). Historical threads, missing links, and future directions in AI in education. *Learning, Media and Technology, 45*(3), 223–235. https://doi.org/10.1080/17439884.2020.1798995

Wimmer, J. J., & Lewis, D. (2023). Who's afraid of ChatGPT? *Phi Delta Kappan, 105*(1), 22–26. https://doi.org/10.1177/00317217231189108

Wong, W. (2023, May 26). ChatGPT is now allowed in schools. *Fortune.* https://fortune.com/2023/05/26/chatgpt-back-in-school-openai-new-york-education/

Wu, D., Wang, S., & Luo, J. (2023). Education in the era of ChatGPT: A critical review of existing literature and implications. *Education and Information Technologies.* Advance online publication. https://doi.org/10.1007/s10639-023-11975-7

Xing, B., & Marwala, T. (2017). Implications of the Fourth Industrial Age on higher education. *The Thinker, 73*(2), 10–15. https://hdl.handle.net/10520/EJC-c9d7689f3

Xu, Y. (2023). Generative AI and writing instruction: Implications for literacy education. *Literacy, 57*(2), 141–150. https://doi.org/10.1111/lit.12293

Yadav, A., & Kumar, A. (2023). Leveraging ChatGPT for programming education: Opportunities and challenges. *Education and Information Technologies, 28,* 11435–11457. https://doi.org/10.1007/s10639-023-11964-w

Zhai, X. (2022). ChatGPT: A new era of AI-powered education. *Education and Information Technologies.* Advance online publication. https://doi.org/10.1007/s10639-023-11757-3

Zhai, X. (2023a). Artificial intelligence writing tools and academic integrity: Concerns and countermeasures. *International Journal of Educational Technology in Higher Education, 20,* Article 46. https://doi.org/10.1186/s41239-023-00414-0

Zhai, X. (2023b). ChatGPT for higher education assessment: A timely tool or Pandora's box? *International Journal of Educational Technology in Higher Education, 20,* Article 12. https://doi.org/10.1186/s41239-023-00398-x

Zhai, X. (2023c). ChatGPT in education: Teachers' attitudes and perceptions. *Education and Information Technologies, 28,* 11219–11237. https://doi.org/10.1007/s10639-023-11935-1

Zhang, D., Yin, X., Wang, D., & Liu, X. (2023). Generative artificial intelligence in education: Opportunities and challenges of ChatGPT. *Smart Learning Environments, 10,* Article 18. https://doi.org/10.1186/s40561-023-00241-y

Zhao, Y. (2023, February 17). What does ChatGPT mean for education? *Education Week.* https://www.edweek.org/technology/opinion-what-does-chatgpt-mean-for-education/2023/02

Zheng, L., & Hu, X. (2023). The impact of ChatGPT on writing: A systematic review. *Journal of Educational Computing Research, 61*(6), 1425–1444. https://doi.org/10.1177/07356331231185945

Zhou, W., Chen, Y., & Li, L. (2023). The promise and peril of ChatGPT in higher education: A case study of AI-augmented essay writing. *Journal of Computing in Higher Education.* Advance online publication. https://doi.org/10.1007/s12528-023-09358-1

Zhu, J., & Xie, Y. (2023). ChatGPT in education: Applications, opportunities, and challenges. *Education and Information Technologies, 28,* 11283–11300. https://doi.org/10.1007/s10639-023-11949-9

Zhu, L. (2023). Ethical considerations of using ChatGPT in academic writing. *Journal of Academic Ethics, 21,* 213–223. https://doi.org/10.1007/s10805-023-09457-9

Ziskin, M., & Harris, A. (2023, February 2). Don't ban ChatGPT in schools. Teach with it. *Harvard Business Review.* https://hbr.org/2023/02/dont-ban-chatgpt-in-schools-teach-with-it

Zuboff, S. (2019). *The age of surveillance capitalism: The fight for a human future at the new frontier of power.* PublicAffairs.

Chapter Nine
Innovative Approaches to Virtual Reality (VR) and Augmented Reality (AR) in Teacher Education

Elhadj Moussa BenMoussa

University of East London, London, UK

Abstract

This chapter explores the integration of Virtual Reality (VR) and Augmented Reality (AR) in teacher education, highlighting their potential to transform traditional training practices. VR and AR offer immersive, interactive simulations that allow teacher candidates to engage in realistic classroom environments, enhancing their practical teaching skills. The chapter discusses key benefits such as personalised learning, increased engagement, and the development of culturally responsive pedagogy. It also addresses significant challenges, including technical barriers, ethical concerns, and the need for pedagogical alignment. Best practices for implementation, such as professional development, collaborative learning, and continuous evaluation, are outlined to guide educators and institutions in leveraging these technologies effectively. The findings emphasise the importance of addressing logistical and ethical challenges while adopting innovative strategies to enhance teacher training.

Keywords: Virtual Reality, Augmented Reality, teacher education, immersive learning, personalised learning, culturally responsive pedagogy.

***

Introduction

Teacher education is entering a new era, driven by advances in technology, particularly with the introduction of Virtual Reality (VR) and Augmented Reality (AR). These immersive technologies are being integrated into teacher training programs to help future educators gain practical experience in a safe, simulated environment. VR and AR create opportunities for teacher candidates to step into virtual classrooms, where they can interact with students, manage different classroom situations, and refine their teaching techniques without the immediate pressures of a real-world setting.

By immersing teacher candidates in realistic, controlled classroom scenarios, VR and AR allow them to practice handling diverse teaching challenges, from managing student behaviour to adjusting instructional strategies based on different learning styles. These simulations provide invaluable hands-on experience, helping candidates develop confidence and competence before entering actual classrooms. In essence, VR and AR serve as a bridge between theoretical knowledge and the practical realities of teaching.

The integration of VR and AR is especially important in today's educational landscape, where classrooms are more diverse and students have varied learning needs. Teachers must be equipped with the skills to create inclusive learning environments, manage technology, and engage students in meaningful ways. VR and AR offer personalised learning experiences for teacher candidates, allowing them to focus on specific areas of improvement and adapt their teaching methods according to the needs of different types of students.

This chapter explores the growing role of VR and AR in teacher education. It highlights how these technologies are transforming traditional approaches to training by providing more dynamic, interactive, and personalised learning experiences. Additionally, it examines some of the challenges associated with implementing VR and AR, such as cost and access, while also considering the future potential of these technologies to reshape teacher preparation programs globally.

Benefits of VR and AR in Teacher Education

Before delving into specific case studies, it's important to outline the key benefits VR and AR bring to teacher education. These technologies can be broadly categorised into three essential areas: enhanced simulation and practice, personalised learning experiences, and the development of culturally responsive pedagogy.

Enhanced Simulation and Practice: A Theoretical and Scholarly Exploration

The integration of Virtual Reality (VR) and Augmented Reality (AR) in teacher education offers a transformative shift in the way educators are trained. One of the most compelling features of these technologies is their ability to create highly immersive and interactive simulations of real-world classroom environments. These enhanced simulations provide teacher candidates with practical experience in managing classrooms, delivering lessons, and interacting with students in a controlled, low-stakes environment. This approach, which contrasts with the traditional methods of classroom placement and observation, gives candidates the chance to engage deeply with teaching scenarios while receiving immediate feedback.

At the heart of this transformative shift is the theoretical framework of experiential learning theory, pioneered by scholars such as David Kolb (1984). Kolb's theory emphasises the importance of experience in the learning process, particularly the cycle of concrete experience, reflective observation, abstract conceptualisation, and active experimentation. VR and AR align perfectly with this framework by allowing teacher candidates to participate in simulated classroom experiences, reflect on their actions, and then try new strategies based on feedback—all within a single training session. The controlled environment offered by VR enhances this cycle, allowing teacher candidates to pause and reflect immediately after encountering challenges, which is often not possible in real classrooms.

Kolb's Experiential Learning and VR

Kolb (1984) posits that "learning is the process whereby knowledge is created through the transformation of experience." In traditional teacher education, this experience often comes from in-person teaching placements, where candidates are placed in classrooms to observe and eventually take on teaching responsibilities. However, the limitations of these placements—such as restricted time, variability in classroom contexts, and delayed feedback—can hinder the learning process. In contrast, VR simulations offer an immersive experience where feedback is immediate, and the scenarios can be customised to target specific teaching skills.

One of the main advantages of using VR and AR is the ability to simulate difficult classroom situations repeatedly. For example, a teacher candidate might face a scenario where a disruptive student interrupts a lesson. In real life, handling such disruptions can be stressful, and feedback from a mentor or supervisor often comes only after the lesson is over. With VR, the candidate can pause the simulation, reflect on their approach, receive feedback, and then re-engage with the scenario to try a different strategy (Dieker et al., 2016). This kind of immediate feedback loop is crucial for developing confidence and competence, particularly in classroom management.

Scholars such as Dieker et al. (2016) argue that VR simulations act as a bridge between theory and practice. They contend that these simulations "offer teacher candidates an opportunity to apply theoretical knowledge in a risk-free, controlled environment, which accelerates the learning process and fosters deeper engagement." This is critical because traditional classroom placements, while valuable, often do not provide the same level of control or repeatability.

Real classrooms are unpredictable, and once a moment has passed, it is difficult to recreate it for further learning.

In VR environments, teacher candidates can practice complex teaching scenarios multiple times, allowing them to refine their techniques. According to Ferdig and Pytash (2020), "one of the key strengths of VR and AR in education is the ability to provide repeatable, immersive experiences that mirror the real world but can be paused, restarted, and altered to suit the learner's needs." This ability to repeat scenarios enables candidates to focus on specific skills, such as classroom management, lesson pacing, or adapting to diverse student needs.

Trial and Error: A Key Feature of Enhanced Practice

The process of trial and error is a crucial component of learning in both the experiential learning model and VR-enhanced teacher education. In traditional classroom placements, teacher candidates often have only one opportunity to handle challenging classroom situations, and their performance is influenced by nerves, unfamiliarity with the environment, or a lack of immediate feedback. Mistakes are inevitable, but the structure of traditional placements limits the ability to learn from them.

VR and AR, by contrast, allow teacher candidates to engage in trial and error without real-world consequences. Huang et al. (2019) emphasise that "the use of immersive simulations in teacher education creates a safe environment for candidates to experiment with different teaching strategies, reflect on their mistakes, and refine their approach without the fear of negatively impacting real students." This is especially important in areas like classroom management, where a single misstep can derail an entire lesson.

The theoretical underpinning for this approach comes from constructivist learning theory, which posits that learners construct knowledge through active engagement with the world around them (Piaget, 1971). In the context of VR-enhanced teacher education, candidates are not passive observers but active participants in the learning process. They are required to make decisions, react to student behaviours, and adapt their teaching strategies in real-time. This active engagement, combined with the ability to learn from mistakes in a controlled environment, creates a rich learning experience that promotes deep understanding and skill development.

John Dewey (1938), a key figure in the development of experiential and constructivist theories, argued that "education is not preparation for life; education is life itself." Dewey's philosophy supports the idea that learning occurs through interaction with one's environment and that experiences should be structured to allow for reflection and growth. VR offers this structure by enabling teacher candidates to interact with virtual classrooms, reflect on their teaching decisions, and grow from each experience.

Realism and Immersion: The Key to Effective Simulation

One of the critical factors that makes VR and AR effective in teacher education is the high level of realism and immersion that these technologies provide. Mikropoulos and Natsis (2011) argue that the "success of virtual environments in educational settings depends largely on the degree of immersion and realism they can achieve." In the context of teacher education, this means that the more realistic the simulated classroom feels, the more effective it is in preparing candidates for real-world teaching.

Immersion in VR creates a sense of presence, where the user feels as though they are truly in the virtual environment. This sense of presence is crucial for teacher candidates, as it allows them to engage fully with the scenarios they are presented with. When the classroom environment feels real, the candidate is more likely to respond authentically to the challenges they face, whether it's managing disruptive students, adjusting lesson plans, or addressing diverse learning needs.

The importance of immersion is further supported by Merchant et al. (2014), who note that "the success of VR-based training in teacher education relies heavily on the realism of the simulations and the degree to which candidates feel immersed in the virtual classroom." In their research, they found that teacher candidates who participated in highly immersive VR simulations were more engaged, showed greater improvement in their teaching skills, and reported higher levels of confidence compared to those who used less immersive technologies.

Social Presence Theory, developed by Short et al. (1976), is also relevant here. The theory suggests that the sense of presence in a mediated environment affects how individuals interact with and respond to the situation. In VR, teacher candidates experience a heightened sense of social presence, which can enhance their emotional and cognitive engagement with the simulated classroom. This level of engagement is essential for fostering the emotional intelligence and relational skills that are key to successful teaching.

Immediate Feedback: A Game-Changer in Teacher Education

One of the most significant advantages of VR in teacher education is the ability to provide immediate, data-driven feedback. In traditional classroom placements, feedback is often delayed and based on memory or subjective impressions. In contrast, VR simulations can offer real-time feedback based on the teacher candidate's actions, allowing for immediate reflection and adjustment.

Anderson et al. (2016) highlight that "the immediacy of feedback in virtual environments is one of the most powerful features of VR in education. Candidates can see the effects of their decisions in real-time and make adjustments on the spot." For example, if a teacher candidate mishandles a classroom disruption, the VR system can provide instant feedback, suggesting alternative approaches or prompting the candidate to reflect on their actions. This feedback loop enables candidates to learn from their mistakes quickly and efficiently, which is often not possible in live classroom settings.

Additionally, the data collected during VR simulations can be used to track a candidate's progress over time. This data can include metrics such as how quickly a candidate responds to student behaviours, how often they adjust their teaching strategies, and how their classroom management skills improve over multiple sessions. According to Ferdig and Pytash (2020), "the use of analytics in VR simulations offers valuable insights into a candidate's development, allowing instructors to provide targeted support and feedback."

Critiques and Limitations of VR in Teacher Education

While VR and AR offer numerous advantages for teacher education, they are not without limitations. One of the primary criticisms is that, no matter how advanced the simulation, VR cannot fully replicate the unpredictability and emotional complexity of a real classroom. Sawyer (2020) argues that "while VR can simulate many aspects of teaching, it cannot capture the full range of emotions, relationships, and spontaneity that occur in real classrooms." In other words, while VR provides a useful tool for developing technical teaching skills, it may fall short in preparing candidates for the relational and emotional demands of the profession.

Moreover, there are concerns about accessibility and equity in the use of VR and AR. The cost of acquiring and maintaining VR equipment can be prohibitive, particularly for underfunded teacher education programs. Gregory et al. (2020) point out that "the high cost of VR technology may limit access for many institutions, particularly those in low-income areas or developing countries." This raises important questions about the equitable distribution of these powerful learning tools and the risk of creating a digital divide in teacher education.

Another limitation is the potential for over-reliance on VR, which could lead to a devaluation of real-world teaching experiences. While VR offers a controlled environment for practice, it cannot replace the complexity and nuance of live classroom interactions. Huang et al. (2019) caution that "VR should be seen as a supplement to, rather than a replacement for, traditional teaching placements. It is essential that teacher candidates also have the opportunity to work with real students in real classrooms."

The use of VR and AR in teacher education offers significant potential for enhancing simulation and practice. These technologies provide teacher candidates with immersive, repeatable experiences that allow for immediate feedback and the opportunity to learn from mistakes without real-world consequences. Supported by theories such as experiential learning, constructivism, and social presence, VR serves as a powerful tool for bridging the gap between theory and practice. However, it is important to recognise the limitations of VR, particularly in its ability to replicate the full emotional and relational dynamics of real classrooms.

As VR and AR technologies continue to evolve, they will likely play an increasingly important role in teacher education. However, their success will depend on careful integration into existing programs, consideration of accessibility issues, and a balance between virtual and real-world teaching experiences. By leveraging the strengths of VR while addressing its limitations, teacher education programs can provide future educators with the skills and confidence they need to succeed in the complex, dynamic environments of modern classrooms.

Personalised Learning Experiences: The Power of VR and AR in Teacher Education

The integration of Virtual Reality (VR) and Augmented Reality (AR) in teacher education represents a significant departure from traditional one-size-fits-all approaches to learning. Personalised learning experiences have gained increasing attention in educational theory and practice, where tailoring education to the specific needs, strengths, and weaknesses of learners is seen as essential to improving outcomes (Tomlinson, 2001). Immersive technologies such as VR and AR provide teacher candidates with personalised learning opportunities that are customised to address their individual challenges and areas for growth, making the learning process more targeted and effective.

In this section, I will explore how VR and AR contribute to personalised learning in teacher education, highlighting relevant theories, key scholars, and empirical research that underscores the significance of this approach. I will also discuss how these technologies align with broader educational philosophies like constructivism and differentiated instruction, and how personalised learning can improve teaching outcomes by providing targeted, practical experiences.

Theoretical Frameworks Supporting Personalised Learning

The concept of personalised learning is deeply rooted in educational theory, particularly in the work of Lev Vygotsky (1978) and his theory of the Zone of Proximal Development (ZPD). Vygotsky argues that learning is most effective when it occurs just beyond the learner's current abilities but is still within reach with the help of guidance or scaffolding. VR and AR technologies offer an ideal platform for implementing Vygotsky's ZPD in teacher education by presenting candidates with scenarios that challenge them at an appropriate level based on their individual progress.

For example, a teacher candidate may struggle with classroom management but excel in instructional planning. Using VR simulations, the candidate can be placed in classroom management scenarios that are tailored to their ZPD—situations that are challenging but manageable with feedback and support. The flexibility of VR and AR allows for this type of individualised learning trajectory, adjusting the complexity of simulations based on the candidate's performance and needs.

John Dely's (1938) theory of experiential learning also plays a significant role in the development of personalised learning experiences through VR and AR. Dewey emphasised that education should be based on the learner's experiences and interactions with their environment. Personalised VR and AR simulations create rich, experiential learning opportunities by immersing teacher candidates in interactive classroom settings where they must actively engage with the material. Dely argued that reflection on these experiences is critical for learning, and VR's ability to offer immediate feedback allows for a reflective process that is continuous and deeply integrated into the learning experience.

In the context of teacher education, differentiated instruction, a concept advanced by Carol Ann Tomlinson (2001), is also closely related to the idea of personalised learning. Differentiated instruction involves adjusting teaching methods and materials to meet the diverse needs of learners. VR and AR serve as powerful tools for differentiated instruction by allowing teacher candidates to practice specific skills, such as classroom management or instructional design, according to their unique learning needs. For instance, a teacher candidate who struggles with using formative assessment techniques can engage in a series of VR simulations that focus on assessing student understanding in real-time, with scenarios that adjust based on their progress and performance.

Tailored Simulations for Individual Learning Needs

One of the most significant advantages of using VR and AR in teacher education is the ability to tailor simulations to meet the specific learning needs of each candidate. Traditional teacher education programs often take a generalised approach, offering limited flexibility to address individual differences in teaching abilities or learning styles (Dieker et al., 2016). Immersive technologies, however, provide a solution by enabling personalised learning paths.

A candidate who struggles with maintaining classroom discipline, for example, can engage in a VR simulation focused on managing disruptive behaviour.

Through repeated practice, they can try different strategies and receive immediate feedback on their performance. If a candidate continues to struggle, the VR platform can adjust the scenario's difficulty, providing more guidance or breaking down the task into smaller, more manageable steps (Ferdig & Pytash, 2020). This adaptive approach aligns with Benjamin Bloom's (1968) mastery learning theory, which holds that all learners can achieve mastery of a subject given the appropriate amount of time and the right instructional conditions. VR and AR facilitate mastery learning by allowing teacher candidates to practice and perfect skills at their own pace, with the opportunity to revisit and refine areas of difficulty.

Similarly, candidates who excel in classroom management but need more experience with instructional design can focus on simulations that challenge them to develop lesson plans and implement them in virtual classrooms. For example, they may be tasked with creating differentiated lesson plans for a diverse group of virtual students, allowing them to experiment with different teaching methods and receive real-time feedback on the effectiveness of their approach.

Research by Huang et al. (2019) highlights the power of personalised learning through VR and AR to enhance engagement and retention. Their study found that teacher candidates who used VR simulations tailored to their learning needs were more engaged with the material and demonstrated greater retention of teaching skills compared to those who participated in traditional teacher training methods. The researchers argue that the ability to tailor simulations to individual learners is one of the key factors driving these outcomes, as it allows candidates to focus on areas where they need the most improvement, rather than following a one-size-fits-all curriculum.

Personalised Feedback and Reflective Practice

One of the core benefits of VR and AR in providing personalised learning experiences is the ability to offer immediate and individualised feedback, which is essential for effective reflective practice. Donald Schön's (1983) theory of reflective practice emphasises that professionals, including teachers, must engage in reflection-in-action—an ongoing process of thinking critically about their actions and decisions while they are engaged in practice. VR and AR environments offer teacher candidates the opportunity to reflect on their actions immediately after completing a task or simulation, rather than waiting for delayed feedback from a mentor or supervisor.

In traditional teacher education programs, feedback is often given after the candidate has completed a lesson, which can delay the learning process and make it harder to apply feedback in real-time situations. By contrast, VR platforms can provide real-time feedback on teaching performance, allowing candidates to immediately adjust their strategies and re-engage with the simulation to test their new approach (Anderson et al., 2016). For example, if a teacher candidate mishandles a student's behavioural outburst in a VR simulation, they can receive instant feedback on their response, reflecting on what went wrong and trying again to improve their classroom management technique.

Schön (1983) argued that reflection-in-action is critical for developing the professional judgment necessary for effective teaching. In the context of VR and AR simulations, this reflective process is facilitated by the fact that candidates can pause the simulation, review their actions, and receive immediate feedback, thus accelerating their development of professional expertise. The personalised nature of this feedback means that it can be tailored to the specific needs of each candidate, focusing on areas where they need the most improvement.

Hattie and Timperley (2007), in their work on feedback, identified three key elements of effective feedback: where the learner is going (the goal), how they are progressing, and what they need to do next to improve. VR and AR environments are particularly Ill-suited to provide this type of feedback because they allow for ongoing assessment of a candidate's performance and continuous adjustment of the learning experience. For instance, a VR simulation might track how quickly a candidate responds to classroom disruptions or how effectively they adjust their lesson plan in response to student feedback. This data can be used to provide specific, actionable feedback on areas for improvement, helping candidates progress toward mastery.

Personalised Learning and Cognitive Load Theory

An important consideration in personalised learning through VR and AR is the concept of cognitive load, a theory developed by John Sweller (1988). Cognitive load theory suggests that the human brain has a limited capacity for processing information, and instructional design must take this into account to avoid overwhelming learners. In the context of teacher education, personalised VR simulations can be designed to manage cognitive load by presenting information and challenges in a way that aligns with the learner's current level of expertise.

For example, a novice teacher candidate might struggle with managing classroom disruptions while simultaneously delivering content. In a traditional classroom setting, this cognitive overload can lead to frustration and reduced learning outcomes. However, VR simulations can be customised to reduce cognitive load by breaking down complex tasks into smaller, more manageable components. A candidate might first practice managing classroom behaviour in isolation, without the added challenge of delivering content, and then gradually integrate these tasks as they build confidence and skill (Sweller, 2010).

This tailored approach ensures that teacher candidates are not overwhelmed by too many demands at once, allowing for more focused and effective learning.

The ability of VR and AR to adapt to the learner's cognitive load is supported by research from Mikropoulos and Natsis (2011), who found that immersive simulations can reduce cognitive overload by offering an intuitive, hands-on learning experience. Their study emphasises the importance of designing VR and AR environments that are cognitively manageable for learners, ensuring that teacher candidates can focus on developing specific skills without becoming overwhelmed by the complexity of the simulation.

Addressing Diverse Learning Styles Through Personalised Learning

Another key benefit of VR and AR in personalised learning is their ability to cater to diverse learning styles. Educational theorists such as Howard Gardner (1983) have argued that learners have multiple intelligences or ways of learning, including linguistic, spatial, kinaesthetic, and interpersonal intelligences. Traditional teacher education programs often prioritise certain types of learning—such as linguistic or logical-mathematical skills—over others, leaving candidates who prefer more hands-on or visual learning experiences at a disadvantage.

VR and AR provide a more inclusive approach to personalised learning by offering multiple modes of engagement. Candidates can interact with simulations in a way that aligns with their preferred learning style. For example, a candidate with strong visual-spatial intelligence might benefit from VR simulations that allow them to visualise the classroom layout and experiment with different seating arrangements to manage student behaviour. A candidate with kinaesthetic intelligence, on the other hand, might prefer to engage with AR tools that allow them to physically interact with lesson materials in a simulated classroom setting (Gardner, 1983).

Moreover, the ability to customise simulations means that candidates can engage with content in a way that feels most intuitive to them. For instance, a candidate who struggles with abstract pedagogical concepts might benefit from a VR simulation that demonstrates these concepts in action, allowing them to see how theory translates into practice in a classroom environment. This flexibility makes VR and AR valuable tools for accommodating diverse learning styles and ensuring that all teacher candidates have access to the types of learning experiences that best suit their needs.

The Future of Personalised Learning in Teacher Education

As VR and AR technologies continue to advance, the potential for further personalisation in teacher education will only grow. Future developments could include even more sophisticated adaptive learning algorithms that track a candidate's performance across multiple simulations and adjust the learning experience accordingly. For example, AI-driven VR platforms could provide real-time insights into a candidate's teaching behaviours, automatically generating new scenarios that target areas for improvement.

Ferdig and Pytash (2020) argue that the future of personalised learning through immersive technologies lies in the integration of artificial intelligence (AI) with VR and AR platforms. AI has the potential to enhance the personalisation of simulations by analysing vast amounts of data on each candidate's performance, learning preferences, and progress. This data can then be used to create even more tailored learning experiences, ensuring that each candidate receives the specific support they need to succeed.

However, the implementation of such advanced technologies will require careful consideration of ethical issues, such as data privacy and equity. As Gibson (2017) points out, the use of AI in personalised learning raises questions about who controls the data generated by these platforms and how it will be used to inform decisions about teacher candidates. Ensuring that personalised learning technologies are accessible to all candidates, regardless of their socio-economic background, will also be essential for promoting equity in teacher education.

Development of Culturally Responsive Pedagogy: Enhancing Teacher Education Through VR and AR

As globalisation and migration reshape societies, classrooms are becoming increasingly diverse, with students hailing from various cultural, linguistic, and socio-economic backgrounds. The growing need for culturally responsive pedagogy—a teaching approach that recognises the importance of including students' cultural references in all aspects of learning—is now paramount in teacher education. For teacher candidates, the ability to navigate these diverse environments and respond sensitively and effectively to the varied cultural needs of their students is a critical skill. Virtual Reality (VR) and Augmented Reality (AR) offer transformative opportunities to train teacher candidates in this area by simulating multicultural classroom settings where they can engage directly with students' diverse experiences and needs.

Culturally Responsive Pedagogy: Theoretical Foundations

Culturally responsive pedagogy stems from the work of educational theorists like Gloria Ladson-Billings (1995), who introduced the concept of culturally relevant teaching. Ladson-Billings emphasised that teaching must affirm students' cultural identities while also promoting academic success and fostering a critical consciousness. This approach requires teachers to be aware of their students' cultural backgrounds and to use this knowledge to enhance learning.

Geneva Gay (2010) further developed these ideas, coining the term culturally responsive teaching. Gay asserts that "culturally responsive teaching uses the cultural knowledge, prior experiences, and performance styles of diverse students to make learning more appropriate and effective for them." This requires that teachers not only acknowledge the diverse cultural backgrounds of their students but also integrate this understanding into their curriculum, instructional strategies, and classroom management techniques. However, teacher candidates often struggle to develop these skills through traditional teacher education methods, which may lack the opportunity for real-time interaction with culturally diverse student populations.

VR and AR offer an immersive and interactive solution to this challenge. These technologies allow teacher candidates to engage with diverse student populations in simulated classroom settings, offering a powerful platform for practising culturally responsive teaching. VR simulations, in particular, provide a safe and controlled environment where candidates can experiment with different strategies for inclusive teaching and receive immediate feedback on their performance. This experiential learning aligns with the ideas of John Dewey (1938), who argued that learning is most effective when it is based on active engagement with real-world problems.

The Role of VR and AR in Developing Cultural Sensitivity

One of the key challenges in developing culturally responsive pedagogy is the lack of direct exposure to diverse classroom environments during teacher training. Many teacher candidates complete their fieldwork in schools that do not fully reflect the diversity they may encounter in their future classrooms. This can leave them unprepared to address the needs of students from different cultural, linguistic, or socio-economic backgrounds. VR and AR can help bridge this gap by allowing teacher candidates to practice teaching in simulated multicultural environments, where they can navigate the complexities of cultural diversity in a structured and supportive setting.

Mutch (2018) emphasises that VR offers a unique opportunity to develop cultural sensitivity in teacher candidates by immersing them in the cultural and linguistic realities of diverse classrooms. Unlike traditional observation or role-playing exercises, where candidates may only passively observe or participate in artificial scenarios, VR places them directly in the middle of a dynamic, interactive environment. In a VR simulation, for example, a teacher candidate might be tasked with managing a classroom where students speak multiple languages, come from different cultural backgrounds, or have varying levels of

English proficiency. The candidate must then adapt their communication strategies, instructional methods, and classroom management techniques to meet the needs of each student.

This kind of immersive experience is critical for developing the skills necessary for culturally responsive teaching. According to Banks (2006), culturally responsive teaching requires teachers to be "cultural mediators" who can bridge the gap between the diverse cultural experiences of their students and the academic content they are teaching. In VR simulations, teacher candidates can practice being cultural mediators by responding to different cultural cues, adjusting their teaching strategies to align with students' cultural practices, and fostering an inclusive learning environment that respects and values cultural diversity.

Simulating Linguistic Diversity and Culturally Relevant Communication

One of the most significant benefits of using VR and AR to develop culturally responsive pedagogy is the ability to simulate linguistic diversity. In many classrooms, teachers are tasked with instructing students who speak multiple languages, some of whom may be English Language Learners (ELLs). Communicating effectively with students who have varying levels of language proficiency can be a daunting task for novice teachers, particularly if they have had limited exposure to linguistically diverse classrooms during their training.

VR simulations can place teacher candidates in classrooms where students speak different languages, allowing them to practice using culturally and linguistically sensitive communication strategies. For example, in a VR simulation, a teacher candidate might encounter a student who primarily speaks Spanish and another student who speaks Mandarin. The candidate must then use a combination of visual aids, gestures, simplified language, and bilingual support to communicate effectively with both students. This process helps teacher candidates develop the flexibility and cultural awareness needed to navigate linguistic diversity in the classroom.

Villegas and Lucas (2002) argue that culturally responsive teaching requires teachers to "acknowledge the legitimacy of the cultural heritages of different ethnic groups, both as legacies that affect students' dispositions toward learning and as resources for teaching and learning." By practising these communication strategies in a VR environment, teacher candidates can learn to recognise and value the cultural and linguistic assets that students bring to the classroom. This not only helps create a more inclusive learning environment but also enhances students' academic outcomes by making the learning experience more relevant and accessible.

Navigating Socio-Economic Contexts

In addition to cultural and linguistic diversity, socio-economic differences among students can present significant challenges for teachers. Students from low-income backgrounds often face unique obstacles that can affect their learning, such as limited access to resources, unstable home environments, or food insecurity. Teachers must be sensitive to these challenges and adjust their teaching strategies accordingly to ensure that all students have the opportunity to succeed.

VR and AR can simulate classrooms where students come from different socio-economic backgrounds, helping teacher candidates develop the skills needed to navigate these complex dynamics. In a VR simulation, for instance, a teacher candidate might encounter a student who lacks access to the internet at home, another who struggles with hunger, and a third who has experienced trauma. The candidate must then adapt their instructional strategies to accommodate these students' needs, perhaps by offering additional support, creating alternative assignments, or providing emotional support.

Research by Darling-Hammond et al. (2005) highlights the importance of teacher training in addressing the socio-economic challenges that students face. Their work suggests that teachers who are trained to recognise and respond to the socio-economic factors affecting their students are better equipped to create supportive and equitable learning environments. VR and AR provide an invaluable opportunity for teacher candidates to practice these skills in a safe and structured environment before they enter the real-world classroom.

Critiques and Limitations of VR and AR in Culturally Responsive Pedagogy

While VR and AR offer significant potential for developing culturally responsive pedagogy, these technologies are not without their limitations. One of the primary critiques is that simulations, no matter how immersive, cannot fully replicate the lived experiences of real students. Paris and Alim (2017) argue that culturally responsive teaching requires a deep understanding of students' cultural backgrounds, which comes from building genuine relationships with students and their families. While VR and AR can simulate diverse classroom environments, they cannot replace the interpersonal connections that are essential for truly understanding and responding to students' cultural experiences.

Additionally, there are concerns about the accessibility of VR and AR technologies, particularly in underfunded teacher education programs. Gregory et al. (2020) point out that the high cost of VR equipment and software may limit the ability of some institutions to adopt these technologies, potentially creating disparities in access to culturally responsive training opportunities. Ensuring that all teacher candidates have the opportunity to benefit from these immersive experiences will require significant investment in both technology and training.

Finally, some critics argue that VR and AR simulations may oversimplify the complexities of cultural diversity. In real classrooms, students' cultural identities are fluid and multifaceted, shaped by a wide range of factors including race, ethnicity, language, gender, and socio-economic status. VR simulations, however, may present a more static and one-dimensional view of culture, which could lead to a superficial understanding of cultural diversity. Sleeter (2011) cautions that teacher candidates must be careful not to reduce culture to a checklist of traits or behaviours but instead engage with students' cultural identities in a nuanced and holistic way.

The Future of Culturally Responsive Pedagogy and Immersive Technologies

Despite these challenges, the future of culturally responsive pedagogy in teacher education is bright, particularly with the continued development of VR and AR technologies. As these tools become more sophisticated, they have the potential to offer even more nuanced and dynamic simulations that reflect the full complexity of cultural diversity in the classroom. Future VR platforms could incorporate artificial intelligence (AI) to create simulations that respond in real-time to the candidate's teaching strategies, offering even more personalised and adaptive learning experiences.

Ferdig and Pytash (2020) suggest that the integration of AI with VR could allow for more sophisticated simulations that adjust based on the candidate's performance, providing targeted feedback on culturally responsive teaching practices. For example, a VR simulation might track how a teacher candidate adapts their instructional strategies to meet the needs of diverse students and offer suggestions for improvement in real-time. This kind of personalised feedback could help candidates develop a deeper understanding of culturally responsive pedagogy and refine their teaching strategies accordingly.

Moreover, as VR and AR technologies become more affordable and accessible, more teacher education programs will be able to integrate these tools into their curricula. This democratisation of access will help ensure that all teacher candidates, regardless of their institution's resources, have the opportunity to develop the skills needed to teach in diverse, multicultural classrooms.

Six Global Case Studies on VR and AR in Teacher Education

To illustrate the global impact and adaptability of VR and AR in teacher education, six case studies from various regions are explored below. These cases highlight the diverse applications of these technologies in different educational systems, revealing both the potential benefits and the challenges associated with their integration.

Case Study 1: The United States – Simulated Classroom Training with TeachLivE

One of the most prominent applications of VR in teacher education is the TeachLivE project in the United States. Developed at the University of Central Florida, TeachLivE uses a VR platform to simulate classroom environments in which teacher candidates can practice their teaching techniques. The simulated students, controlled by human actors and artificial intelligence, respond to the candidate's actions in real-time, creating a highly realistic classroom experience (Dieker et al., 2014).

TeachLivE is particularly effective in helping candidates improve their classroom management skills. The program allows for repeated practice, giving candidates the opportunity to handle challenging student behaviours without the risks associated with live classroom placements. According to Dieker et al. (2016), teacher candidates using TeachLivE showed significant improvements in their ability to manage student behaviour, plan lessons, and engage students.

While TeachLivE has shown great promise, it also has its critics. Some argue that because the simulated students are controlled by actors, the program lacks true unpredictability. Real-world classrooms are often more chaotic and complex than any simulation can replicate. Moreover, there is concern that over-reliance on VR might lead to a detachment from the realities of teaching, where emotional intelligence and relationship-building are key components of success (Sawyer, 2020).

Case Study 2: The United Kingdom – Using AR to Support Special Education Needs (SEN) Teachers

In the United Kingdom, AR has been integrated into teacher training programs to support Special Education Needs (SEN) teachers. The SENscape project, led by researchers at the University of Leeds, uses AR to create interactive learning environments for students with special needs, such as autism or sensory processing disorders. These environments are also used to train SEN teachers, giving them a chance to experience how students with special needs perceive the world and how best to support them (Sutherland & Good, 2019).

By using AR, teacher candidates can simulate the sensory overload experienced by autistic students or the challenges faced by students with visual or auditory impairments. This gives candidates a deeper understanding of their students' experiences, helping them develop more effective teaching strategies.

While the SENscape project has been lauded for its innovative approach to training SEN teachers, it also faces limitations. Some critics argue that AR simulations may oversimplify the complexities of special education. Real-life students with special needs exhibit a wide range of behaviours and challenges, and AR may not capture this diversity adequately. There is also a concern that AR could create an artificial distance between teachers and their students, reducing opportunities for the kind of deep, empathetic understanding that is crucial in special education (Sutherland & Good, 2019).

Case Study 3: Australia – Preparing Teachers for Rural and Remote Classrooms

In Australia, where rural and remote communities often face a shortage of qualified teachers, VR is being used to prepare teacher candidates for the unique challenges of working in these areas. The Remote Classroom project, developed by researchers at Queensland University of Technology (QUT), simulates the experience of teaching in a rural classroom with limited resources, multi-age groups, and significant cultural diversity (Gregory et al., 2020).

The simulations expose teacher candidates to the logistical challenges of rural teaching, such as managing classrooms with students of varying ages and abilities, integrating Indigenous knowledge into the curriculum, and building relationships with students and parents in tight-knit communities. Candidates who completed the program reported feeling better prepared for rural placements and more confident in their ability to adapt to these environments.

Although the Remote Classroom project addresses an important need in Australian education, it has limitations. Critics argue that VR simulations cannot fully prepare candidates for the emotional and psychological challenges of rural teaching, such as isolation, limited professional support, and the pressure of working in close-knit communities where teachers often play multiple roles (Willems & Bossu, 2012). Moreover, the cost of VR technology may be prohibitive for smaller, rural teacher education programs, limiting access to these valuable training tools.

Case Study 4: China – Training Teachers for Large, High-Pressure Classrooms

In China, where class sizes often exceed 50 students, VR is being used to train teachers to manage large, high-pressure classrooms. The ClassSim project, developed by East China Normal University, uses VR to simulate crowded classrooms where candidates must practice keeping all students engaged while maintaining discipline and ensuring learning outcomes are met (Jin & Zhang, 2020).

ClassSim is designed to replicate the challenges of managing large groups of students, where attention can easily drift and classroom management becomes a major issue. The simulations allow candidates to practice strategies for keeping all students on task, even in a noisy and chaotic environment.

Although ClassSim offers valuable practice for managing large classrooms, it has limitations. Critics argue that the highly controlled nature of the VR environment cannot fully replicate the unpredictability and emotional complexity of real-world classrooms. In addition, some educators worry that focusing too much on classroom management in VR simulations may lead candidates to adopt overly authoritarian teaching styles, which can stifle student creativity and engagement (Jin & Zhang, 2020).

Case Study 5: South Africa – Addressing Teacher Shortages through VR Training

South Africa faces a chronic shortage of qualified teachers, particularly in rural and disadvantaged areas. To address this challenge, the Virtual Teacher Training project was launched by the University of Cape Town in collaboration with the South African Department of Education. This initiative uses VR to fast-track the training of new teachers, providing them with the skills they need to manage classrooms effectively and deliver the curriculum in under-resourced schools (Makhanya, 2021).

The project focuses on classroom management, curriculum delivery, and inclusive teaching practices. It allows candidates to practice these skills in a virtual environment, receiving feedback on their performance and improving their techniques over time.

While the Virtual Teacher Training project has shown promise, it also faces significant challenges. Critics argue that VR cannot fully prepare candidates for the reality of teaching in under-resourced schools, where teachers must deal with issues such as poverty, violence, and lack of infrastructure (Jansen, 2017). Moreover, there are concerns that VR training could exacerbate existing inequalities, as wealthier teacher education programs may have better access to these technologies than underfunded schools and institutions.

Case Study 6: Brazil – Using VR to Promote Culturally Responsive Teaching

In Brazil, where classrooms are often highly diverse, VR is being used to train teachers in culturally responsive pedagogy. The EduVR project, developed by the Federal University of São Carlos, simulates classrooms with students from different cultural and linguistic backgrounds, allowing teacher candidates to practice inclusive teaching strategies (Gatti et al., 2021).

EduVR places teacher candidates in classrooms where students speak different languages and come from different cultural backgrounds, challenging them to adapt their teaching to meet the needs of all students. The program also includes simulations of working with Indigenous students, many of whom face unique challenges in the Brazilian education system.

EduVR has been praised for its focus on inclusivity and cultural sensitivity, but it also faces challenges. Critics argue that VR simulations, while valuable, cannot fully capture the complexities of real-world classrooms, where teachers must build relationships with students and adapt to ever-changing classroom dynamics (Gatti et al., 2021). There are also concerns about the accessibility of these technologies in Brazil's underfunded public schools, where many teachers lack access to even basic teaching resources.

The integration of VR and AR in teacher education presents both exciting opportunities and significant challenges. As the six case studies illustrate, these technologies are being used around the world to address a wide range of educational needs, from managing large classrooms in China to promoting culturally responsive teaching in Brazil. While the benefits of VR and AR are clear—enhanced simulation, personalised learning, and the development of culturally responsive pedagogy—there are also important questions that need to be addressed. How can the impact of these technologies be measured? How can they be made accessible to all teacher candidates, regardless of where they live or the resources available to them?

As VR and AR continue to evolve, they hold the potential to reshape teacher education in profound ways. However, for this potential to be fully realised, educators, policymakers, and researchers must work together to address the limitations and challenges associated with these technologies. Only then can VR and AR truly transform the way teachers are trained, ensuring that they are well-prepared to meet the diverse needs of their students in an ever-changing world.

Increased Engagement and Motivation: The Role of VR and AR in Teacher Education

The integration of Virtual Reality (VR) and Augmented Reality (AR) into teacher education offers a new dimension of interactivity and engagement, which is crucial for enhancing learning outcomes. The immersive, interactive nature of these technologies not only makes the learning process more dynamic and enjoyable but also significantly increases motivation among teacher candidates. This heightened engagement translates into better retention of knowledge and improved development of teaching skills, which are essential for success in the classroom. Scholars such as Freina and Ott (2015) have emphasised that the unique ability of VR and AR to create immersive learning environments can lead to deeper involvement in learning activities, enhancing both short-term and long-term educational outcomes.

In this section, I explore how the interactive features of VR and AR foster increased engagement and motivation in teacher education. Drawing on key educational theories, empirical research, and examples from the field, I examine why these technologies are so effective in capturing and maintaining learners' attention and how this engagement directly contributes to improved performance and skill retention.

Theories of Engagement and Motivation in Learning

To understand why VR and AR are so effective at increasing engagement and motivation, it is helpful to consider key theories of motivation in education. Self-Determination Theory (SDT), developed by Deci and Ryan (1985), provides a valuable framework for understanding how these technologies can enhance motivation. SDT posits that intrinsic motivation—the internal drive to engage in an activity for its inherent satisfaction—arises when three basic psychological needs are met: autonomy, competence, and relatedness.

1. Autonomy refers to the sense of control or ownership over one's learning. VR and AR provide teacher candidates with autonomy by allowing them to explore, interact, and make decisions within the virtual classroom. They are not passive recipients of information; instead, they actively engage in learning scenarios where they must make choices about how to manage a classroom, deliver a lesson, or interact with students.
2. Competence involves the feeling of effectiveness in one's actions. Through repeated practice in VR simulations, teacher candidates can develop and refine their teaching skills, gaining a sense of mastery as they see their performance improve. Real-time feedback in these simulations further enhances the candidate's sense of competence by providing clear indicators of progress and areas for improvement (Freina & Ott, 2015).
3. Relatedness is the need to feel connected to others. While VR may not replicate the real-life relationships that develop in a traditional classroom, the use of avatars and social interaction tools within these virtual environments can still foster a sense of connection. Teacher candidates interact with virtual students, colleagues, and mentors, building a form of relational engagement that simulates the real-world teaching experience.

By satisfying these psychological needs, VR and AR increase intrinsic motivation, making teacher candidates more eager to participate in learning activities and more committed to their development as educators.

Immersive Learning and Flow Theory

Another important theory that explains the increased engagement associated with VR and AR is Flow Theory, developed by Mihaly Csikszentmihalyi (1990). According to this theory, individuals experience a state of "flow" when they are deeply immersed in an activity that challenges their skills without overwhelming them. This state is characterised by intense focus, a sense of control, and a feeling of enjoyment.

VR and AR environments are uniquely suited to promote flow because they create immersive, interactive experiences that are both engaging and challenging. In a VR classroom simulation, for example, a teacher candidate may be tasked with managing a disruptive class while delivering a lesson on a complex subject. The candidate must balance multiple demands, adapt their teaching strategies, and respond to student needs—all while remaining fully engaged with the task at hand. This level of engagement, combined with the sense of control that comes from interacting directly with the virtual environment, fosters the flow state, which Csikszentmihalyi argues is key to enhancing learning and motivation.

Gibson (2017) highlights that one of the reasons VR and AR are so effective in maintaining engagement is their ability to "immerse learners in environments that demand their full attention and cognitive resources, resulting in sustained focus and deeper learning." The interactivity of these technologies ensures that learners are constantly active participants in the learning process, rather than passive observers, which significantly enhances engagement and motivation.

Gamification and Engagement in Teacher Training

The concept of gamification, or the application of game-like elements to non-game contexts, is another key factor in the increased engagement associated with VR and AR in teacher education. By incorporating elements such as challenges, levels, rewards, and feedback, VR simulations can transform the learning process into an engaging, game-like experience. Teacher candidates may progress through different levels of difficulty in a classroom management simulation, earn badges or points for successfully handling disruptive behaviours, or receive instant feedback that reinforces their learning.

Deterding et al. (2011) argue that gamification enhances motivation by appealing to the human desire for achievement, competition, and recognition. In the context of teacher education, gamified VR simulations can increase candidates' motivation to engage with the material by providing clear goals, immediate feedback, and a sense of accomplishment. For example, a candidate who successfully manages a difficult classroom scenario in a VR simulation may receive a reward in the form of a virtual badge or a congratulatory message, reinforcing their sense of competence and encouraging them to continue practising.

Additionally, the use of gamification can help reduce the fear of failure that often hinders engagement in traditional teacher training. In a VR environment, candidates can experiment with different teaching strategies and make mistakes without the fear of negative consequences. This freedom to fail encourages experimentation and risk-taking, which are essential for developing the creative and adaptive thinking skills required for effective teaching (Salen & Zimmerman, 2004).

Engagement, Motivation, and Knowledge Retention

One of the most important outcomes of increased engagement and motivation in learning is improved retention of knowledge and skills. Research shows that learners who are more engaged with the material are more likely to retain information and apply it effectively in future situations. Freina and Ott (2015) conducted a study on the impact of VR on learning outcomes and found that learners who participated in VR-based training programs demonstrated significantly higher levels of engagement and knowledge retention compared to those who underwent traditional training methods.

In teacher education, this improved retention is particularly valuable because it ensures that candidates not only acquire theoretical knowledge but are also able to apply it in practice. By immersing candidates in interactive simulations, VR and AR create a learning experience that is memorable and impactful, leading to deeper cognitive processing and long-term retention of teaching skills.

Mikropoulos and Natsis (2011) support this view, arguing that "immersive learning environments like VR and AR promote active engagement, which is critical for the encoding and retrieval of information." When teacher candidates are actively involved in managing a classroom or delivering a lesson in a virtual environment, they are more likely to remember the strategies they used and the feedback they received, enabling them to transfer these skills to real-world teaching contexts.

Real-World Examples of Increased Engagement through VR and AR

The use of VR and AR in teacher education has already shown promising results in increasing engagement and motivation. For example, at Stanford University's Graduate School of Education, teacher candidates use VR simulations to practice classroom management, lesson planning, and instructional strategies in a highly engaging, interactive environment. According to Brown (2018), "the use ofVR in teacher education at Stanford has significantly increased candidate engagement, with many reporting that the simulations make them feel more prepared and confident to handle real-world teaching challenges."

Similarly, the University of Central Florida's TeachLivE program, which uses VR to simulate classroom environments, has been shown to boost both engagement and motivation among teacher candidates. Dieker et al. (2016) found that teacher candidates who participated in TeachLivE reported higher levels of motivation to practice teaching strategies and greater confidence in their abilities to manage real classrooms. The immersive nature of the VR simulations allowed candidates to feel fully engaged in the learning process, leading to better retention of the skills they developed during the training.

Limitations and Criticisms

Despite the numerous benefits of increased engagement and motivation through VR and AR, there are also some limitations and criticisms to consider. One concern is that the novelty of these technologies may lead to a temporary increase in engagement, which could diminish over time as candidates become more familiar with the virtual environment. Sawyer (2020) argues that "while VR and AR offer a unique and engaging learning experience, there is a risk that the initial excitement will wear off, leading to a decline in motivation." To mitigate this, teacher education programs must ensure that VR and AR are integrated into the curriculum in meaningful ways that maintain long-term engagement, rather than relying solely on the novelty of the technology.

Another criticism is the potential for over-reliance on gamification, which may encourage candidates to focus on earning rewards rather than developing a deep understanding of the material. While gamification can be an effective tool for enhancing engagement, it is important that teacher education programs strike a balance between motivating candidates through game-like elements and ensuring that they remain focused on the underlying educational goals (Deterding et al., 2011).

Finally, there is the issue of accessibility. Not all teacher education programs have the resources to implement VR and AR technology, which may lead to disparities in engagement and motivation among candidates from different institutions. Gregory et al. (2020) highlight the need for greater investment in technology infrastructure to ensure that all teacher candidates have access to the benefits of immersive learning environments.

The interactive nature of VR and AR in teacher education significantly enhances engagement and motivation among teacher candidates. By creating immersive, dynamic learning environments that promote autonomy, competence, and relatedness, these technologies align with key motivational theories such as Self-Determination Theory and Flow Theory. Furthermore, the gamification of VR and

AR simulations add an additional layer of motivation, encouraging candidates to engage deeply with the material and experiment with new teaching strategies. This increased engagement leads to better retention of knowledge and skills, ultimately preparing candidates for the challenges of real-world teaching. However, it is important to recognise the limitations of VR and AR in maintaining long-term engagement and ensuring equitable access to these technologies. As immersive technologies continue to evolve, teacher education programs must find ways to integrate them meaningfully into the curriculum, ensuring that they enhance, rather than replace, traditional teaching methods.

Challenges and Considerations: VR and AR in Teacher Education

While the potential of Virtual Reality (VR) and Augmented Reality (AR) in teacher education is vast, their successful implementation is not without significant challenges. These challenges span both technical and pedagogical dimensions, requiring thoughtful planning and strategic investment. This section explores two major areas of concern: the technical and logistical challenges associated with implementing VR and AR in educational settings, and the importance of pedagogical integration to ensure that these technologies support, rather than detract from, established learning goals and practices.

Technical and Logistical Challenges

One of the most pressing challenges in integrating VR and AR into teacher education programs is the substantial technical and logistical investment required. To create immersive learning experiences, institutions must invest in specialised hardware, such as VR headsets, AR-enabled devices, and compatible computers. Additionally, there are considerable costs associated with the development or licensing of VR and AR software, as well as the infrastructure required to support these technologies.

Radu (2014) points out that "the successful implementation of AR-based learning experiences requires robust technical support and significant investment in hardware and software." This is particularly true in large-scale teacher education programs, where institutions must ensure that the technology is available to all teacher candidates, not just a select few. The need for institutional investment is often a barrier, especially for under-resourced universities or teacher preparation programs that may not have the financial capacity to purchase and maintain VR and AR systems. Furthermore, institutions may face additional costs related to training staff on how to use the technology effectively, as well as ongoing maintenance and updates.

Technical issues related to device compatibility and software integration also present challenges. As Radu (2014) notes, "the heterogeneity of AR hardware and software poses significant challenges for educational institutions, which must ensure that their systems are interoperable and compatible with other educational technologies." This is particularly true in teacher education, where programs often use a range of digital platforms for instruction, assessment, and communication. Ensuring that VR and AR technologies integrate seamlessly with these existing systems can be difficult, especially if the technology is new and not yet standardised across the industry.

Accessibility is another critical concern. Not all teacher candidates may have equal access to the devices required for VR and AR experiences, particularly those from low-income or rural backgrounds. This digital divide raises important questions about equity in teacher education: How can institutions ensure that all candidates have access to the benefits of VR and AR, regardless of their socio-economic status? Gregory et al. (2020) argue that "without adequate infrastructure and support, the use of immersive technologies in teacher education may exacerbate existing disparities in access to educational resources."

Moreover, scalability is a logistical challenge. While some teacher education programs have successfully piloted VR and AR technologies, scaling these initiatives to larger cohorts can be problematic. Dede (2010) emphasises that "one of the key challenges in scaling up educational technologies is ensuring that they remain cost-effective, sustainable, and accessible across different institutional contexts." For smaller institutions or programs with limited resources, the cost of implementing VR and AR on a large scale may be prohibitive, making it difficult to ensure widespread adoption.

Additionally, bandwidth and connectivity issues can further complicate the implementation of these technologies. VR, in particular, requires significant processing power and high-speed internet to deliver immersive experiences without lag or interruption. For institutions in regions with limited internet infrastructure, these technical challenges can be a major obstacle to integrating VR and AR into teacher education.

Pedagogical Integration

While overcoming technical challenges is essential, it is equally important to ensure that VR and AR are pedagogically sound and meaningfully integrated into teacher education programs. One of the most common pitfalls in adopting new technologies in education is the tendency to use them for their novelty value, without fully considering how they align with instructional goals. To avoid this, educators must ensure that VR and AR activities complement, rather than replace, traditional teaching methods, and that they are grounded in established pedagogical principles.

Dunleavy and Dede (2014) argue that "the integration of VR and AR into educational settings must be driven by pedagogical considerations, not by the technology itself." In other words, the use of these immersive technologies should be purposeful, designed to enhance the learning process rather than merely introduce novelty. One of the key challenges in achieving this is ensuring that VR and AR activities align with clearly defined learning objectives. For teacher candidates, this means that the simulations and experiences created through VR and AR must directly support the development of essential teaching skills, such as classroom management, lesson planning, and differentiated instruction.

For example, a VR simulation in which teacher candidates practice managing a classroom should not only provide an immersive experience but also align with specific curriculum standards and assessment criteria. Without this alignment, there is a risk that VR and AR activities could become disconnected from the broader educational goals of the teacher preparation program. Dede (2010) emphasises the importance of ensuring that "educational technology must be tightly integrated with curriculum standards, pedagogical goals, and assessment practices if it is to have a meaningful impact on learning outcomes."

One challenge in aligning VR and AR with pedagogical goals is ensuring that these technologies support active learning. Kolb's (1984) experiential learning theory, for example, posits that learning is most effective when it involves active engagement, reflection, and the application of knowledge in real-world contexts. VR and AR have the potential to provide such active learning experiences, but only if they are designed with this in mind. A classroom management simulation, for instance, should not only immerse teacher candidates in a virtual classroom but also prompt them to reflect on their actions, receive feedback, and adjust their strategies in future iterations.

Another consideration is the role of assessment in VR and AR activities. Ferdig and Pytash (2020) argue that "effective pedagogical integration of VR and AR requires clear mechanisms for assessing learning outcomes." This raises important questions about how educators can assess the skills that teacher candidates develop through immersive experiences. Traditional forms of assessment, such as written exams or reflective essays, may not fully capture the competencies that teacher candidates develop in VR environments. Instead, educators may need to adopt new forms of assessment that are more aligned with the interactive nature of VR and AR, such as performance-based assessments or digital portfolios that track candidates' progress through simulations.

Additionally, educators must be mindful of how VR and AR align with constructivist pedagogical principles. Constructivism, which emphasises the importance of learners actively constructing their own knowledge through experience and reflection, is particularly relevant to the use of VR and AR in teacher education. According to Vygotsky's (1978) social constructivist theory, learning is a social process that is mediated through interaction with others and the environment. In the context of VR and AR, this means that teacher candidates should not be passive participants in simulations but should actively engage with the virtual classroom, collaborate with peers, and reflect on their experiences to construct new understanding.

However, one of the challenges in achieving this level of engagement is the potential for over-reliance on the technology itself, rather than on the pedagogical principles that should guide its use. Salomon (1992) warns of the "seductive allure" of educational technologies, which can sometimes lead educators to prioritise the technology over the learning outcomes it is meant to support. To avoid this, educators must remain focused on the core competencies that teacher candidates need to develop and use VR and AR as tools to support, rather than replace, these pedagogical goals.

Teacher Training and Professional Development

Another critical aspect of pedagogical integration is ensuring that educators themselves are adequately trained to use VR and AR effectively in teacher education. Dunleavy and Dede (2014) argue that "professional development is essential for ensuring that educators understand how to integrate VR and AR into their teaching in ways that align with instructional goals." Without proper training, there is a risk that educators may either underutilise the technology or use it in ways that are not pedagogically sound.

Professional development programs should focus not only on the technical aspects of using VR and AR but also on how to design and implement immersive learning experiences that support active learning, reflection, and assessment. Educators need to be equipped with the skills to design VR and AR activities that are aligned with curriculum standards and learning objectives, as well as the ability to assess the learning outcomes that result from these activities. Moreover, ongoing support and collaboration are essential for ensuring the long-term success of VR and AR integration. Gibson (2017) highlights the importance of "creating a community of practice" where educators can share their experiences, challenges, and successes in using immersive technologies. By fostering collaboration and knowledge-sharing, institutions can ensure that best practices for using VR and AR in teacher education are continually refined and improved.

Ethical and Privacy Concerns: Navigating the Complexities of VR and AR in Teacher Education

While Virtual Reality (VR) and Augmented Reality (AR) technologies offer exciting opportunities for innovation in teacher education, they also present significant ethical and privacy challenges. These issues stem primarily from the way these immersive technologies collect, store, and use personal data. The sensitive nature of this data, combined with the interactive and highly personalised nature of VR and AR environments, raises important questions about user privacy, consent, and the ethical use of data. As VR and AR become more integrated into teacher education, institutions must establish clear guidelines and policies to ensure that these technologies adhere to ethical standards and protect the privacy of users.

In this section, I will explore the key ethical and privacy concerns associated with the use of VR and AR in teacher education. Drawing on scholarly research and case studies, I will discuss the potential risks of data collection, the importance of informed consent, and the role of institutions in safeguarding student and teacher candidate privacy.

Data Collection and Privacy in Immersive Technologies

One of the primary ethical concerns with VR and AR technologies is the vast amount of personal data they collect from users. Unlike traditional educational tools, VR and AR platforms gather a range of data points, including biometric information, eye-tracking data, facial expressions, movement patterns, and voice recordings. This data can provide valuable insights into user behaviour and learning outcomes, but it also poses significant risks if not handled properly. Slater (2018) notes that "the immersive nature of VR means that more personal and potentially sensitive data is collected than in traditional digital environments, raising the stakes for data privacy and security."

In the context of teacher education, this data can be used to track teacher candidates' progress, assess their performance in simulations, and provide personalised feedback. For example, a VR platform might track how long a teacher candidate spends engaging with virtual students or how quickly they respond to classroom disruptions. While this data can enhance the learning experience by providing real-time analytics and feedback, it also raises important questions about how this data is stored, who has access to it, and how it might be used beyond the immediate educational context.

A major concern is the potential for misuse of this data. Slater (2018) warns that "without proper safeguards, the data collected in VR environments could be exploited for purposes beyond education, such as marketing, surveillance, or even discrimination." In the case of teacher education, the data collected from VR simulations could be used to create detailed profiles of teacher candidates, which might influence decisions about their future employment or advancement. Institutions must therefore take steps to ensure that data collected through VR and AR platforms is used solely for educational purposes and is protected from unauthorised access.

Informed Consent and Transparency

Another ethical issue related to the use of VR and AR in education is the need for informed consent. Teacher candidates and students who use these technologies must be fully informed about what data is being collected, how it will be used, and who will have access to it. Bailenson (2018) argues that "transparency and informed consent are essential components of ethical VR and AR use, particularly in educational settings where users may not fully understand the extent of data collection." This is especially important when the data being collected goes beyond standard academic performance metrics and includes more intimate details such as biometric information.

In traditional classroom settings, data collection is generally limited to academic records, attendance, and basic personal information. However, in VR environments, the scope of data collection expands dramatically. For instance, many VR platforms track users' physical movements, monitor their eye gaze to determine where they are looking, and record their speech for later analysis. This level of data collection requires a higher standard of informed consent, as users may not be aware of how much personal information they are sharing simply by participating in a VR simulation.

Institutions must therefore develop clear and comprehensive consent forms that outline exactly what data is being collected, how it will be used, and what measures are in place to protect it. These consent forms should be written in accessible language and provided to teacher candidates and students before they engage with VR or AR technologies. Additionally, institutions should offer users the option to opt out of certain types of data collection if they are uncomfortable with sharing specific information. Ensuring that users have control over their personal data is a critical component of ethical VR and AR use in education.

The Role of Institutions in Safeguarding Privacy

To address privacy concerns, institutions must take proactive steps to protect the data collected through VR and AR platforms. This includes developing and implementing robust data protection policies, investing in secure data storage solutions, and training staff on best practices for managing sensitive information. Gregory et al. (2020) emphasise that "the responsibility for protecting student privacy lies not only with the technology providers but also with the institutions that implement these technologies."

One of the most effective ways institutions can safeguard privacy is by adhering to established data protection standards, such as the General Data Protection Regulation (GDPR) in the European Union. The GDPR provides a comprehensive framework for managing personal data, including guidelines for data collection, storage, and processing. By following these guidelines, institutions can ensure that they are meeting the highest standards for data privacy and security. In regions where GDPR does not apply, similar national or regional data protection regulations should be followed to ensure compliance with legal and ethical standards.

Additionally, institutions should conduct regular audits of their VR and AR systems to ensure that data is being handled securely and in accordance with privacy policies. These audits should include a review of how data is collected, stored, and accessed, as well as an assessment of the risks associated with data breaches or unauthorised access. Anderson and Rainie (2018) argue that "regular audits and transparency reports can help build trust between institutions and users by demonstrating a commitment to privacy and accountability."

Another important consideration is the potential for data breaches. As VR and AR technologies become more widespread, they may become targets for cyberattacks, particularly if the data they collect is valuable. Institutions must therefore invest in cybersecurity measures to protect against data breaches and unauthorised access. This includes encrypting data, securing networks, and implementing multi-factor authentication for users accessing sensitive information.

Ethical Use of VR and AR in Simulations

In addition to privacy concerns, the use of VR and AR in teacher education raises ethical questions about the nature of the simulations themselves. For example, in a VR simulation designed to help teacher candidates practice classroom management, the virtual students may display behaviours that reflect real-world challenges such as disruptive behaviour, language barriers, or socio-economic disadvantage. While these simulations can provide valuable learning experiences, they must be designed carefully to avoid reinforcing harmful stereotypes or oversimplifying complex social issues.

Sleeter (2011) warns that "there is a risk that VR simulations, if not designed thoughtfully, could perpetuate stereotypes by presenting diverse students in reductive or one-dimensional ways." For example, a VR simulation that depicts students from low-income backgrounds as consistently disruptive or underperforming could unintentionally reinforce negative biases among teacher candidates. To mitigate this risk, VR simulations should be developed with input from educators, sociologists, and cultural experts to ensure that they present students from diverse backgrounds in a nuanced and respectful manner.

Moreover, the ethical use of VR and AR in simulations requires that teacher candidates engage critically with the content they encounter. Rather than passively accepting the scenarios presented to them, teacher candidates should be encouraged to reflect on the ethical implications of their interactions with virtual students and consider how these interactions may inform their real-world teaching practices. This reflective process aligns with Schön's (1983) theory of reflective practice, which emphasises the importance of critical reflection in professional development.

Addressing Bias and Inclusivity in VR and AR

Another important ethical consideration is the potential for bias in the design and implementation of VR and AR technologies. Just as traditional educational materials can reflect biases related to race, gender, class, and ability, VR and AR simulations can also be influenced by the perspectives of the developers who create them. Noble (2018) highlights the risks of "algorithmic bias," where the algorithms that power VR and AR experiences may reflect the biases of their creators, leading to inequitable outcomes for users.

In the context of teacher education, this means that VR and AR simulations must be designed with inclusivity in mind, ensuring that they accurately reflect the diversity of real-world classrooms. Simulations should not only include students from a variety of cultural, linguistic, and socio-economic backgrounds but should also avoid reinforcing harmful stereotypes or biases. For example, a simulation that includes students with disabilities should present these students in a way that emphasises their strengths and capabilities, rather than focusing solely on their challenges.

To address these concerns, institutions should work with diverse teams of developers, educators, and cultural experts when designing VR and AR simulations. This collaborative approach can help ensure that the simulations are inclusive, equitable, and representative of the diverse student populations that teacher candidates will encounter in their professional careers. The use of VR and AR in teacher education presents exciting possibilities for innovation, but it also raises significant ethical and privacy concerns. Institutions must take proactive steps to protect the data collected through these technologies, ensuring that it is used responsibly and securely. Informed consent, transparency, and adherence to data protection standards are critical for safeguarding user privacy. Moreover, the ethical use of VR and AR in simulations requires careful attention to bias, inclusivity, and the potential for reinforcing harmful stereotypes. By addressing these challenges, institutions can ensure that VR and AR technologies are used in ways that are both educationally valuable and ethically sound.

Best Practices for Implementation of VR and AR in Teacher Education

The successful integration of Virtual Reality (VR) and Augmented Reality (AR) in teacher education hinges not only on the availability of technology but also on the effective and strategic use of these tools in educational settings. While VR and AR offer significant potential to enhance the learning experiences of teacher candidates, their implementation requires thoughtful planning, ongoing support, and a commitment to continuous improvement. Below, I outline three best practices for the successful implementation of VR and AR in teacher education: professional development for educators, fostering collaborative learning, and establishing continuous evaluation and improvement processes.

Professional Development

Ongoing professional development for educators is fundamental to the successful integration of VR and AR in teacher education. Without adequate training, even the most advanced technology will fail to meet its educational potential. Many educators may be unfamiliar with immersive technologies and need support in learning how to effectively incorporate them into their teaching practices. Professional development programs must focus not only on the technical aspects of VR and AR but also on how these tools can be used to meet pedagogical goals and enhance the overall learning experience.

Ertmer and Ottenbreit-Leftwich (2010) argue that "teachers' beliefs, knowledge, and skills are key factors influencing their adoption of educational technologies." For educators to feel confident in using VR and AR, they need hands-on experience with these technologies, coupled with clear guidance on how to integrate them into their instructional methods. This professional development should be ongoing, offering opportunities for educators to refine their skills as the technology evolves.

Key components of professional development for VR and AR implementation include:

- Technical training: Educators must become familiar with how to use VR and AR devices, such as VR headsets, AR-enabled smartphones, and the associated software. This training should cover not only basic functions but also more advanced features, such as customising simulations or creating content using VR and AR platforms.
- Pedagogical integration: Educators need to understand how to align VR and AR activities with curriculum standards, learning objectives,

and assessment practices. Professional development should provide examples of how these technologies can complement traditional teaching methods, such as using VR simulations to enhance experiential learning or AR overlays to visualise complex concepts.

- Overcoming challenges: Educators must be prepared to address potential challenges, such as technical difficulties, student discomfort with the technology, or issues with accessibility. Professional development programs should include strategies for troubleshooting common problems and ensuring that all students can participate fully in VR and AR experiences.

As part of the professional development process, educators should also be encouraged to participate in communities of practice, where they can share their experiences, challenges, and successes in using VR and AR in the classroom. According to Gibson (2017), "creating a network of educators who can collaborate and learn from one another is essential for sustaining the effective use of educational technology." By fostering a culture of collaboration and continuous learning, institutions can ensure that educators remain engaged with and committed to the successful implementation of VR and AR.

Collaborative Learning

Encouraging collaborative learning within VR and AR environments is another best practice for maximising the benefits of these technologies in teacher education. While VR and AR are often associated with individual, immersive experiences, they also provide opportunities for peer interaction and teamwork. Incorporating collaborative learning activities into VR and AR-based training programs helps teacher candidates develop essential skills such as communication, problem-solving, and teamwork—skills that are critical for effective teaching.

Johnson et al. (2016) emphasizes the importance of collaborative learning in virtual environments, stating that "VR and AR provide unique opportunities for co-teaching simulations, group problem-solving, and peer feedback, which promote deeper learning and the development of interpersonal skills." In teacher education, these collaborative activities can take many forms, such as:

- Co-teaching simulations: In a VR simulation, two or more teacher candidates can co-teach a virtual classroom, allowing them to practice collaborative planning, coordination, and classroom management. This type of simulation mimics real-world teaching scenarios where teachers often work together to manage a class or lead interdisciplinary lessons.
- Group problem-solving scenarios: VR and AR environments can be designed to include complex, real-world challenges that require

teacher candidates to work together to find solutions. For example, a VR simulation might place candidates in a classroom where they must collaboratively manage student behaviour, implement a differentiated lesson plan, or address a cultural misunderstanding. These activities help candidates develop their problem-solving abilities while learning to work effectively as part of a team.

- Peer feedback and reflection: After participating in a collaborative VR or AR activity, teacher candidates can engage in peer feedback sessions, where they reflect on their experiences and provide constructive feedback to one another. These sessions can be facilitated either within the VR environment or in a traditional classroom setting, depending on the institution's resources. Reflecting on their experiences helps candidates consolidate their learning and identify areas for improvement.

Collaborative learning within VR and AR environments also fosters a sense of community among teacher candidates, helping them build relationships with their peers and develop their communication skills. This is particularly important in teacher education, where the ability to work well with colleagues, students, and parents is essential for professional success. By integrating collaborative activities into VR and AR-based training programs, institutions can enhance the overall training experience and prepare teacher candidates for the realities of working in diverse, dynamic school environments.

Continuous Evaluation and Improvement

A continuous evaluation process is critical for assessing the effectiveness of VR and AR in teacher education and for identifying areas where these technologies can be refined and improved. Without regular evaluation, it is difficult to determine whether VR and AR-based training programs are meeting their intended goals, such as improving teaching skills, enhancing engagement, or supporting learning retention. Hew and Brush (2007) argue that "ongoing evaluation is essential for ensuring that educational technology initiatives are aligned with institutional goals and responsive to the needs of students and educators."

Effective evaluation involves collecting feedback from a variety of stakeholders, including teacher candidates, educators, and administrators. Each group brings a unique perspective to the evaluation process:

- Teacher candidates: Gathering feedback from teacher candidates is crucial for understanding how they perceive the value of VR and AR in their training. Candidates can provide insights into which aspects of the technology are most helpful, which simulations are most realistic, and what challenges they face when using the technology. Surveys, interviews, and focus groups are useful methods for collecting this feedback.

- Educators: Educators who facilitate VR and AR-based activities can offer valuable insights into how these technologies align with curriculum standards and learning objectives. They can also identify areas where additional professional development or technical support is needed to improve the implementation process.
- Administrators and stakeholders: Collecting feedback from administrators and other stakeholders helps ensure that VR and AR initiatives are aligned with broader institutional goals, such as improving student outcomes, increasing engagement, or enhancing access to innovative learning tools.

Key elements of a continuous evaluation process include

- Formative and summative assessments: Formative assessments, such as regular feedback from teacher candidates and educators, can be used to make real-time adjustments to VR and AR-based activities. Summative assessments, such as final evaluations or end-of-program surveys, provide a comprehensive view of the overall effectiveness of the technology.
- Data-driven decision-making: VR and AR platforms often include built-in analytics tools that track user interactions, progress, and outcomes. By analysing this data, institutions can gain insights into how candidates are engaging with the technology, which activities are most effective, and where improvements are needed. Ferdig and Pytash (2020) note that "the use of data analytics is essential for identifying patterns in user behaviour and ensuring that VR and AR technologies are being used effectively."
- Iterative improvement: Based on the feedback and data collected through continuous evaluation, institutions should regularly update and refine their VR and AR-based training programs. This iterative approach ensures that the technology remains responsive to the evolving needs of teacher candidates and educators. It also allows institutions to address any technical or pedagogical challenges that arise over time.

By adopting a continuous evaluation process, institutions can ensure that VR and AR are not only implemented effectively but also continuously improved to maximise their impact on teacher education. This commitment to ongoing assessment and improvement helps create a dynamic, responsive training environment that adapts to the needs of both teacher candidates and educators.

The successful implementation of VR and AR in teacher education requires more than just access to cutting-edge technology—it also requires thoughtful planning, ongoing professional development, collaborative learning opportunities, and a commitment to continuous evaluation and improvement. By investing in professional development, encouraging peer interaction and teamwork within VR and AR environments, and regularly assessing the effectiveness of these tools, institutions can ensure that VR and AR enhance the training experience for teacher candidates. These best practices help ensure that immersive technologies are used in ways that support educational goals, improve teaching skills, and prepare candidates for the diverse challenges they will face in real-world classrooms.

Conclusion

The integration of Virtual Reality (VR) and Augmented Reality (AR) into teacher education presents transformative potential, revolutionising how future educators are trained. By providing immersive, interactive, and highly practical training experiences, VR and AR bridge the gap between theoretical learning and real-world application. These technologies offer teacher candidates the opportunity to engage in simulated classroom environments, practice teaching strategies, and develop the critical skills needed to manage diverse student populations—all within a safe, controlled setting.

However, realising the full potential of VR and AR in teacher education requires addressing several key challenges. Technical and logistical hurdles, such as the cost of hardware, software integration, and issues of accessibility, must be overcome to ensure that these technologies are available to all teacher candidates. Institutions must also address the pedagogical challenges of aligning VR and AR with curriculum standards and learning objectives, ensuring that these tools enhance rather than replace traditional teaching methods.

Equally important are the ethical and privacy concerns surrounding the use of VR and AR, particularly regarding the collection and use of personal data. Clear guidelines and policies must be established to protect the privacy of teacher candidates and students while ensuring that the data collected through immersive technologies is used responsibly and transparently.

By adopting best practices, such as ongoing professional development for educators, fostering collaborative learning within VR and AR environments, and implementing continuous evaluation processes, educational institutions can successfully integrate these technologies into their teacher education programs. These practices not only ensure that VR and AR are used effectively but also help create a dynamic, adaptive learning environment that meets the evolving needs of both teacher candidates and educators.

In conclusion, VR and AR offer powerful tools for enhancing teacher training, providing opportunities for experiential learning, personalised instruction, and culturally responsive pedagogy. This chapter has aimed to guide educators and institutions in leveraging these technologies to innovate and improve their teacher education programs. By addressing the technical, pedagogical, and ethical challenges associated with VR and AR, and committing to best practices, institutions can prepare the next generation of educators to thrive in increasingly complex and diverse classroom environments.

References

Anderson, C. and Rainie, L., 2018. The Future of Ill-Being in a Tech-Saturated World. *Pew Research Center.* Available at: <https://www.pewresearch.org> [Accessed 28th 2024].

Anderson, E., Imms, W. and Maher, D., 2016. Immersive Learning: The Real World, the Virtual World, and the Bridge Betlen. *Australian Journal of Education,* 60(3), 253-265.

Bailenson, J.N., 2018. *Experience on Demand: What Virtual Reality Is, How It Works, and What It Can Do.* New York: W.W. Norton & Company.

Banks, J.A., 2006. *Cultural Diversity and Education: Foundations, Curriculum, and Teaching.* 5th ed. Boston: Pearson Education.

Bloom, B.S., 1968. *Learning for Mastery.* UCLA: CSEIP.

Brown, A., 2018. Engaging with Virtual Reality in Teacher Education: The Stanford Experience. *Journal of Educational Technology,* 54(2), 101-113.

Csikszentmihalyi, M., 1990. *Flow: The Psychology of Optimal Experience.* New York: Harper & Row.

Darling-Hammond, L., Ii, R.C., Andree, A., Richardson, N. and Orphanos, S., 2009. Professional Learning in the Learning *Profession: A Status Report on Teacher Development in the United States and Abroad.* Washington, DC: National Staff Development Council.

Dede, C., 2010. *Emerging Technologies for Learning: Impact on Cognition and Pedagogy.* Boston: Harvard University Press.

Deterding, S., Dixon, D., Khaled, R. and Nacke, L., 2011. From Game Design Elements to Gamefulness: Defining Gamification. In*: Proceedings of the 15th International Academic MindTrek Conference: Envisioning Future Media Environments.* ACM, pp. 9-15. https://doi.org/10.1145/2181037.2181040

Dieker, L.A., Rodriguez, J.A., Lignugaris-Kraft, B., Hynes, M.C. and Hughes, C.E., 2016. The Potential of Simulated Environments in Teacher Education: *Current and Future Possibilities. Teacher Education and Special Education,* 39(1), 21-33. https://doi.org/10.1177/0888406413512683

Dunleavy, M. and Dede, C., 2014. Augmented Reality Teaching and Learning. In: M. J. Bishop and J. Elen, eds. *Handbook of Research on Educational Communications and Technology.* New York: Springer, 735-745. https://doi.org/10.1007/978-1-4614-3185-5_59

Ertmer, P.A. and Ottenbreit-Leftwich, A.T., 2010. Teacher Technology Change: How Knowledge, Confidence, Beliefs, and Culture Intersect. *Journal of*

Research on Technology in Education, 42(3), 255-284. https://doi.org/10.1080/15391523.2010.10782551

Ferdig, R.E. and Pytash, K.E., 2020. Exploring the Potential of Virtual Reality in Teacher Education. *International Journal of Educational Technology in Higher Education*, 17(1), 1-17.

Freina, L. and Ott, M., 2015. A Literature Review on Immersive Virtual Reality in Education: State of the Art and Perspectives. In: *The International Scientific Conference eLearning and Software for Education* (eLSE). Bucharest, Romania, pp. 133-141. https://doi.org/10.12753/2066-026X-15-020

Gardner, H., 1983. *Frames of Mind: The Theory of Multiple Intelligences.* New York: Basic Books.

Gay, G., 2010. *Culturally Responsive Teaching: Theory, Research, and Practice.* 2nd ed. New York: Teachers College Press.

Gibson, S., 2017. The Evolution of Teacher Training in the Digital Age: From Chalkboards to VR Classrooms. *Educational Technology & Society*, 20(3), 123-137.

Gregory, S., Wood, D. and Parker, C., 2020. Virtual Reality in Education: Exploring Its Potential for Training Teachers in Rural and Remote Classrooms. *Australian Journal of Teacher Education*, 45(7), 1-18.

Hattie, J. and Timperley, H., 2007. The PoIr of Feedback. *Review of Educational Research*, 77(1), 81-112. https://doi.org/10.3102/003465430298487

Hew, K.F. and Brush, T., 2007. Integrating Technology into K-12 Teaching and Learning: Current Knowledge Gaps and Recommendations for Future Research. *Educational Technology Research and Development*, 55(3), 223-252. https://doi.org/10.1007/s11423-006-9022-5

Huang, H.M., Rauch, U. and Liaw, S.S., 2019. Investigating Learners' Attitudes Toward Virtual Reality Learning Environments: Based on Constructivist Learning Theories. *Computers & Education*, 55(3), 1171-1182. https://doi.org/10.1016/j.compedu.2010.05.014

Jacobson, M.J., Taylor, S.M. and Richards, D., 2021. Teaching the Teachers: Exploring the Potential of VR for Ongoing Teacher Professional Development. *Educational Media International*, 58(2), 101-118.

Johnson, D., Adams Becker, S., Estrada, V. and Freeman, A., 2016. *NMC Horizon Report: 2016 Higher Education Edition. Austin*, TX: The New Media Consortium.

Kolb, D.A., 1984. *Experiential Learning: Experience as the Source of Learning and Development.* Englewood Cliffs: Prentice-Hall.

Ladson-Billings, G., 1995. Toward a Theory of Culturally Relevant Pedagogy. *American Educational Research Journal*, 32(3), 465-491. https://doi.org/10.3102/00028312032003465

Merchant, Z., Goetz, E.T., Cifuentes, L., Keeney-Kennicutt, W. and Davis, T.J., 2014. Effectiveness of Virtual Reality-Based Instruction on Students' Learning Outcomes in K-12 and Higher Education: A Meta-Analysis. *Computers & Education*, 70, 29-40. https://doi.org/10.1016/j.compedu.2013.07.033

Mikropoulos, T.A. and Natsis, A., 2011. Educational Virtual Environments: A Ten-Year Review of Empirical Research (1999–2009). *Computers & Education*, 56(3), 769-780. https://doi.org/10.1016/j.compedu.2010.10.020

Mutch, C., 2018. Virtual Reality and Cultural Sensitivity: Engaging Teacher Candidates in Multicultural Classrooms. *International Journal of Educational Research*, 92(4), 45-56.

Noble, S.U., 2018. *Algorithms of Oppression: How Search Engines Reinforce Racism.* New York: NYU Press.

Paris, D. and Alim, H.S., 2017. *Culturally Sustaining Pedagogies: Teaching and Learning for Justice in a Changing World.* New York: Teachers College Press.

Piaget, J., 1971. *The Theory of Stages in Cognitive Development.* New York: McGraw-Hill.

Radu, I., 2014. Augmented Reality in Education: A Meta-Review and Cross-Media Analysis. *Personal and Ubiquitous Computing,* 18(6), 1533-1543. https://doi.org/10.1007/s00779-013-0747-y

Salomon, G., 1992. The Technology is in the Mind of the Learner: The Influence of Social and Cognitive Factors on Students' Use of Technology. *Educational Technology,* 32(1), 11-20.

Salen, K. and Zimmerman, E., 2004. *Rules of Play: Game Design Fundamentals.* Cambridge, MA: MIT Press.

Schön, D.A., 1983. *The Reflective Practitioner: How Professionals Think in Action.* New York: Basic Books.

Short, J., Williams, E. and Christie, B., 1976. *The Social Psychology of Telecommunications.* London: Wiley.

Slater, M., 2018. Immersive Virtual Reality: A Short History of a Breakthrough. *Frontiers in Robotics and AI,* 5(112), 1-7.

Sleeter, C., 2011. *The Academic and Social Value of Ethnic Studies: A Research Review.* Washington, DC: National Education Association.

SIller, J., 1988. Cognitive Load During Problem Solving: Effects on Learning. *Cognitive Science,* 12(2), 257-285. https://doi.org/10.1016/0364-0213(88)90023-7

SIller, J., 2010. Cognitive Load Theory: Recent Theoretical Advances. In: J.L. Plass, R. Moreno and R. Brünken, eds. *Cognitive Load Theory. Cambridge*: Cambridge University Press, pp. 29-47. https://doi.org/10.1017/CBO9780511844744.004

Tomlinson, C.A., 2001. *How to Differentiate Instruction in Mixed-Ability Classrooms.* 2nd ed. Alexandria, VA: ASCD.

Villegas, A.M. and Lucas, T., 2002. Preparing Culturally Responsive Teachers: Rethinking the Curriculum. *Journal of Teacher Education,* 53(1), 20-32. https://doi.org/10.1177/0022487102053001003

Vygotsky, L.S., 1978. Mind in Society: *The Development of Higher Psychological Processes.* Cambridge, MA: Harvard University Press.

Willems, J. and Bossu, C., 2012. Equity Considerations for Open Educational Resources in the Glocalization of Education. *Distance Education,* 33(2), 185-199. https://doi.org/10.1080/01587919.2012.692051

Yoshida, K., 2020. Virtual Reality Simulations for Large Classrooms in Japan: A Case Study of Teacher Preparation. *Asian Journal of Education and e-Learning,* 8(4), 102-114.

Chapter Ten
Use of Artificial Intelligence in Advance Academic Course Development

Abirlal Mukherjee

GITAM (Deemed to be University), Bengaluru, India

Abstract

The research focuses on the potential of artificial intelligence in academic course development, eliminating the imagination or "great expectations." The research method tries to analyse different elements of AI and data available to identify its potential in academic course development. The results penetrate into the areas of applicability in using AI in state-of-the-art education when it does not focus on the footprints, challenges, and disadvantages of AI explicitly. The functionalities of AI, types, and specific areas of improvement are elaborated in this chapter, which will give a comprehensive view of how AI can enhance human abilities, skills, and learning in an online teaching and learning environment.

Keywords: artificial intelligence, advance course development, AI and education, AI policy, AI research and development.

***

Introduction

Artificial intelligence and content creation are two of the most used and popular integrations. The present is a time of mass consumption of content created by artificial intelligence. When AI has gained a bad reputation due to technological developments like Deepfake and FaceSwap to fake the face in audiovisual contents, and is called a high alarm situation in the developing countries and underdeveloped countries where there are limited opportunities and laws to tackle the AI challenges, there are several utility sides to explore and several to explore in the field of AI studies. AI has come up with different solutions and perfection that previously needed much time, human labour, and intelligence.

Educational research and development is one of the biggest sectors that can consume AI as a giant fasting for it. The biggest reason behind the consumption of artificial intelligence in the educational sector is taking place: the data, the reasoning, creative, inferential, calculative, and analytical autonomous abilities of artificial intelligence. But the newborn baby is yet to grow. After the launch of ChatGPT in late 2022, the fabrications and aspirations regarding the roles of AI in human affairs, especially in education, seemed foolish. There are associated risks. Incorporating AI in course development represents a pivotal juncture in education. The synergy between advanced technology and pedagogy promises a metamorphosis that places personalisation, engagement, and efficiency at the forefront of modern learning experiences. It is a delicate balance where AI's capabilities coalesce with the human touch, forging a new era in education that transcends boundaries and empowers learners like never before. The research addresses all the possible uses of artificial intelligence in developing dependent, independent, instant, and on-demand courses using different functionalities of the same. The paper discusses how AI can help in academic course development without creating teaching machines. What are the gaps to bridge in policymaking and ethical monitoring to keep AI as a worthy contributor?

What is artificial intelligence (AI)?

There is a significant evolution in the meaning of the term "teaching." While a few decades ago, teaching was strict and limited to certain strategies and planning, the past years have seen liberation in teaching. The term teaching is not as heavy as it sounded in the recent past; it is now friendly, more of learning about the students and the way they learn, playful, and research-driven. The incorporation of digital technology in teaching and learning paved the path for research and development opportunities in the field of teaching and learning. After the mass awareness of AI, there is a huge concern: "Will AI replace human jobs?" and in teaching, "Can AI replace human teachers?" Artificial intelligence is a branch of computer science created as a substitution for human intelligence to aid in reasoning, calculation, prediction, and autonomous learning systems. AI needs a large amount of data to function autonomously.

The history of artificial intelligence has its roots in history, but in the mid-twentieth century, broadly, the advancement of the study and its use came into the picture. Humans tend to create comfortable and less effortful thinking mechanisms; in ancient Greek mythology, there are mentions of "automatons," and there were efforts to create effective mechanisms to work independently. For example, in Greek mythology, "Talos" or "Talus" or "Talon" was a giant bronze automaton created by Hephaestus, the god of blacksmiths and metalworking, and given to King Minos of Crete to protect the island. Pandora is another example; she was created out of clay as a punishment to Prometheus by Hephaestus and the other gods and given to Athena. Pandora was given a box and instructed not to open it, but she opened it out of curiosity.

In 1936, the British mathematician and computer scientist Alan Turing came up with his paper "On Computable Numbers with the Theoretical Foundation of Computing and Artificial Intelligence." His concept was the universal machine, and he also introduced the idea of a theoretical computing device that can simulate any algorithmic process. Around 1943, Warren McCulloch, a neurophysiologist, and Walter Pitts, a logician, came up with the mathematical model of an artificial neuron, laying the groundwork for the neural network theory. In 1956, the term "artificial intelligence" was coined at the Dartmouth Conference, which was under the leadership of John McCarthy, Marvin Minsky, Nathaniel Rochester, and Claude Shannon. The delegates of this conference were primarily scientists interested in creating machines that can simulate human intelligence. The same decade saw the development of The Logic Theorist (1955) and the General Problem Solver (1957). The following decades saw a focus on the expert systems that encoded human knowledge to solve specific problems.

The late 1980s and late 1990s are known as "AI Winter." This period saw a significant downfall in the field of AI research. The first decade of the twenty-first century saw rapid growth in the field of deep learning. The contemporary period is the most effective and popular in the history of AI studies. AI came into mass notice and interest, and, for the first time, launched to help the commoners through their smart devices to reduce their mental loads. There is data-based and knowledge-based AI:

- **Data-based AI:** "Artificial neural networks" have the mapping ability and computation connectivity that lead to the output of the prompt or input. The study of behaviour is old, and the very advanced technology also studies the input pattern and the output satisfaction. Data-driven AI is trained before it is made public. When there is an input, the system first predicts and minimizes the loss throughout the operations. The training period of a data-driven system is rigorous, and until the rate of error is very low, it keeps on learning.
- **Knowledge-based AI:** Data-driven AI has taken the popular market, but there is knowledge-based AI, which may be critical in educational use. Data-driven AI generates and represents the preexisting human data with a model of manipulation. Knowledge-based AI is the formulas, forms, patterns of behaviour, understanding, systems, or judgment that are reproduced. For knowledge, conceptual clarity is important, and concepts are determined by context. And determining a concept from a concept and giving an output is a complex process. In a changing or dynamic world where concepts and forms are evolving

and contexts are ever-expanding, it is difficult to model knowledge-based AI. Therefore, knowledge-based intelligence has made much progress in arithmetic and physics rather than in the other disciplines.

AI and Education

One decade ago, we imagined what would happen if a machine got a brain? The imagination is now reality. In the realm of education, the incorporation of innovative technologies has sparked a remarkable transformation in research and development, with artificial intelligence (AI) leading the charge. This cutting-edge convergence has birthed a plethora of groundbreaking applications that revolutionise the educational landscape. Here, we delve into the inventive utilisation of AI in course development, unveiling the ways it is reshaping pedagogical paradigms. The advent of AI has ushered in an era of hyper-personalisation, where algorithms meticulously dissect students' learning behaviours and preferences. This data-driven approach results in the creation of bespoke learning paths that cater to individual strengths and weaknesses, optimally fostering comprehension and engagement.

AI has taken on the role of a discerning curator, swimming in the vast reservoirs of educational content to unearth the most relevant and apt resources. Moreover, it actively contributes to content creation through Natural Language Generation (NLG), offering concise explanations and educational materials. This technological leap has bestowed the power of automation on the assessment process. AI-powered tools seamlessly evaluate assignments, quizzes, and exams. The immediate feedback loop nurtures real-time learning, rectifying misconceptions promptly. The dynamic landscape of education now boasts AI-fuelled adaptive learning platforms. These platforms continually monitor student progress, modulating the course's complexity and content based on individual advancements. The result is a tailored education experience that optimises growth.

AI-driven language translation tools have transcended linguistic barriers. By translating course materials into diverse languages, AI ensures that educational content becomes universally accessible, heralding an era of global learning connectivity. AI's analytical prowess shines through in its ability to decipher student interaction data. This resource is pivotal in identifying patterns, illuminating trends, and steering course refinements that elevate learning outcomes. AI avatars and virtual assistants stand at the forefront of student support systems. They cater to queries, provide insights, and guide learners through course materials, fostering an environment of constant guidance. AI-engineered gamification elements infuse courses with interactivity and motivation. Through automated progress tracking, point allocation, and competitive dynamics, students are propelled towards active involvement.

The AI-driven creation of interactive simulations and virtual labs marks a watershed moment in STEM education. These experiential learning tools deepen comprehension by allowing students to immerse themselves in hands-on exploration. AI's predictive capabilities empower educators to forecast students' prospects of success. By scrutinising behavioural cues, attendance records, and performance metrics, instructors can intervene proactively to bolster struggling learners. Two revolutionary models, Artificial Intelligence (AI) and Machine Learning (ML), showed a sense of the research and development possibilities to shift some of the major problem areas in education models and strategy, an article published on Global Market, Wadhwani and Loomba ("Artificial Intelligence (AI) in Education Market Size to Exceed $30 Bn by 2032"), Insight says that there is a high possibility in terms of components that Artificial Intelligence (AI) in education market from the service segment may cross the value of 10 billion USD by 2032. The education market size by different components is given in Table 10.1.

Table 10.1 Wadhwani and Loomba, "Artificial Intelligence (AI) in Education Market Size to Exceed $30 Bn by 2032".

By component	By development	By technology	By application	By end-use
- Solution - Services Professional Services Managed Services	- On-Premise Cloud	- Machine Learning - Deep Learning - Natural Language Processing (NLP)	- Learning Platform and Virtual Facilitators - Intelligent Tutoring System (ITS) - Smart Content - Fraud and Risk Management	- Higher Education - K-12 Education - Corporate Learning

As AI weaves into education, ethical considerations are paramount. Ensuring bias-free content and safeguarding data privacy are essential aspects that need diligent oversight. AI's prowess extends to fostering collaboration by suggesting ideal group compositions based on complementing skills. It also provides a technological canvas for seamless virtual teamwork. After the launch of ChatGPT in late 2022, there has been a huge mass consumption and befooling interpretation of the possibilities of AI in the near future, especially in education. The Assistant Director-General of Education at UNESCO, Stefania Giannini, wrote that "The speed at which generative AI technologies are being integrated into education systems in the absence of checks, rules, or regulations, is astonishing" (UNESCO). UNESCO shows concern in a press release that, while ChatGPT reached a Himalayan audience size of 100 million active users monthly in the first month of 2023, there is only one country that has released regulations for generative artificial intelligence ("Guidance for

Generative AI in Education and Research"). The Digital Futures Commission of the United Kingdom has developed a proposal for a certificate course to upskill educators regarding the appropriate and ethical use of artificial intelligence. The training on the use of artificial intelligence is highly absent, which is a concern for proper monitoring. In recent years, there has been a growth of online and blended degree courses in many of the leading universities. These universities even take handwritten submissions for evaluation. There is a high chance of students taking advantage of artificial intelligence to write the answers or reproduce answers generated by AI. The data reflected in response to the prompts in ChatGPT are the manipulated data from other individuals used without any proper reference or citations, which is highly unethical to use in serious research work. When prompted with, "Should we rely upon ChatGPT in serious research?" The ChatGPT-generated text indicated the following concerns:

Figure 10.1 Chat GPT's response on "Should we Rely Upon ChatGPT in Serious Research?" (GPT 4.0).

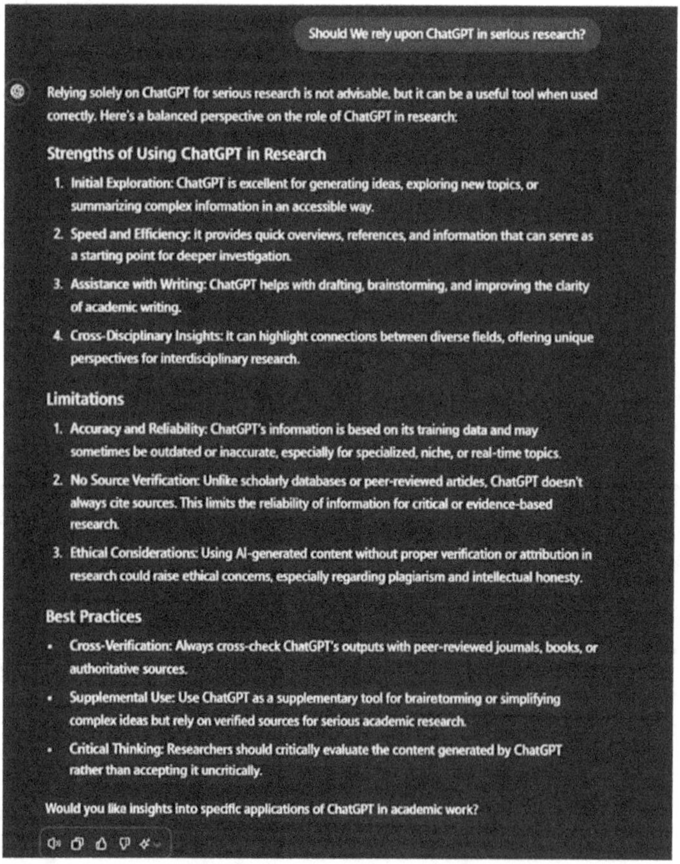

Should We rely upon ChatGPT in serious research?

Relying solely on ChatGPT for serious research is not advisable. but it can be a useful tool when used correctly. Here's a balanced perspective on the role of ChatGPT in research:

Strengths of Using ChatGPT in Research

1. **Initial Exploration:** ChatGPT is excellent for generating ideas, exploring new topics, or summarizing complex information in an accessible way.
2. **Speed and Efficiency:** It provides quick overviews, references, and information that can serve as a starting point for deeper investigation.
3. **Assistance with Writing:** ChatGPT helps with drafting, brainstorming, and improving the clarity of academic writing.
4. **Cross-Disciplinary Insights:** It can highlight connections between diverse fields, offering unique perspectives for interdisciplinary research.

Limitations

1. **Accuracy and Reliability:** ChatGPT's information is based on its training data and may sometimes be outdated or inaccurate, especially for specialized, niche, or real-time topics.
2. **No Source Verification:** Unlike scholarly databases or peer-reviewed articles, ChatGPT doesn't always cite sources. This limits the reliability of information for critical or evidence-based research.
3. **Ethical Considerations:** Using AI-generated content without proper verification or attribution in research could raise ethical concerns, especially regarding plagiarism and intellectual honesty.

Best Practices

- **Cross-Verification:** Always cross-check ChatGPT's outputs with peer-reviewed journals, books, or authoritative sources.
- **Supplemental Use:** Use ChatGPT as a supplementary tool for brairetorming or simplifying complex ideas but rely on verified sources for serious academic research.
- **Critical Thinking:** Researchers should critically evaluate the content generated by ChatGPT rather than accepting it uncritically.

Would you like insights into specific applications of ChatGPT in academic work?

While ChatGPT and similar language models can be valuable tools for various tasks, including generating ideas, drafting content, and providing information on a wide range of topics, it's important to exercise caution when relying on them for serious research. Here are some considerations:

1. Accuracy of information: ChatGPT's responses are generated based on patterns learned from diverse data sources on the internet. It doesn't have real-time access to databases or the ability to verify the latest information. Therefore, it's crucial to independently verify information obtained from ChatGPT in serious research.
2. Bias and limitations: The model may inadvertently produce biased or inaccurate information, as it reflects the biases present in its training data. It might not be aware of recent events or advancements, and its responses may lack nuance or context.
3. Lack of critical thinking: ChatGPT doesn't possess critical thinking skills and might provide information that sounds plausible but is incorrect. Researchers need to critically evaluate the information and cross-reference it with reliable sources.
4. No guarantee of quality: The quality of responses can vary, and there's no guarantee that every response will be accurate or suitable for research purposes. It's important to use the model as a supplement rather than a sole source of information.
5. Ethical considerations: If using ChatGPT for research, researchers should be aware of potential ethical concerns, such as unintentional biases in responses and the responsible use of AI technology.
6. Human expertise: While ChatGPT can provide information on a wide range of topics, human expertise is essential for interpreting, contextualising, and validating the information. Researchers should collaborate with domain experts to ensure the accuracy and reliability of their findings.

While ChatGPT can be a helpful tool in the research process, it should be used with caution. It's crucial to corroborate information with reliable sources, be aware of the model's limitations, and exercise critical thinking. Combining AI capabilities with human expertise is often the most effective approach in serious research (OpenAI, 2024).

Collecting resources is challenging, and maintaining regularly updated data is the biggest challenge for traditional research; therefore, it takes even years to update one ongoing research project. AI can easily collect data from free-to-access sources. AI can access all the websites and library resources open to access, but artificial intelligence, as of now, does not have access to the huge academic copyrights and research materials. The paid journals and publishing houses have been taking the copyright and access to these resources, which, if freed, can make the AI database oceanic. Research work can be revolutionised by moderating the AI access to these data. A study on the website of Words Rated, Curcic (2023), says that the share of open-access articles increased by 58 percent in the last decade, and in 2020, 56.8 percent of the articles published were somewhat open access. AI can access this data, and with the help of its structuring and data manipulating intelligence, it can easily frame and reframe the contents of the academic courses.

ChatGPT responds almost in the same language pattern to almost all the users on the same prompt, and you can regenerate the language also multiple times, again, they are the same for everyone, but every human writing is different in a unique way, every human writing is the reflection of the complex cognitive development of the person, the experiences since post birth. A subject expert has their philosophies and style of communication, and that is even more prominent in research writings. ChatGPT cannot assume a person's individuality and produce text; it produces in a general way, therefore maintaining the high-quality research along with innovative and critical thinking that is achieved through long academic thinking and life experience. The instant served manipulated plagiarized data cannot be used as of now, until all these qualities are prominent in ChatGPT. ChatGPT, like most of the other AI technologies, learns from the inputs given by the users, but the learning is limited as people also learn how to give technical inputs that best match the responses. Otherwise, the learning of ChatGPT from human users can not address the individual communication style of a person. For example, though ChatGPT learns the prompts given by people, when someone gets a response from ChatGPT, it is not personalised to the person who has given the prompt, but it is a result of all the learning of ChatGPT in general.

The Borgesian revelation of understanding has not and will not—and, we submit, cannot—occur if machine learning programs like ChatGPT continue to dominate the field of AI. However useful these programs may be in some narrow domains (they can be helpful in computer programming, for example, or in suggesting rhymes for light verse), we know from the science of linguistics and the philosophy of knowledge that they differ profoundly from how humans reason and use language. These differences place significant limitations on what these programs can do, encoding them with ineradicable defects.

UNESCO has proposed that all governments quickly regulate generative AI in schools. There are some points highlighted by the leading organisation:

1. The suggested minimum age is 13 years to use artificial intelligence in the classrooms.

2. There must be proper adaptation of data protection, privacy standards, and organisation of teacher training to properly manage AI in schools.
3. The government should monitor the schools using generative AI with a human-centred approach.
4. Governments should introduce policies to regulate and guide the ethical use of artificial intelligence in education.

Challenges in Advance Course Development and Contribution of AI

Cost and Time

Developing an advanced academic course involves a range of expenses. This encompasses the creation of a detailed curriculum, including course outlines, learning objectives, and assessments, which may necessitate input from subject matter experts, instructional designers, and curriculum developers. Content creation extends to generating lectures, assignments, quizzes, and various multimedia elements like videos and interactive content. Technology costs encompass the acquisition of a Learning Management System (LMS), potentially requiring expenses for platforms such as Moodle, Canvas, or Blackboard. Additionally, if custom platforms are developed, this can lead to significant costs. Faculty or instructor salaries or honorariums constitute another key expenditure. Marketing and promotion expenses include advertising through channels like social media or paid advertising, as well as the design of promotional materials or websites. Administrative costs involve tools for managing enrolments and registrations, along with support and help desk expenses for assisting students. There are also assessment-related costs, such as grading tools or services, as well as the design and administration of exams or quizzes. Compliance and accreditation expenses involve ensuring the course complies with legal and regulatory standards, along with potential fees for seeking accreditation from educational bodies.

Research and development costs may include pilot programs for testing the course with a small group of students before a full-scale launch. Miscellaneous expenses could include travel and accommodation for in-person components or conferences, as well as licenses for specialised software or tools. Ongoing maintenance and updates involve costs for periodically updating and improving course content, along with technical maintenance for the platform or LMS. Traditional higher education costs a huge amount, either for the government or the learner. Online and distant courses are cheaper than offline courses, and MOOCs are more cost-effective, but if we dream of making higher education more available and eliminating the financial barrier in higher education, we need to think about something that may cost very little, and the quality is not compromised. A study published by Athabasca University, Canada, gives data regarding MOOC costs in the US: "We find costs ranging from $38,980 to $325,330 per MOOC, and costs per completer of $74-$272, substantially lower than costs per completer of regular online courses, by merit of scalability" (Hollands & Tirthali, 2014).

Education is a sector where knowledge and intelligence play significant roles, along with innovative teaching styles. Teaching is a humane activity; it somewhere needs the understanding of the learner. For the last few years, machines could not independently teach, but it is time to introduce the technology to understand human behaviour and actions, human character traits, and human feelings. Through the use of AI, we can bring the cost of creating effective academic courses to a minimal level. The existing knowledge can be manipulated every time on demand by the machine to create and recreate content; it can more easily learn, unlearn, and relearn than human beings. Through effective data records, analysis, and progression statistics, AI can reduce the huge number of student management costs and eliminate the payment to the content developers and the teachers.

Data Collection and Quality Content

Collecting resources is challenging, and maintaining regularly updated data is the biggest challenge for traditional courses; therefore, it takes even years to update one ongoing course curriculum. AI can easily collect data from free-to-access sources. AI can access all the websites and library resources open to access, but artificial intelligence, as of now, does not have access to the huge academic copyright and research materials. The paid journals and publishing houses have been taking the copyright and access to these resources, which, if freed, can make the AI database oceanic. Research work can be revolutionised by moderating the AI access to these data. A study on the website of Words Rated, Curcic (2023), says that the share of open-access articles increased by 58 percent in the last decade, and in 2020, 56.8 percent of the articles published were somewhat open access. AI can access this data, and with the help of its structuring and data manipulating intelligence, it can easily frame and reframe the contents of the academic courses.

Maintaining the Data Updates

After spending a huge amount of time and money when a course is developed, the biggest challenge is updating and maintaining the content. The traditional data findings and citations are a lot of work to redo. It is very difficult for the big platforms containing thousands of MOOCs to update the content and lectures from time to time. It sometimes needs to reframe and revise the content structures. AI also needs a huge time and cost to produce the output and maintain the data; initially, it needs rigorous training to reduce the output error, and once the level of error is very low, it should be given continuous training to meet the changes and demands of the time. Therefore, data is the fuel for AI. AI can be modelled on the secondary sources of data, and rely on the big data storage houses, but for specific needs, AI may need specific data storage to store the base and user data. But there is a huge possibility that once multiple AI technologies are developed, there can be an integration between them to get knowledge and a database, which may change the pathways of education.

Student engagement and satisfaction

One of the biggest challenges for traditional MOOCs is engaging students. There are common discussion forums for interaction, but they are not effective enough to engage the students. The uniform nature of the courses often makes them uninteresting to different learners. The learner's expectations and prior knowledge levels are not always addressed. Online and blended courses are new to most universities, and they do not have a structured plan and effective pedagogy to address the student engagement and satisfaction challenges. Clear learning objectives, free interaction, varied content delivery, peer learning and collaboration, feedback loops, professional development for the instructors, and reflecting and adapting are the basic pedagogical elements of online and blended learning.

AI-powered adaptive learning systems can adjust the pace and difficulty of the course material based on the learner's performance. Adaptive learning has been shown to enhance and improve retention, boost learner engagement, reduce training time, and provide rich analytics that track the learner's performance over time (10 Ways Artificial Intelligence Is Transforming Instructional Design, n.d.).

Accessibility

Due to huge costs and human engagements, the popular MOOC platforms like Coursera and Edx cannot provide degree-level courses cheaply; they cost huge amounts. Financial burden is one of the biggest challenges in higher education. There are millions of students who cannot enter the big universities for fear of huge tuition fees. If AI can reduce the cost of running degree courses in different innovative ways, like reduction of resource costs and manipulation, instructors' salaries, data management, and assessment, courses can be delivered more cheaply, and everybody can have access to higher degrees. Apart from the course elements, AI can also make accessible the student support system, assessment data, scores, performance of the faculty, student feedback, and reflection more comprehensible and researchable (Yu et al., 2017).

Personalisation

The present MOOC models are based on recorded lectures and predetermined questionnaires, forums, and checks for understanding. The assignments are peer reviewed by the co-learners. AI can bring a drastic change in the sense:

- There can be an inductive interaction between the AI and the learner; this can be a formative evolution or mapping of the characteristics of the student response.
- The videos can be replaced by an interactive interface.
- The instruction language complexity level may be set according to the reading comprehension of the student.
- The questions or checks for understanding can be formed based on the student's reading level and comprehension style (which can be assessed through the type of questions the person asks and the type of clarification the person needs).
- Translation can be done as per the requirements of the student. Minimize the translation errors.
- 3D or AI characters can act as teachers; this needs rigorous training to align with human behaviour.

A mixed AI model that runs on the data and knowledge base can train the student behaviours for one single subject. Through rigorous training, the student behaviours can be categorized. The popular MOOC platforms like Coursera and Edx already have and can collect the student behavioural data for understanding how the students respond to particular instructions, teaching styles, activities, question answers, and behaviours. Since online learning gives a wide range of one-to-one learning possibilities, there is a huge scope to map the background, personal traits for effective learning. Students in advanced online courses interact with the intelligence system, which is a great opportunity for the artificial intelligence to get trained. Recently, a platform like Coursera has added a groundbreaking feature called "Coach," which is an AI bot to explain, summarize, bullet point, and suggest questions for practice to the users or students.

This "coach" AI can be very useful if it is used to learn the students' behaviour. For now, the bot itself suggests some questions to the students, like:

- Explain this topic in simple terms.
- Give me a summary.
- Give me practice questions.
- Give me real-life examples.

Standardizing the questions can help in a controlled study of the students' instincts and also lead in nature, but since it is a new feature, it is good to begin with a controlled setting and gradually train the AI. Though there is a question bar for customized questions, which may help the learners to query the questions not in the suggested list.

Assessment

Assessment as learning is a groundbreaking philosophy in academia that looks into assessing the students as a process of learning itself; assessment should not be a separate work. Assessment as an integrated constituent of learning can be improved in accuracy, efficiency, and personalization with the assistance of artificial intelligence. As discussed before, AI can take a more comprehensive, interactive, and effective placement evaluation to manipulate the entire course content on several parameters: the content rigor, the language rigor, the speed and pace, the questions rephrasing, the form of presentations, and more.

Objective tests like multiple choice questions, true-false questions, and other objective types of questions can be presented as more interactive, animated, and humanized and easily graded to save the time of the educator.

Subjective assessment needs an advanced natural language processing (NLP) tool to grade the descriptive questions ranging from short answer questions, small descriptions or essays, reports, or academic writings to long essays and other open-ended responses by analysing sentence structure. To value and grade based on organization, language quality, and creativity, the artificial intelligence needs long training and very little detection error in its knowledge map.

The concept of "assessment as learning" subscribes to the idea of formative assessment rather than summative assessment. AI is capable of real-time feedback, which generally in traditional courses needs time; on advanced platforms like Coursera and Edx, they follow peer review, which is time-consuming, and at the same time, there are several cases when the peer review is not properly done or done at all (*Help Articles*, n.d.). The real-time feedback can be on assignments, projects, quizzes, and presentations, but that must be more than just giving the data-driven figures and graphs; AI should elaborate and develop the possibilities of development in areas of improvement for a student. The unconscious or subjective bias in grading is a traditional problem in assessment; AI can minimize the personal beliefs and bias in academic grading.

As a part of personalization, AI can adjust the difficulty level of the assessment or test based on the placement evaluation and formative feedback.

Prediction and inference can be one of the major qualities of AI; it can analyse the pattern of the student behaviour and suggest remedial actions and conduct for them for better future outcomes.

Plagiarism and similarity are usually checked at the end of the written work. With the help of AI data stores and real-time detection, if the author or the student is given the real-time source of the data origin, it can help in restructuring and citing the work, which will lead to a finer academic writing or research outcome.

Conclusion

Academia has endless possibilities to learn from each conduct of it, all the institutions of bodies, both online and offline, have huge data to constantly learn the student behaviour and patterns, in general and in specific contexts. It is not that AI has all the possibilities to take charge of the entire academia, but it can significantly minimize unnecessary teacher-learner load and time. Personalization, assessment, interaction, skill development, and dishonesty in research can be significantly taken under control. Also, personal biases can be eliminated.

References

ChatGPT. chat.openai.com/share/615ea9ec-09f4-4506-a2c4-d11e57abc374.

Chomsky, Noam (2023, March 8). et al. "Opinion | Noam Chomsky: The False Promise of ChatGPT." *The New York Times.* www.nytimes.com/2023/03/08/opinion/noam-chomsky-chatgpt-ai.html

Curcic, D. (2023, June 2). *Open access publishing statistics.* https://wordsrated.com/open-access-publishing-statistics/#:~:text=Over%2056.86%25%20of%20academic%20articles,third%20year%20in%20a%20row.

Guidance for generative AI in education and research. (2024, June 19). UNESCO. https://www.unesco.org/en/articles/guidance-generative-ai-education-and-research#:~:text=UNESCO's%20first%20global%20guidance%20on,visión%20of%20these%20new%20technologies

Help articles. (n.d.). Coursera. https://www.coursera.support/s/article/learner-208279966-Solve-problems-with-peer-graded-assignments?language=en_US

UNESCO (2023). "Education in the Age of Artificial Intelligence." *The UNESCO Courier,* vol. 7257. unesdoc.unesco.org/ark:/48223/pf0000387029_eng

UNESCO: Governments Must Quickly Regulate Generative AI in Schools. *UNESCO,* 8 Sept. 2023, www.unesco.org/en/articles/unesco-governments-must-quickly-regulate-generative-ai-schools?hub=83250

Wadhwani, P., & Loomba, S. (2023, February 2). Artificial Intelligence (AI) in Education Market size to exceed $30 bn by 2032. *Global Market Insights Inc.* https://www.gminsights.com/pressrelease/artificial-intelligence-ai-in-education-market#:~:text=Artificial%20Intelligence%20(AI)%20in%20Education%20Market%20size%20is%20predicted%20to,by%20Global%20Market%20Insights%20Inc

Yu, H., Miao, C., Leung, C., & White, T. J. (2017). Towards AI-powered personalization in MOOC learning. *Npj Science of Learning, 2*(1). https://doi.org/10.1038/s41539-017-0016-3

Chapter Eleven
Navigating the EdTech and AI Revolution: Integrating Technology into Education with Ethical Responsibility

Sunita Maharaj-Landaeta

University of Trinidad and Tobago, Republic of Trinidad and Tobago

Alfredo Landaeta

Government of Montserrat

Abstract

This chapter explores the integration of educational technology (EdTech) and artificial intelligence (AI) in education, particularly focusing on their evolution, impact, and ethical considerations. It examines both the promise and challenges of these technologies in transforming learning environments, enhancing personalized education, and streamlining administrative tasks. Notable programs, such as Jamaica's "Tablets in Schools" and Trinidad and Tobago's "SmarterTT" initiative, highlight the Caribbean's efforts in implementing EdTech, with successes and limitations in areas like internet connectivity and teacher training. The chapter also delves into adaptive learning platforms, such as Knewton Alta and DreamBox, demonstrating their potential to tailor educational experiences to individual students. However, concerns related to data privacy, algorithmic bias, and the reduction of human interaction in education are raised. Furthermore, the importance of responsible practices in AI integration, including issues of equity, accessibility, and financial considerations, is emphasized. The chapter concludes with frameworks for the effective selection and implementation of EdTech, suggesting that successful technology integration in education hinges on a balanced approach that combines technological innovation with ethical and human-centred practices.

Keywords: Artificial Intelligence, AI and education, AI in education, ethical responsibility, personalized learning.

***

Introduction

The rapid advancement of educational technology (EdTech) and artificial intelligence (AI) has dramatically transformed the educational landscape, reshaping how knowledge is delivered and experienced in classrooms across the globe. These trends have picked up quickly in the Caribbean as well. Today's educational environments are increasingly infused with a diverse array of digital tools that enhance learning, engage students, and facilitate innovative teaching methods that were once beyond reach. AI, in particular, promises to revolutionize teaching by offering personalized educational experiences, automating and creating more efficient modalities for carrying out cumbersome administrative tasks, and providing data-driven insights to enhance student outcomes. However, these benefits come with significant ethical concerns, potential drawbacks, and financial implications that must be critically examined to ensure AI and EdTech are used responsibly and equitably in education.

In the Caribbean, there have been significant efforts to integrate technology into education. For instance, in Jamaica, the Ministry of Education, Youth and Information launched the "Tablets in Schools" program in 2014. This initiative aimed to improve digital literacy and access to educational resources by providing tablets to students in primary and secondary schools. However, the program also faced challenges, including limited internet connectivity and the need for extensive teacher training to effectively use these new tools in the classroom (Ministry of Education, Youth and Information, Jamaica, 2024). Another example is the eConnect and Learn (eCAL) program implemented by the Trinidad and Tobago Ministry of Education (MOETT) from 2010 to 2013, when it was replaced by the SmarterTT initiative (Martin, 2021). The program consisted of a "one laptop per child" initiative. "However, after investing large sums of money in the program, the initiative was abandoned as unsuccessful" (Martin, 2021).

Historical Context and Current Developments

The integration of technology in education began gaining traction in the early 2010s with the introduction of adaptive learning platforms and AI-driven tools. For instance, eBooks and interactive whiteboards started gaining traction in the US K-12 school system, offering an alternative to traditional textbooks (McCrea, 2009). While AI-driven tools were not as widespread in the early 2010s, there was a growing interest in using technology to assess and track individual student performance. The developments in this area laid the groundwork for more sophisticated AI-driven tools that would follow in later years. One of these examples is a company in India, *Think & Learn*, founded in 2011 by Byju Raveendran; in 2019, his company was valued at USD 5.7 billion, serving more than 35 million students (Kunju, 2019).

Nowadays, platforms like Knewton Alta (John Wiley & Sons Inc., 2024) and DreamBox (Discovery Education Inc., 2024) offer personalized learning experiences by adapting to individual student needs through real-time data analysis. Knewton Alta utilizes adaptive algorithms to tailor assignments in real-time, adjusting to each student's performance and knowledge level to ensure targeted learning and effective gap closure. By 2024, the landscape had evolved dramatically, with AI tools such as ChatGPT, Perplexity, Claude, Gemini, and Co-Pilot becoming integral to modern classrooms, offering new ways to engage students and streamline educational processes. However, their rapid adoption has also raised concerns about potential impacts on students' critical thinking skills and the ethical implications of heavily relying on AI.

A Look at Some Technological Platforms Used in Education: Pros and Cons

Knewton Alta is a platform that operates by analysing student performance data as they complete assignments, using adaptive algorithms to customize the learning experience. These algorithms assess which areas a student struggles with and adjust the content accordingly, providing additional practice or instructional support in those specific areas. This approach helps ensure that each learner receives personalized learning tailored to their individual needs, promoting deeper understanding and mastery of the subject matter (John Wiley & Sons Inc., 2024).

However, the use of adaptive algorithms also has disadvantages. Research has shown that these algorithms, while effective in identifying gaps in knowledge, rely heavily on data inputs that may not always capture the full complexity of a student's learning needs. This can lead to a narrow focus on measurable outcomes, potentially at the expense of fostering a holistic understanding of the subject matter (O'Neil, 2016). Studies have raised concerns about the potential of these systems to reinforce existing biases present in the data, which could disproportionately impact certain groups of students, exacerbating educational inequities (Eubanks, 2018).

The increased reliance on technology and algorithms can reduce the role of human intuition and teacher-student interaction. Educators may become more dependent on the algorithm's recommendations, which could lead to less nuanced decision-making. This shift might limit opportunities for personalized emotional and social support, aspects of learning that are crucial but difficult for algorithms to replicate (Selwyn, 2019). This reduction in direct interaction could potentially diminish the overall quality of the educational experience, as the human element in teaching—vital for addressing the diverse needs of students—becomes less prominent.

DreamBox Learning is another adaptive learning platform focused on mathematics education, designed to tailor instruction to individual student needs by analysing their performance data in real time. The platform adjusts the difficulty and type of content presented to each student, offering targeted support in areas where they struggle while advancing them through concepts they have already mastered. This adaptive approach aims to promote a deepened understanding of mathematical concepts and improve student outcomes by providing personalized learning experiences. While the advantages of DreamBox Learning are clear in its ability to provide customized instruction that meets students where they are, it is important to recognize that there are also some considerations to keep in mind. Research suggests that while adaptive algorithms are effective in addressing specific knowledge gaps, they may sometimes prioritize measurable outcomes, which could limit opportunities for students to explore broader, more conceptual aspects of mathematics (Luckin, 2017). However, this structured approach can be highly beneficial for reinforcing foundational skills and ensuring mastery of key concepts before moving on to more complex material.

Another consideration is the potential for existing biases in the data to influence the adaptive process, though this is a concern that many educational technologies are actively working to mitigate (Noble, 2018). It's also worth noting that the use of such platforms can lead to a shift in the teaching dynamic, where the role of the teacher may become more focused on facilitating and interpreting data-driven insights rather than directly guiding every aspect of student learning (Williamson, 2020). This can be a positive development, as it allows teachers to focus on more personalized interaction and support, although it also underscores the need to maintain a balance between technological and human elements in the learning process.

More recently, ChatGPT has become quite popular among students, teachers and other educational stakeholders (Fütterer et al., 2023). It is an AI-driven language model designed to assist with a wide range of tasks, from answering questions and providing explanations to generating content and facilitating conversations. By leveraging vast amounts of data and advanced algorithms, ChatGPT can understand and respond to user inputs in a way that feels natural and conversational. This makes it a versatile tool for learning, content creation, and even brainstorming, offering users quick access to information and ideas tailored to their needs. While the strengths of ChatGPT lie in its ability to generate coherent, contextually relevant responses across a broad spectrum of topics, it's important to consider some of the challenges associated with AI-driven models. One of the key benefits is the immediacy and breadth of information it can provide, which can be particularly useful in educational settings where quick access to information is crucial. However, because

ChatGPT relies on patterns in data, it may occasionally produce content that is either overly generic or lacks the nuanced understanding that a human expert might provide. This can sometimes result in responses that are informative but not deeply insightful or contextually rich (Bender et al., 2021).

Another consideration is the model's reliance on existing data, which means that its responses are shaped by the information it has been trained on. While this allows for a wide range of knowledge, it also means that any biases present in the training data can be reflected in its outputs. Ongoing efforts in AI research are aimed at mitigating these biases, but it's a reminder that human oversight is still necessary when interpreting and applying the information provided by AI (Broussard, 2018).

Moreover, while ChatGPT offers significant potential for enhancing productivity and learning, it is essential to maintain a balance between its use and human interaction. AI tools like ChatGPT can efficiently handle routine tasks and provide quick answers, freeing up time for deeper, more personalized engagement. However, it's crucial to ensure that the human element remains central, especially in areas requiring critical thinking, empathy, and ethical judgment (Floridi & Cowls, 2019).

Perplexity is a key metric in natural language processing, especially when evaluating models like ChatGPT. It gauges the uncertainty a model has in predicting the next word in a sequence, with a lower perplexity indicating greater accuracy and confidence in its predictions. In educational contexts, a model with low perplexity is more likely to produce coherent, contextually appropriate responses, which enhances the learning experience by delivering clear and relevant information. However, it's important to note that while low perplexity is desirable for generating accurate text, it doesn't always guarantee the depth or quality of content, particularly when dealing with complex concepts or nuanced human emotions (Radford et al., 2019).

In the broader field of educational technology, several other AI platforms are gaining prominence for their contributions to personalized learning. Socratic by Google, for example, is an AI-powered app that assists students with homework by analysing questions through text or photos and offering step-by-step explanations (Google, n.d.). By leveraging machine learning, Socratic provides quick and clear solutions, making it a valuable tool for helping students understand challenging concepts (Google AI, 2020). Similarly, Carnegie Learning specializes in personalized math instruction using AI-driven tools that adapt to each student's pace and needs. This platform combines cognitive science with adaptive learning technologies to offer a tailored educational experience, helping students build a strong foundation in mathematics (Pane et al., 2014).

Gradescope is another innovative AI platform that streamlines the grading process by using machine learning to group similar responses. This approach makes grading more efficient and consistent, allowing educators to focus more on delivering meaningful feedback rather than on the mechanical aspects of grading, which ultimately enhances the educational experience for students (Singh et al., 2017). Century Tech, on the other hand, integrates AI with neuroscience insights to create personalized learning pathways. The platform continuously analyses individual learning behaviours and adapts content to match each student's strengths and weaknesses, promoting more effective learning outcomes (Luckin, 2017).

These AI platforms demonstrate the transformative potential of artificial intelligence in education. By offering personalized, adaptive learning experiences and automating routine tasks, they enable educators to concentrate on more impactful interactions with students. Nonetheless, as with any technology, it's crucial to maintain a balance between AI use and human oversight to ensure that learning remains comprehensive and inclusive.

Building a Foundation for Responsible Technology Use-Defining Responsible EdTech and AI Practices

Defining responsible EdTech and AI practices is essential for enhancing learning while addressing the ethical and practical challenges that come with integrating technology into education. As AI platforms like ChatGPT, Socratic by Google, Carnegie Learning, Gradescope, and Century Tech become increasingly prevalent in educational settings, it is crucial to consider how these technologies impact key aspects such as data privacy, accessibility, and equity.

Responsible EdTech and AI use is predicated on understanding both the benefits and risks associated with these technologies. One of the most critical aspects is ensuring data privacy (Taxila Business School, 2020). Platforms like ChatGPT and Gradescope, which rely on vast amounts of student data to function effectively, must manage and protect this data in strict compliance with legal standards, such as the Family Educational Rights and Privacy Act (FERPA) in the United States. Educators and institutions must be vigilant in safeguarding student information from unauthorized access and misuse, as the integrity of personal data is central to building trust in these technologies (Regan & Jesse 2019). A particularly sensitive area within data privacy is the protection of children and their families. The use of AI in education often involves the collection of detailed and sensitive information about students, including minors, which can lead to specific child protection issues. For example, data breaches or unauthorized access to educational platforms could expose children to risks such as identity theft, online predators, or the misuse of their personal information. Moreover, the aggregation of data over time could lead to profiling, where children are unfairly categorized or judged based on their digital footprints. This not only threatens their privacy but could also have long-term implications for their future opportunities (Barassi, 2020). Protecting the privacy of children and their families is thus paramount, requiring that any data collected is handled with the highest levels of security and in accordance with legal and ethical guidelines designed to protect vulnerable populations.

Accessibility is another fundamental consideration in responsible EdTech practices. AI platforms like Socratic and Century Tech offer significant potential to personalize learning experiences, but it is imperative that these tools are designed and implemented in a way that is accessible to all students, including those with disabilities. Ensuring compliance with accessibility standards and providing alternative means of engagement—such as tools that incorporate screen readers, alternative text descriptions, and adaptable user interfaces—are critical steps toward creating an inclusive learning environment that supports diverse student needs (Seale, 2013).

Equity in education is equally vital. As platforms like Carnegie Learning and Gradescope are integrated into classrooms, it is essential to ensure that all students, regardless of their socio-economic background, have equal access to these technological tools and resources. Addressing the digital divide by providing adequate infrastructure, such as reliable internet access and necessary devices, as well as offering training and support to both students and educators in using these technologies effectively, is crucial (Selwyn, 2019). Moreover, equity extends to the algorithms themselves. There is a need for continuous evaluation and mitigation of potential biases that could disproportionately affect marginalized groups, ensuring that all students benefit equally from AI-driven educational tools (Eubanks, 2018).

Beyond these considerations, the ethical implications of AI integration in education are profound. AI systems, by their nature, require vast amounts of student data to function effectively, which raises concerns about potential privacy violations. Furthermore, AI algorithms, if trained on biased datasets, may inadvertently perpetuate inequalities, disadvantageous to certain student groups. As AI takes on responsibilities traditionally held by educators, there is a risk that the role of teachers could be diminished, leading to a depersonalized educational environment where human interaction is reduced.

Overall, responsible EdTech and AI practices require a balanced approach that enhances learning while addressing ethical and practical concerns. By prioritizing data privacy, especially in relation to child protection, accessibility, and equity, educators can ensure that these powerful technologies are used in a way that is not only effective but also just and inclusive. As AI continues to shape the future of education, it is essential to remain mindful of these principles to create a learning environment that benefits all students.

Realizing technological integration in education responsibly requires more than an understanding of ethical challenges—it necessitates a structured and technical approach to embedding these principles into the decision-making process. The ethical concerns around data privacy, accessibility, and equity must inform every stage of technology selection, deployment, and use. This is where a robust framework for effective technology selection and implementation plays a critical role. Such a framework enables institutions to operationalize responsible practices by incorporating technical and evaluative processes.

Frameworks for Effective Technology Selection and Implementation

Realizing technological integration in education responsibly requires a comprehensive approach that moves beyond simply addressing ethical concerns. It demands a structured and technical framework to embed principles like data privacy, accessibility, and equity into every stage of technology selection, implementation, and evaluation. This is where a robust framework for effective technology selection becomes essential—not only to mitigate risks but to ensure these tools genuinely enhance educational outcomes while adhering to ethical standards.

A well-designed framework operationalizes these responsible practices by incorporating technical evaluations and continuous monitoring. For instance, during the needs assessment phase, institutions can identify potential risks—such as data privacy vulnerabilities or accessibility gaps—using benchmarks like GDPR compliance or Section 508 accessibility standards. In the vendor selection phase, schools can enforce accountability by requiring transparent reporting on security protocols and algorithmic biases, ensuring alignment with ethical principles.

In the implementation phase, rigorous testing environments can evaluate AI tools like ChatGPT or Gradescope for equity impacts and accessibility, using methods like WCAG compliance testing for screen readers and adaptive interfaces. Scalability assessments ensure tools function consistently across diverse classroom settings, addressing digital divide concerns. Finally, the monitoring and evaluation phase ensures ongoing refinement by gathering actionable data on tool performance and unintended consequences, enabling institutions to address algorithmic bias or implementation flaws effectively.

This structured framework not only helps institutions realize the principles of responsible EdTech and AI integration but also ensures that these technologies directly contribute to inclusive, equitable, and impactful learning experiences. By embedding these technical processes, educators can bridge the gap between ethical considerations and practical implementation, creating a system that benefits all students while fostering trust and accountability.

Needs Assessment and Goal Setting

The first step in any effective technology selection framework is conducting a thorough needs assessment. This involves identifying the specific challenges or gaps that the technology is intended to address. For example, if a school district is considering adopting a new digital assessment tool, the process might begin by surveying teachers and administrators to determine the current limitations of existing assessment methods. This could include identifying issues such as the time required to administer assessments, the ability to provide timely feedback, and the need for tools that accommodate diverse student learning styles. Once the needs are clearly understood, the next step is to set specific, measurable goals for what the technology should achieve. These goals should align with the broader educational objectives of the institution, such as improving student engagement, increasing assessment efficiency, or enhancing personalized learning opportunities. Clear goal setting provides a benchmark against which the effectiveness of the selected technology can be measured, ensuring that it meets the institution's needs.

Research and Evaluation of Tools

With goals in place, the next step is to research and evaluate the available technology options. This involves reviewing the features, usability, scalability, and cost of various tools to determine which best meets the identified needs and goals. For instance, a school district might explore several digital assessment platforms, comparing them based on their ease of use for both students and teachers, their ability to scale across multiple classrooms or schools, and their cost-effectiveness. In this stage, it is also important to consider the alignment of the technology with the curriculum and instructional strategies. Tools that seamlessly integrate with existing systems and support the delivery of curriculum content are more likely to be effective. Patricia Ertmer and Anne Ottenbreit-Leftwich's work highlights that technology platforms aligning closely with instructional goals and being easy for teachers to incorporate into their daily routines are more likely to be adopted and used effectively (Ertmer & Ottenbreit-Leftwich, 2013).

Pilot Testing

After selecting a potential tool, the next step in the framework is pilot testing. This involves implementing the technology on a small scale to gather data on its effectiveness in a real-world educational setting. For example, a school might pilot a new digital assessment tool in a few classrooms before rolling it out across the entire district. During this pilot phase, teachers and students can provide feedback on the tool's usability, effectiveness, and any challenges they encounter. Pilot testing allows for the identification of potential issues before a full-scale implementation, reducing the risk of widespread problems. It also provides an opportunity to assess whether the technology meets the goals set during the needs assessment phase. Data gathered during the pilot can be used to make informed decisions about whether to proceed with a broader implementation, adjust, or consider alternative options.

Professional Development

Effective technology implementation requires more than just selecting the right tool; it also involves ensuring that educators are equipped to use it effectively. Professional development is a critical component of the framework, providing teachers with the training and support they need to integrate the technology into their instructional practices. This could include workshops, online tutorials, and ongoing coaching to help teachers become proficient with the new tool. Research by Lawless and Pellegrino emphasizes the impact of professional development on the success of technology integration in schools. Their findings indicate that when teachers receive adequate training and support, they are more likely to use the technology effectively and to its full potential (Lawless & Pellegrino, 2007). Therefore, any framework for technology selection and implementation should include a robust professional development plan.

Reviewing and Iteration

Finally, effective technology implementation is an ongoing process that requires continuous reviewing and iteration. After the technology has been implemented, it is important to gather feedback from all stakeholders, including teachers, students, and administrators. This review can be used to identify any issues, assess the impact of the technology on learning outcomes, and make necessary adjustments to improve its effectiveness (The Pennsylvania State University, 2024). For instance, if a school district finds that the new digital assessment tool is not being used as intended or is not improving student outcomes, it might need to revisit the training provided, make changes to the tool's configuration, or even reconsider its selection. The iterative process of reviewing and refinement ensures that the technology remains aligned with the school's educational goals and continues to meet the needs of its users. Therefore, technology selection and implementation in education requires a structured framework that includes needs assessment, research and evaluation, pilot testing, professional development, and continuous feedback. By following this framework, schools can ensure that the technologies they adopt are practical, scalable, cost-effective, and aligned with their educational goals. This systematic approach not only increases the likelihood of successful technology integration but also ensures that the tools selected truly enhance learning outcomes.

Seamless Classroom Integration Strategies

Integrating technology into the classroom should enhance the learning process by making it more interactive, personalized, and engaging. A practical approach to achieving this is through blended learning, which combines traditional face-to-face instruction with online activities. For example, a science teacher might use both hands-on experiments and virtual labs, allowing students to explore scientific concepts in multiple ways. This method caters to diverse learning styles, offering a more comprehensive understanding of the material. AI-powered platforms like Khan Academy further enhance this approach by adapting content and pacing to individual students' needs, ensuring that each learner receives instruction at the appropriate level. Research has shown that such personalized learning approaches can lead to improved learning outcomes, particularly in subjects like mathematics and science (Pane et al., 2017).

Engaging Students and Fostering Collaboration

EdTech and AI tools are also instrumental in enhancing student engagement and fostering collaboration. Digital tools like interactive whiteboards and collaborative software, such as Google Docs, enable students to work together seamlessly, regardless of their physical location. For instance, tools like Padlet create collaborative learning environments by allowing real-time interaction, which promotes a sense of community and shared learning. This collaborative approach is particularly effective in promoting critical thinking and problem-solving skills, as students are encouraged to work together to explore ideas and solutions (Dillenbourg, 1999).

Ensuring Effective Learning Outcomes

To ensure that technology integration leads to effective learning outcomes, it is essential to align these tools with the educational objectives of the curriculum. Setting clear goals helps educators evaluate the success of the technology, while regular assessments and feedback mechanisms are crucial for continuous improvement. For example, a history teacher using an interactive timeline tool can measure its effectiveness by assessing student engagement and understanding of historical events. Continuous improvement involves regularly updating and refining the technology based on feedback and observed outcomes, ensuring that it remains relevant and effective in achieving educational goals (Ertmer & Ottenbreit-Leftwich, 2013). Similarly, in Grenada, the government initiated the "Smart Classroom Project" in 2018, which equipped classrooms with interactive whiteboards, high-speed internet, and multimedia projectors. This project transformed traditional classrooms into tech-enabled, interactive learning environments. While this initiative has been successful in promoting digital learning, it also highlighted the importance of continuous professional development for teachers to maximize the benefits of these technologies (Government of Grenada, Ministry of Education, 2024).

Global Perspectives on Teacher Training

Teacher training is a critical component of successful technology integration, and innovative approaches are being adopted worldwide to address this need. In Finland, for example, teacher training programs emphasize the creative integration of technology into digital pedagogy, encouraging educators to use digital tools in ways that enhance creativity and critical thinking. Meanwhile, in rural India, where internet access is often limited, teacher training focuses on leveraging mobile technology and offline resources to ensure that students still benefit from technology-enhanced learning. These diverse approaches highlight the importance of tailoring technology integration strategies to the unique contexts of different educational environments, providing valuable insights into overcoming challenges and maximizing the potential of EdTech and AI (Sahlberg, 2011; UNESCO, 2020). Effective technology integration in the classroom, therefore, requires thoughtful strategies that enhance learning, engage students, and align with educational goals. By adopting approaches such as blended learning, fostering collaboration, ensuring continuous feedback, and considering global perspectives on teacher training, educators can successfully integrate technology in ways that improve learning outcomes and prepare students for the future.

Empowering Teachers and Students in the Digital Age

In the digital age, empowering teachers and students requires a concerted effort to develop digital competence, transform traditional roles, and cultivate essential skills such as critical thinking and lifelong learning. Effective technology integration hinges on equipping educators with the knowledge and confidence to harness digital tools in ways that enhance the learning experience (Tzafilkou et al., 2023).

Developing Digital Competence and Confidence

Building digital competence among teachers is crucial for the successful integration of technology in the classroom. Professional development programs should focus on both technical skills and pedagogical knowledge, ensuring that teachers are not only comfortable with the tools they are using but also understand how to apply them effectively in their teaching. For instance, workshops that train teachers to use digital tools like interactive whiteboards can be followed by collaborative opportunities where educators share best practices and learn from one another. This ongoing support, whether through online communities or mentoring, helps sustain and deepen digital literacy over time. Research indicates that such comprehensive professional development, which includes technical training and ongoing support, significantly enhances teachers' confidence and competence in using digital tools (Lawless & Pellegrino, 2007).

Transforming Teacher Identity and Roles

The role of teachers is evolving in the digital age, shifting from being primarily transmitters of knowledge to becoming facilitators of learning. The introduction of AI in education plays a significant role in this transformation, allowing for personalized instruction that caters to individual student needs. With AI handling more routine tasks such as grading and delivering basic instruction, teachers can focus on facilitating discussions, providing individualized support, and guiding students in developing critical thinking skills. This shift not only transforms the learning experience into a more student-centred approach but also empowers teachers to engage more deeply with their students' learning processes, fostering a more interactive and supportive classroom environment (Wiliam, 2011).

Cultivating Critical Thinking and Lifelong Learning Skills

Integrating technology in the classroom should also aim to develop students' critical thinking and lifelong learning skills. Digital tools offer unique opportunities for students to engage in complex tasks that require critical analysis and independent thought. For example, students using digital research tools can explore multiple perspectives on a historical event, compare sources, and develop their interpretations. Additionally, encouraging students to participate in online communities or work on digital projects can foster a culture of lifelong learning. This approach not only prepares students for the demands of the digital world but also instils a mindset of continuous learning and intellectual curiosity, which are crucial skills for success in the twenty-first century (Jenkins et al., 2009).

Empowering teachers and students in the digital age requires a multifaceted approach that includes developing digital competence, transforming traditional educational roles, and cultivating critical thinking and lifelong learning skills. By focusing on these areas, educators can effectively integrate technology into their teaching practices, creating a more dynamic, student-centred learning environment that prepares students for the challenges and opportunities of the digital world.

Ensuring Accountability in Tech-Enhanced Learning

Accountability in tech-enhanced learning is essential to ensure that educational technologies are integrated effectively and contribute meaningfully to student outcomes. This requires a systematic approach, beginning with the establishment of clear goals and metrics and extending to the utilization of data to measure the impact of these technologies. When implemented correctly, accountability helps educators and administrators make informed decisions about the use of technology in the classroom, ensuring that it enhances, rather than detracts from, the learning experience.

Establishing Clear Goals and Metrics

The foundation of accountability in tech-enhanced learning lies in setting clear, measurable goals. These goals should align with the overall educational objectives of the institution and be specific enough to allow for accurate assessment. For instance, a school might implement virtual labs with the goal of increasing student engagement in science subjects. Success metrics for this initiative could include participation rates in the virtual labs, feedback surveys that gauge student interest and understanding, and performance on assessments that measure knowledge retention and application. An example of this approach is seen in the implementation of flipped classrooms, where the goal might be to improve student comprehension of complex topics by allowing students to review materials at their own pace before participating in classroom activities. Metrics to assess the success of this initiative could include the frequency of student logins to the learning platform, completion rates of pre-class assignments, and improvements in test scores following the implementation of the flipped classroom model (Bergmann & Sams, 2012). Establishing these goals and metrics ensures that the technology is being used effectively to enhance learning.

Utilizing Data to Measure Impact

Data plays a crucial role in measuring the impact of technology on learning outcomes. The ability to collect, analyse, and interpret data allows educators to assess the effectiveness of tech-enhanced learning initiatives and make necessary adjustments. Online platforms often come equipped with analytics tools that track student progress, such as time spent on tasks, completion rates, and quiz scores. These data points provide valuable insights into how students interact with the content and where they may be struggling.

A specific example of data utilization can be seen in adaptive learning platforms like DreamBox Learning, which adjust the difficulty of tasks based on real-time student performance data. Teachers can use the data generated by such platforms to identify struggling students and tailor their instruction to meet individual needs. This approach not only helps in personalizing learning but also in holding the technology accountable for delivering the promised educational outcomes (Pane et al., 2014). Moreover, data-driven accountability extends to assessing the broader impacts of technology in education. For instance, in the United Kingdom, a significant study conducted by Louis-Philippe Beland and Richard Murphy at the London School of Economics and Political Science (LSE) analysed the effects of banning mobile phones in secondary schools. Their research, titled "Ill Communication: Technology, Distraction & Student Performance" (2016), found that banning mobile phones led to a significant improvement in student test scores, particularly among low-achieving students. This study highlights the importance of using data to measure the unintended consequences of technology use in educational settings. By tracking academic performance before and after the mobile phone ban, the researchers demonstrated a clear link between reduced distractions and improved learning outcomes (Beland & Murphy, 2016). Additionally, studies on the impact of AI in education further underscore the importance of data-driven accountability. For example, the RAND Corporation's study titled "Promising Evidence on Personalized Learning" (Pane et al., 2017) evaluated the effectiveness of AI-driven personalized learning systems. The study found that students using AI-powered adaptive learning platforms demonstrated modest improvements in mathematics performance, highlighting the potential of AI to enhance educational outcomes. However, the study also emphasized the need for proper teacher training and careful integration of AI systems into existing curricula to ensure these technologies are used effectively. Furthermore, the OECD's report "Artificial Intelligence in Education: Challenges and Opportunities for Sustainable Development" (2019) explored the broader implications of AI in education, including the ethical concerns related to data privacy, bias, and equity. The report stressed the importance of ongoing evaluation and mitigation of potential biases in AI algorithms to prevent reinforcing existing inequalities and ensure that all students benefit from AI-driven educational tools.

Ensuring accountability in tech-enhanced learning requires a comprehensive approach that begins with establishing clear goals and metrics and is supported by the rigorous use of data to measure impact. By doing so, educators can make informed decisions about the integration of technology in the classroom, ensuring that it serves as a tool for enhancing student engagement, comprehension, and overall academic success. As technology continues to evolve, so too must the strategies for holding it accountable, ensuring that it fulfils its potential to positively transform education.

Challenges Faced by Mobile Populations

Mobile populations, like refugees and migrant workers, face barriers to accessing education, including language differences and limited connectivity. Innovative solutions are needed to ensure quality education for these groups. For instance, Syrian refugee children in Lebanon often face challenges in continuing their education because the instruction is primarily in English or French, which they may not speak fluently. Additionally, limited internet access in many refugee camps restricts their ability to engage in online learning, making it difficult for them to keep up with their peers (UNHCR, 2022).

Leveraging Technology for Mobile Learning

Technology can provide education to mobile populations through mobile learning platforms, offline resources, and language translation tools. For example, Duolingo offers accessible language learning through a mobile app, which can be particularly useful for refugees needing to learn a new language quickly (Duolingo, 2023). Additionally, Kolibri, an offline platform, has been deployed in various refugee camps, including those in Lebanon, to provide curriculum-aligned educational content without the need for internet access. This ensures that even in connectivity-challenged environments, children can continue their education (Learning Equality, 2023).

Cultural Impacts of Technology in Education

Positive Cultural Impacts

Technology can enrich education by incorporating diverse cultural perspectives. Online platforms enable global connections, fostering awareness and cultural exchange. For instance, platforms like ePals and Global Classroom allow students from different countries to collaborate on projects, promoting cross-cultural understanding (ePals, 2023). Virtual field trips, such as those offered by Google Arts & Culture, expose students to historical and cultural landmarks worldwide, broadening their horizons (Google Arts & Culture, 2023).

Addressing Cultural Sensitivity and Inclusivity

Educators must ensure digital tools and content are culturally sensitive and inclusive, selecting resources that represent diverse cultures accurately. For example, resources like the Smithsonian Learning Lab provide culturally diverse content that educators can use to create inclusive lesson plans (Smithsonian

Learning Lab, 2023). Technology can also support language preservation efforts in indigenous communities, maintaining cultural heritage. For instance, platforms like FirstVoices are used to preserve and teach indigenous languages, ensuring they are passed on to future generations (FirstVoices, 2023).

Financial Implications of AI and EdTech in Education

Reducing Operational Costs

AI can reduce operational costs by automating tasks like grading and scheduling, freeing up time and resources. For instance, platforms like Gradescope use AI to streamline the grading process, allowing educators to focus more on instruction (Gradescope, 2023). However, the initial investment in acquiring AI tools, training educators, and maintaining systems can be substantial, often requiring significant upfront capital and ongoing financial commitments (EdTech Review, 2023).

Challenges for Underfunded Schools

The financial burden of AI implementation is particularly challenging for underfunded schools, potentially exacerbating existing inequalities. Schools with limited budgets may struggle to afford the necessary infrastructure, such as high-speed internet and updated hardware, which are essential for effective AI integration (Digital Divide Council, 2023). Ensuring compliance with legal and ethical standards adds to the financial considerations, particularly for institutions operating on a global scale, where different regions may have varying requirements (OECD, 2023).

Balancing Creativity with Established Pedagogical Standards

While embracing technological innovations, balancing creativity with established pedagogical standards ensures that new approaches enhance learning without compromising academic rigor. For example, incorporating digital storytelling in an English class can enhance creativity while meeting writing and communication standards. Tools like Storybird allow students to create visually engaging stories that align with curriculum goals, blending creativity with educational outcomes (Storybird, 2023).

Conclusion

The EdTech and AI revolution in education is not merely a wave of technological advancement; it represents a profound transformation in how we approach learning and teaching. As we navigate this new landscape, it is imperative to remember that technology is a tool—powerful, yes, but only as effective as the principles guiding its use. The promise of AI and EdTech lies in their ability to create more personalized, efficient, and engaging learning environments. However, realizing this potential requires a steadfast commitment to ethical responsibility, inclusivity, and equity.

In the Caribbean and globally, the integration of these technologies must be approached with a careful balance between innovation and the preservation of educational integrity. The success of initiatives like Jamaica's "Tablets in Schools" and Grenada's "Smart Classroom Project" underscores the transformative potential of EdTech, yet also highlights the challenges that come with such large-scale implementations. The financial implications, cultural impacts, and ethical considerations outlined in this discourse are not peripheral concerns; they are central to ensuring that technology serves as a bridge to greater educational equity, rather than a barrier.

As educators, policymakers, and stakeholders, our role is to ensure that the deployment of AI and EdTech is aligned with the fundamental goal of education: to empower learners with the knowledge, skills, and values they need to navigate an increasingly complex world. This involves not only selecting and implementing technology with care but also continuously reflecting on its impact, learning from global best practices, and being vigilant against the unintended consequences that may arise.

The future of education is undoubtedly intertwined with technology, but it is the human element—our creativity, empathy, and ethical judgment—that will determine whether this future is one of opportunity for all. As we continue to integrate AI and EdTech into our classrooms, let us do so with a clear vision: to create a learning environment that is not only technologically advanced but also profoundly human, where every student can thrive.

References

Angwin, J., Larson, J., Mattu, S., & Kirchner, L. (2016, May 23). Machine bias: There's software used across the country to predict future criminals. And it's biased against Blacks. *ProPublica.* https://www.propublica.org/article/machine-bias-risk-assessments-in-criminal-sentencing

Barassi, V. (2020). *Child data citizen: How tech companies are profiling us from before birth. MIT Press.* https://doi.org/10.7551/mitpress/12415.001.0001

Bélanger, L.-P., & Murphy, R. (2016). Ill communication: Technology, distraction & student performance (CEP Discussion Paper No. 1350). London School of Economics. https://cep.lse.ac.uk/pubs/download/dp1350.pdf

Bergmann, J., & Sams, A. (2012). *Flip your classroom: Reach every student in every class every day. International Society for Technology in Education.*

Broussard, M. (2018). *Artificial unintelligence: How computers misunderstand the world.* MIT Press. https://doi.org/10.7551/mitpress/11022.001.0001

Cadwalladr, C., & Graham-Harrison, E. (2018, March 17). Revealed: 50 million Facebook profiles harvested for Cambridge Analytica in major data breach. *The Guardian.* https://www.theguardian.com/news/2018/mar/17/cambridge-analytica-facebook-influence-us-election

Digital Divide Council. (2023). *Addressing the digital divide: Challenges for underfunded schools.* Accessed August 29, 2024. https://www.digitaldividecouncil.com

Dillenbourg, P. (1999). What do you mean by collaborative learning? In P. Dillenbourg (Ed.), *Collaborative learning: Cognitive and computational approaches* (pp. 1–19). Elsevier.

Discovery Education Inc. (2024). *DreamBox.* Accessed July 15, 2024. https://www.dreambox.com/

Duolingo. (2023). *Language learning for refugees: The impact of mobile apps on language acquisition.* Accessed August 29, 2024. https://www.duolingo.com/research

EdTech Review. (2023). *The financial costs of implementing AI in schools.* Accessed August 29, 2024. https://www.edtechreview.in

ePals. (2023). *Global classroom projects: Fostering cross-cultural understanding.* Accessed August 29, 2024. https://www.epals.com

Eubanks, V. (2018). *Automating inequality: How high-tech tools profile, police, and punish the poor.* St. Martin's Press.

FirstVoices. (2023). *Preserving indigenous languages through technology.* Accessed August 29, 2024. https://www.firstvoices.com

Google AI. (2020, August 12). Socratic by Google: Helping students with homework. Google Blog. https://blog.google/products/assistant/socratic-google-helping-students-homework/

Google Arts and Culture. (2023). *Virtual field trips: Exploring cultural landmarks around the world.* Accessed August 29, 2024. https://artsandculture.google.com

Google. (n.d.). Socratic. Accessed August 10, 2024. https://socratic.org/

Gradescope. (2023). *Streamlining the grading process with AI: Benefits for educators.* Accessed August 29, 2024. https://www.gradescope.com

Holmes, W. (2019). *AI and education: Adaptive learning, smart assessment, and the future of learning.* Routledge.

Jenkins, H., Clinton, K., Purushotma, R., Robison, A. J., & Weigel, M. (2009). *Confronting the challenges of participatory culture: Media education for the 21st century.* MIT Press. https://doi.org/10.7551/mitpress/8435.001.0001

John Wiley & Sons, Inc. (2024). *Knewton Alta.* Accessed March 29, 2024. https://support.knewton.com/s/article/Knewton-Adaptive-Learning-and-How-it-s-Unique

Kunju, S. (2019, December 21). How education technology has evolved in 2010s. *NDTV.* https://www.ndtv.com/education/how-education-technology-edtech-has-evolved-in-2010s-2152433

Learning Equality. (2023). *Kolibri in refugee camps: Offline learning for connectivity-challenged populations.* Accessed August 29, 2024. https://learningequality.org/kolibri

Luckin, R. (2017). *Enhancing learning and teaching with technology: What the research says.* UCL Institute of Education Press.

Martin, V. L. (2021). Trinidad and Tobago secondary school principals' perspective of implementing a laptop initiative [Doctoral dissertation, Walden University]. Walden Dissertations and Doctoral Studies Collection. https://scholarworks.waldenu.edu/dissertations

McCrea, B. (2009, December 10). 5 K-12 technology trends for 2010. *The Journal.* https://thejournal.com/articles/2009/12/10/5-k12-technology-trends-for-2010.aspx

Ministry of Education, Youth and Information, Jamaica. (n.d.). Tablets in schools programme. Accessed August 29, 2024. https://moey.gov.jm/tablets-in-schools-programme

Noble, S. U. (2018). *Algorithms of oppression: How search engines reinforce racism.* NYU Press.

Noddings, N. (2013). *Caring: A relational approach to ethics and moral education.* University of California Press.

O'Neil, C. (2016). *Weapons of math destruction: How big data increases inequality and threatens democracy.* Crown Publishers.

Organisation for Economic Co-operation and Development. (2019). Artificial intelligence in education: Challenges and opportunities for sustainable development. OECD Publishing. https://doi.org/10.1787/9f000646-en

Organisation for Economic Co-operation and Development. (2023). Legal and ethical standards for AI in education: A global perspective. Accessed August 29, 2024. https://www.oecd.org

Pane, J. F., Steiner, E. D., Baird, M. D., & Hamilton, L. S. (2017). *Promising evidence on personalized learning.* RAND Corporation. https://doi.org/10.7249/RR2042

Radford, A., Wu, J., Child, R., Luan, D., Amodei, D., & Sutskever, I. (2019). Language models are unsupervised multitask learners. OpenAI. https://openai.com/research/language-models-are-unsupervised-multitask-learners

Sahlberg, P. (2011). *Finnish lessons: What can the world learn from educational change in Finland?* Teachers College Press.

Seale, J. (2013). *E-learning and disability in higher education: Accessibility research and practice.* Routledge. https://doi.org/10.4324/9780203095942

Selwyn, N. (2019). Should robots replace teachers? AI and the future of education. Polity Press.

Singh, A., Aghazadeh, S., Fusco, M., Kochenderfer, M. J., & Liang, P. (2017). *Gradescope: A fast, flexible, and fair system for scalable assessment of handwritten work.* Proceedings of the Fourth ACM Conference on Learning @ Scale, 81–88. https://doi.org/10.1145/3051457.3051467

Smithsonian Learning Lab. (2023). *Creating inclusive lesson plans: Using culturally diverse content.* Accessed August 29, 2024. https://learninglab.si.edu

Storybird. (2023). *Enhancing creativity in the classroom: Digital storytelling with Storybird.* Accessed August 29, 2024. https://www.storybird.com

Student Survey. (2024). *Secondary students' perspectives on AI in education.* [Unpublished raw data].

Taxila Business School. (2020). *The ethics of AI in education: Why it matters and how to address it.* Accessed July 15, 2024. https://taxila.in/blog/the-ethics-of-ai-in-education/

The Pennsylvania State University. (2024). *Framework for stakeholder inclusion in the technology planning process.* Accessed July 23, 2024. https://coaction.psu.edu/inclusion-framework-phases/

UNESCO. (2020). Leveraging technology to improve education in rural India. United Nations Educational, Scientific and Cultural Organization.

UNHCR. (2022). *Education for Syrian refugees in Lebanon: Overcoming language barriers.* Accessed August 29, 2024. https://www.unhcr.org/education

Wiliam, D. (2011). Embedded formative assessment. Solution Tree Press.

Williamson, B. (2017). *Big data in education: The digital future of learning, policy, and practice.* SAGE. https://doi.org/10.4135/9781529714920

PART III: OVERCOMING CHALLENGES IN DIVERSE CONTEXTS

Chapter Twelve

The Challenge of the Digital Divide in the Classroom: Insights, Interventions, and Innovations in the Indian Context

Alka Ranjan

JAIN (Deemed-to-be-University), Bengaluru, India

Abstract

The digital divide produces inequity in society and challenges the globalized world. This chapter aims to understand the phenomenon of the digital divide in the Indian education system by appraising the determinants and evaluating its psychological, social, and educational impact on students. It highlights the existing ethical issues and suggests a systemic approach to alleviate them. It explains certain interventions and innovations to handle the challenges of the digital divide in the Indian education system.

Keywords: digital divide, education, classroom, interventions, India.

***

Introduction

Digitalization touches all aspects of our day-to-day life in the globalized world. The field of education has also become digitized with information and communication technologies (ICT). The impact is so substantial that it is called the "fourth industrial revolution" (Schwab, 2016). Undeniably, information and communication technologies have become an inherent part of the education field to such an extent that it is challenging the existing definition of literacy.

The "fourth industrial revolution" (Schwab, 2016) has led to certain challenges that this world faces today. One such challenge is the "digital divide." The chapter's primary objective is to examine the digital divide phenomenon in the Indian educational context. By conducting a literature review, the study aims to identify the determinants, assess the impact, appraise the ethical implications, and evaluate the interventions used to mitigate the digital divide in education in India.

Determinants of Digital Divide

Researchers (Dewan & Riggins, 2005; Wei et al., 2011) have used a conceptual theoretical outline to understand the digital divide phenomenon. They have given the "ICT Adoption Cycle" to understand the three levels of the digital divide across the three levels of analysis. The three levels of the digital divide are "access – usage – skills" occurring across the three levels of analysis, i.e., "individual – organizational - global." There is an interaction across the levels.

The first level is the lack of physical access to ICT. It points to the gap between those who have access and those who do not have access to the ICT tools. According to the International Telecommunication Union (ITU, 2023), there is a 66% gap between developed and developing countries. This viewpoint is considered a "narrow" focus (Friedman, 2001).

The second level of the digital divide is digital usage, which emphasizes digital skills. Individual differences in the skills to use the ICT are found despite having physical access to the technology. Motivation plays a crucial role in learning digital skills (van Dijk, 2005). ITU (2023) identifies five areas of digital skills: "communication/ collaboration; problem-solving; safety; content creation; and information/data literacy." Data shows that the "information/ data literacy" skill is the predominant skill identified (ITU, 2023, p. 15).

The third level of the digital divide is the differences in digital outcomes regarding learning, job readiness, employment, monetary benefits, etc. Less research is available in this area. Research must also be conducted to understand the dynamics and the effect of the interaction between the three levels.

The above framework has been applied to the educational digital divide in the Indian context and is represented diagrammatically in the following figure.

Figure 12.1 Digital Divide in the Indian Education System.

	"ICT Adoption Stages" (Wei *et al*., 2011) in India		
Student	Level 1 digital divide ⇨	Level 2 digital divide ⇨	Level 3 digital divide
School			
State / Country			

Thus, various factors determine the digital divide in education (Figure 12.1). The determinants at level 1, "Digital access divide," are discussed below.

Gender: A gender disparity exists in internet use, as only one-third of Indian women use the internet (UNICEF, 2017), which accounts largely for the gendered digital divide globally. Figure 12.2 shows that only 35% of females use the internet nationally. This disparity is more acute in rural India, where only 31% of the females use the internet compared to 40% in urban India. Social norms restrict girls to household responsibilities, which puts physical and economic constraints on them, thus limiting their participation in the digital space. Figure 12.3 shows the gender disparity in access to digital devices in adolescent boys and girls in 10 states of India. The largest disparity exists in the Haryana state, followed by Maharashtra. The Centre for Catalysing Change and Digital Empowerment Foundation (2020) has identified five key dimensions that should be assessed to understand this crisis. These dimensions pertain to access, literacy skills, content, family attitude, and connectivity to digital devices. Thus, these dimensions can be assessed to understand the skewed gender gap in digital access in the country.

Figure 12.2 Internet Use according to Gender in %.

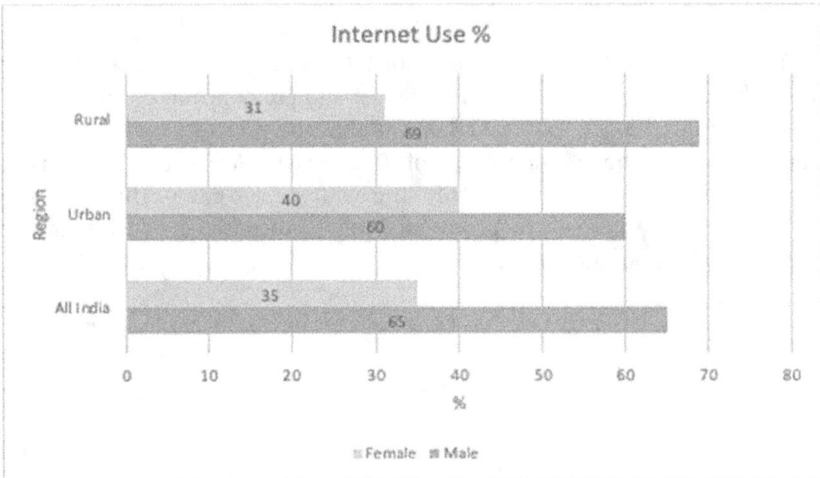

Source: Internet and Mobile Association of India (IAMAI) & Nielsen (2019).

Figure 12.3 Access to digital devices in Adolescent girls and boys.

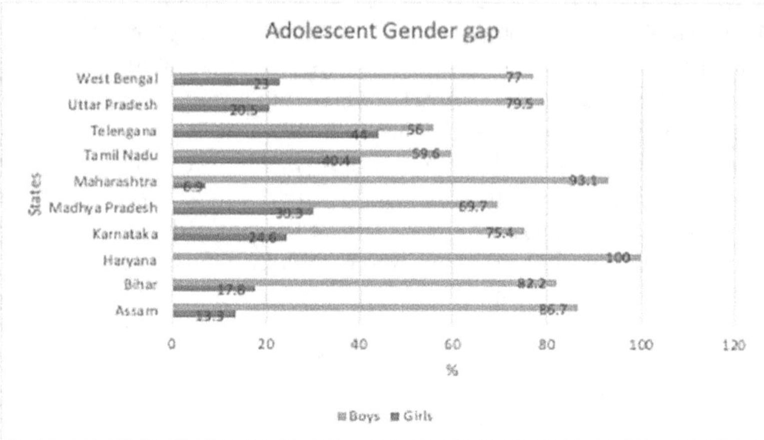

Source: C3-DEF Survey (2020); Centre for Catalysing Change & Digital Empowerment Foundation (2020), p. 13.

Disability: According to the Centre for Internet and Society, the differently abled have difficulty accessing the Internet and are deprived of information, thus violating the law. Figure 12.4 shows that the prevalence of disability is highest between the productive ages of 10 to 49 years. Access to education and employment is the highest priority for this younger age group. Thus, difficulty in accessing the internet will further increase the disparity in educational and job skills of this population.

Figure 12.4 Percentage (%) of the Disabled Population according to age group.

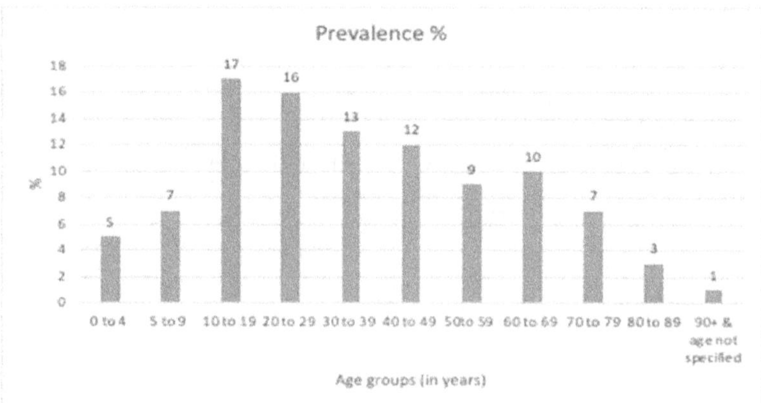

Source: Census 2011, Government of India.

Age: There is a generational gap in internet usage. ITU (2023) data shows a global 14%-point difference in internet use between youth (15-24 years) and the remaining population, and this generation gap has been consistent across the regions over the last four years.

A similar trend is observed in the Indian context. According to IAMAI and Nielsen (2019), "2/3 of the internet users in India are in the age group of 12-29 years and this age group corresponds to more than 70% of internet users in rural areas." The following figure shows this trend.

Figure 12.5 Internet users in India according to Age (in %).

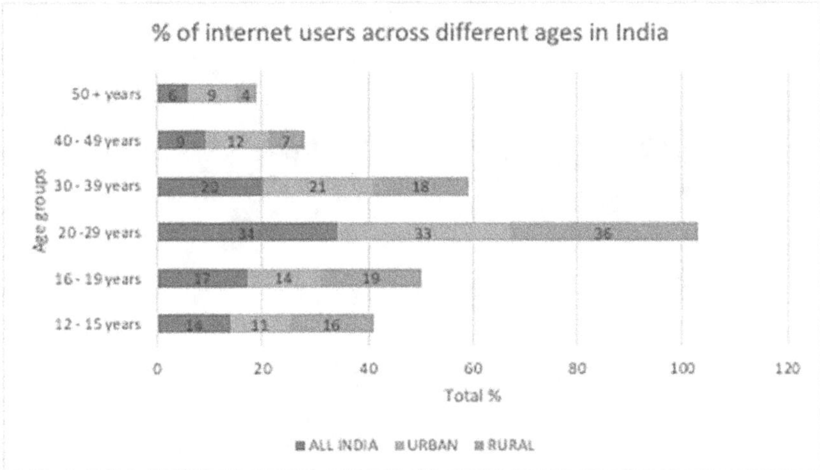

Source: IAMAI & Nielsen (2019).

Socio-economic status: According to IAMAI & Nielsen (2019), those belonging to the New Consumer Classification System (NCCS: A - high, B – upper-middle, and C – middle) have greater Internet access (about 25% at the national level) due to their accessibility to more educational and financial resources. The population from lower-middle and low socioeconomic status has 14% and 4% internet access, respectively, at the national level. The disparity among the students of various caste groups' access to digital infrastructure, i.e., computers with internet facilities (Inequality report, Oxfam, 2022), is also found. According to the report, only 4% of the scheduled caste (SC) and scheduled tribe (ST) have access to digital infrastructure as compared to 7% in other backward castes (OBC) and 21% of the remaining caste groups. Thus, socio-economic status is a major determinant of differences in internet access.

The rural-urban difference in computer and internet use: Though the percentage of households having computer and internet facilities (Figure 12.6) is higher in urban areas, the growth rate of internet use is double in rural areas, as shown in Figure 12.7. The data (Figure 12.8) shows that presently more than half (54%) of internet users live in rural India (KANTAR ICUBE & IAMAI, 2023).

Figure 12.6 Households with a Computer and Internet Facilities in %.

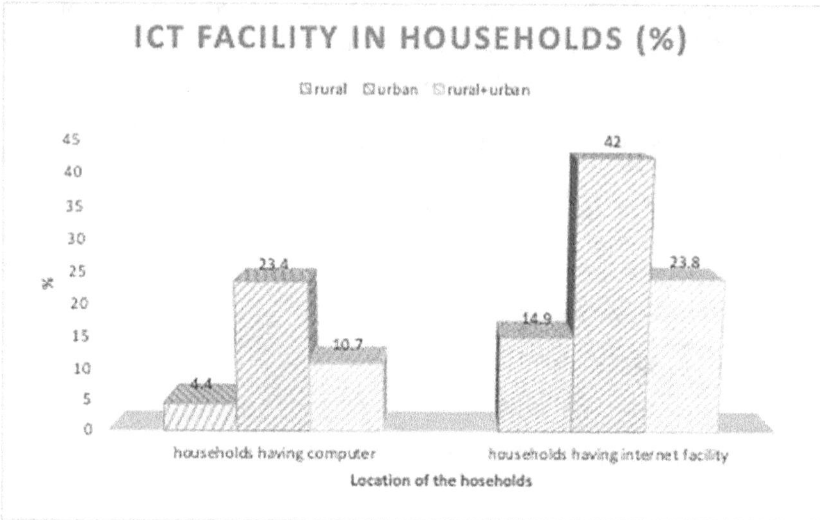

Source: NSS Report No. 585, Government of India (July 2017 – June 2018).

Figure 12.7 Growth Rate in Urban and Rural India.

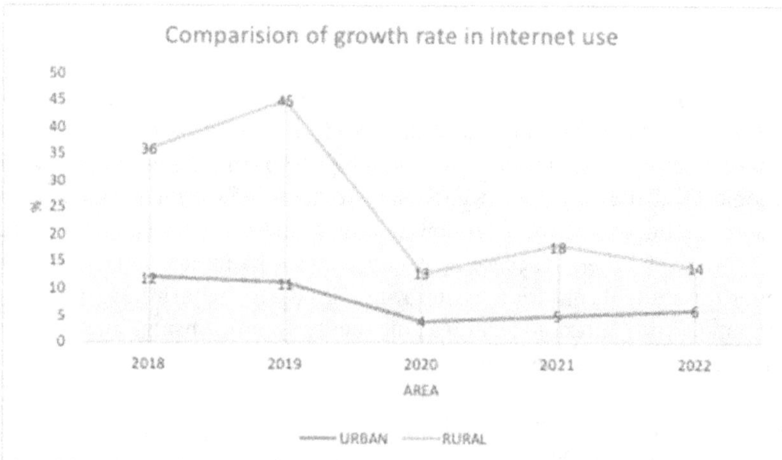

Source: ICUBE 2022 - Kantar & IAMAI (2023).

Figure 12.8 Internet Users in Urban and Rural India (in %).

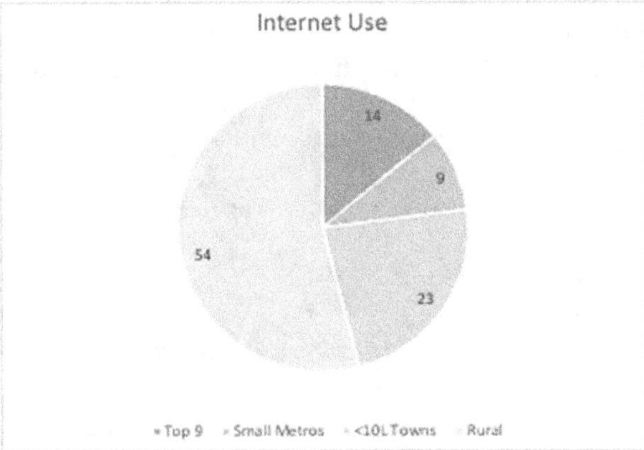

Source: ICUBE 2023 – Kantar & IAMAI (2023).

Electricity: According to the Government of India (2017), electricity infrastructure in India is almost universal but lacks quality and reliability of supply across the country. The Ministry of Rural Development Mission Antyodaya (2019) further states that below 50% of households receive electricity for less than 12 hours on average (Figure 12.9).

Figure 12.9 Electricity Power Supply in Indian Households.

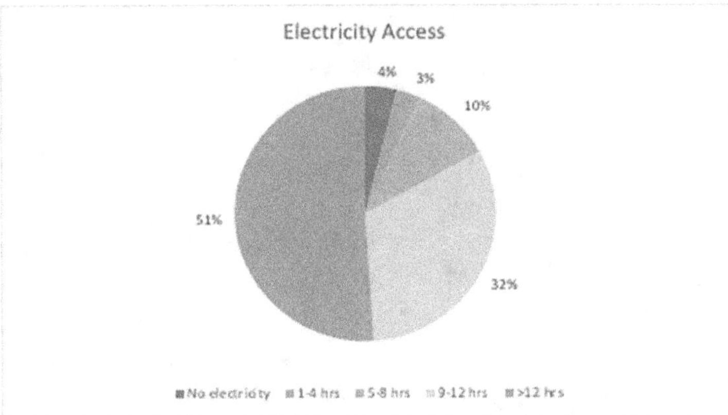

Source: Ministry of Rural Development Mission Antyodaya (2019).

Lack of infrastructure: Figure 12.10 shows the lack of digital infrastructure in schools. Over 60% of government and private schools lack digital infrastructure. The difference is about 8% between the types of schools. Even if computers are available, they may not have internet access. Very few schools have access to both a computer and the Internet. There is not much difference in the availability of digital infrastructure in either of the schools. Figure 12.11 shows that children and adolescents use WhatsApp most frequently for studying.

Figure 12.10 Comparison of digital infrastructure in government and private schools.

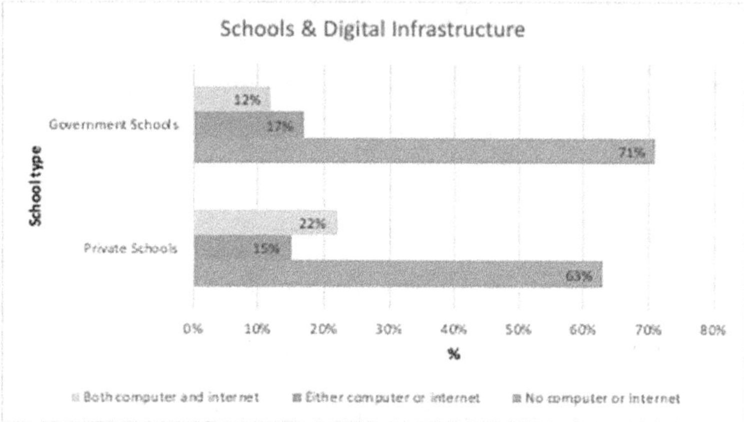

Source: Government of India, Ministry of Education (2020-2021).

Figure 12.11 Tech-based tools for studying and learning.

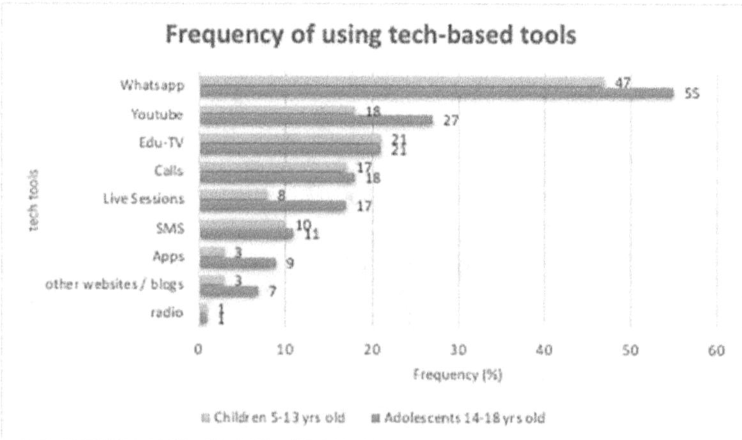

Source: UNICEF India, 2021.

The determinants at level 2, "Digital usage divide," are discussed below.

- Language: The majority of digital content is available in English. This poses challenges for students who know only regional or vernacular languages. Digital content should continue to be translated into other languages to reach a wider audience.
- Digital literacy: The rate of digital illiteracy is high in the Indian population (Internet Literacy Index across India, 2021), but the ability to use the Internet is low in India.
- Digital Safety: It is important to ensure the digital safety of students. The National Commission for Protection of Child Rights (NCPCR) and the Central Board of Secondary Education (CBSE) have published on digital safety and security in children.

Thus, the determinants of level 1, i.e., digital access, and level 2, digital usage will impact digital outcome – level 3 of the digital divide. This will influence the attainment of sustainable development goals (SDGs).

Impact of Digital Divide on Indian Students

It is important to understand the impact of the digital divide on students. Digitalization has changed the nature of traditional classrooms, which have been replaced by online and hybrid classrooms conducted in the virtual space.

Due to this transition, students and teachers are required to equip and upgrade their digital skills. However, individual differences in access and usage of ICT among students have led to the digital divide. Thus, the digital divide is further exacerbating the already existing inequality gap among the different sections of students, which can have a psychological, social, economic, and educational effect on them.

Psychological Impact

The digital divide can have a psychological and cognitive effect on the students. There will be individual differences in computer familiarity and experience among the students (Abdulkareem & Lennon, 2023). There will also be a difference in the digital capability determining computer self-efficacy (Wei et al., 2011). Mathrani et al. (2021) state that "individuals affected by this divide lack belief and self-efficacy, resulting in lower digital productivity." Self-efficacy will also shape the students' attitudes toward the access and usage of computers. Lower familiarity and experience with computers can lead to computer-based anxiety (Abdulkareem & Lennon, 2023).

Motivational access-based digital divide has been found among Nepali students (Joshi, 2023). The study found differences in motivational levels in using ICT in education based on socio-demographic variables, such as gender, ethnicity, origination, school type, and parents' employment.

Teachers can use effective methods to engage the students in ICT-based education; otherwise, no learning will happen (Reeve, 2012). Bergdahl et al. (2018) found that just using technology in teaching does not engage the students. Opportunities should be given to the students to engage. In their results, they found two factors that could engage the students. These factors were "noticing the students' participation" by teachers and "orchestrating the technologies to gain insight into students' knowledge and learning processes." Håkansson-Lindqvist (2015) further stated that a gap exists in ICT use in classrooms and suggested that teachers should be trained to integrate technologies into learning.

The digital divide can also produce inadequate socio-emotional development in the students and have cognitive effects. Most students in India use mobile phones, particularly WhatsApp, to study (UNICEF India, 2021). Studies have shown that WhatsApp positively impacts learning (Cetinkaya, 2017), though more in-depth studies need to be done in the Indian context. Studies have shown that those who frequently use smartphones have shorter attention spans, which can impact learning (Lui & Wong, 2012). Excessive use can strain their eyes and lead to poor attention (De-Sola Gutiérrez et al., 2016). Most children check their smartphones while studying (National Commission for Protection of Child Rights, 2021). This can lead to distraction and a loss of concentration. Hence, effective measures and best practices need to be introduced in schools for mobile phone use.

Social Impact

The digital divide can further exacerbate the existing socio-economic disparities. It also contributes to economic burdens, making it financially challenging for some families to purchase ICT tools for their children. The Knowledge Gap theory further states that knowledge gaps can widen between the socio-economic groups due to ICTs, and groups having greater access to information and technology will acquire and apply knowledge more effectively (Tichenor et al., 1970).

The digital divide has also created a generation gap between "digital natives," i.e., students born in the late 1990s, and "digital immigrants," i.e., their parents. This generation gap can affect the mentoring and guidance from parents and teachers, which will act as an impediment to the attainment of the "zone of proximal development (ZPD)" as described by Vygotsky (1978). Consequently, the digital divide negatively impacts the well-being of the student.

Impact on Education

The impact on education can be understood using the access, usage, and outcome framework (Dewan & Riggins, 2005). The following table 12.1 depicts these factors across the continuum.

Table 12.1 Factors Contributing to the Digital Divide and Their Impact on Educational Access, Usage, and Outcomes.

Framework		
Access	**Usage**	**Outcome**
- Limited access to online educational resources - Lack of access to technology and internet at home - Inability to participate in online classes	- Lack of motivation - Lack of digital engagement (Warschauer, 2003)	- Incomplete assignments - Lower grades and test scores - Limited educational opportunities - Lack of opportunity to develop digital literacy skills (Baturay & Demiray, 2018) - Widening achievement gap (Chinn & Fairlie, 2010) - Decreased peer interaction and communication with teachers outside of the classroom (Kirschner & Karpinski, 2010). - Inability to complete assessments (Warschauer & Matuchniak, 2010). - High drop-out rates - Inability to develop digital citizenship skills, such as responsible online behaviour and digital security (Baturay & Demiray, 2018).

As a result, limited access to online educational content leads to numerous adverse educational outcomes that further exacerbate the existing problems of unemployment and socio-economic disparity in society.

Table 12.1 also shows that literacy needs to be redefined. Literacy is currently defined as the ability to read and write. The digital skill component should also be added in this digitalization era.

Ethical Implications of the Digital Divide in Education

The digital divide violates human rights by fostering social exclusion, vulnerability, and inequality. The Constitution of India protects the "Right to Education" as a "Fundamental Right, under Directive Principles of State Policy and Fundamental Duties." However, the digital divide deprives children, particularly girls, those living in rural and remote areas, and poor and disabled children of the right to education. This has long-term consequences of producing marginalization of these groups, thus further widening the already existing digital and socio-economic inequality. It also threatens the data security and privacy of students by exposing them to cyberbullying and cybercrime. This impedes meeting the target of sustainable development goals (SDGs) by 2030, particularly SDG 9, which aims at building a sustainable digital infrastructure.

Hence, social responsibility should be fostered at all levels. School authorities and parents should impart adequate awareness and digital knowledge to the students to safeguard their privacy. Responsible digital habits need to be promoted in society. The United Nations General Assembly has taken a positive step and declared access to the Internet a basic human right. However, the digital divide still impacts society's moral obligations to ensure universal, equal, and safe access to education.

Interventions and Strategies to Mitigate Digital Divide in Classrooms

A multi-pronged systemic approach can reduce the digital divide in education. This approach can be adapted to focus on complex interrelationships between students and their socio-politico-economic contextual factors. Interventions can be implemented at the individual, community, and social levels. The constituent factors at each level are shown in the figure below.

Figure 12.12 Levels of the Systemic Approach.

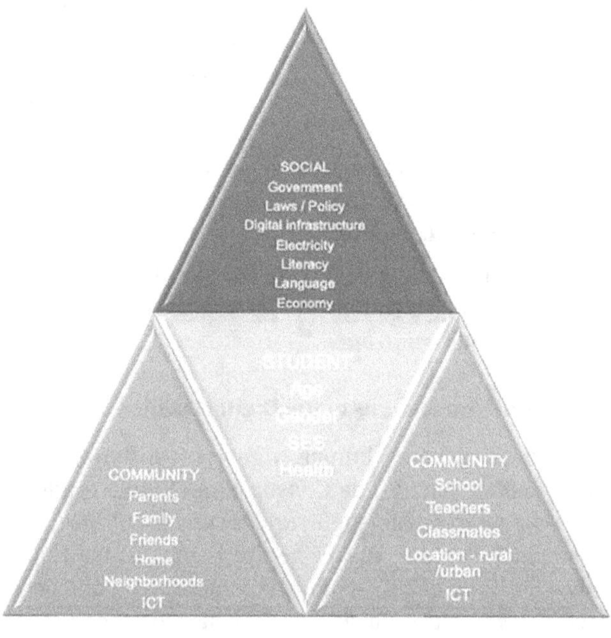

The interventions and strategies at each level have been discussed.

Individual-level interventions: Students are the target of micro-level interventions. They should be given guidance and training in digital access and usage for education purposes. Hobbs (2010) states that this practice can imbibe responsible digital use and digital habits. Training modules should be designed according to the age and education level of the student.

Healthy digital practices will familiarize them with computers and help them develop favourable attitudes toward using ICT. This will also reduce computer-based anxiety during exams. Introducing healthy lifestyle practices such as digital detox and mindful technology can help them manage distractions and effectively use the ICT tools. Effective steps, such as changing public attitudes, can be taken by increasing awareness of gender disparity. The digital divide among the differently-abled has not been given much attention. More research is required in this area.

Community-level interventions: A Needs assessment should be conducted to evaluate the pros and cons in a particular school setting. Figure 12.10 shows the digital infrastructure of the government and private schools in India. Community interventions such as organizing digital camps and introducing Edtech in schools can be conducted through government financial aid. Infrastructural arrangements should be made to give ICT access to children in their neighbourhoods. Digital infrastructure can be upgraded so that the medium for taking exams and conducting arts/crafts and laboratory classes is improved virtually.

Educators can be empowered through continuous training in digital innovations, expertise, and applications. Workshops and peer learning are some of the methods that can be used. Training modules can be designed separately for urban and rural teachers. Parents can be trained and provided with psycho-educational resources to become aware of the importance of digital tools in education. Banks can give subsidies and loans to buy ICT tools that will reduce the economic burden on each household.

Social-level interventions: The Indian government has implemented a subsidized network and has introduced programs such as the National Rural Internet Mission. Free Wi-fi has been given to some educational institutions, and public wi-fi hotspots have been made available to the citizens. Some policies on digital education have been outlined in the National Education Policy (NEP, 2020). Digital platforms like "Study Webs of Active-learning for Young Aspiring Minds" (SWAYAM) have been introduced to make learning accessible to the larger population through Massive Online Open Courses (MOOC). Open sources have been provided, and programs are broadcast through DIKSHA and television for the rural population. The government has also introduced a digital library and repository. Further improvement can be made by translating the online educational resources into various regional languages. Digital literacy programs have also been introduced, and the government has formed the "Monitoring Committee on Digital India." These initiatives will certainly make digital education accessible to the public.

Conclusion

Thus, ICT tools have become an integral part of the Education system in the "fourth industrial revolution" era. The Indian education system is slowly adapting to this monumental change. The Indian government has introduced many initiatives, yet these steps should be integrated across the systemic levels.

Effective steps should be taken to implement government policies at the grassroots level. A program similar to UGC's 'NEP SAARTHI' can be initiated at the school level. Teachers should be adequately trained to use ICT in various applied aspects of education by making it an integral part of undergraduate and postgraduate curricula. ICT knowledge and skills should be a criterion for their promotion. The effectiveness of an ICT application should be determined through research and students' feedback. Conferences, symposia, workshops, and exhibitions can demonstrate effective ICT applications in education. Thus, academia can play an effective role.

Schools can organize programs for parents, such as operating parental control and restricting content on ICT. Volunteers can be recruited from the parents and grandparents' community to spearhead the programs further. The use of ICT in teaching subjects can be discussed during parent-teacher meetings held at schools. Rules can be made in schools on the use of mobile phones on-premises. Drama and scientific research can popularize safe ICT use among students. Comic books and stories can be written on the topic. The media can play an effective role. Effective messages can be communicated through radio, television, and social media. Research grants can be given for cost-effective and user-friendly ICT use in education at government schools.

A developing country like India has a large student population and restricted economic resources. Thus, a systemic approach being effectively implemented at the grassroots level will help the country to meet the challenges and adapt to digitalization.

Recommendations

The term 'literacy' ought to be redefined to include the digital skill component. The minimum criteria for digital access and skills need to be officially determined. The ICT content should continue to be translated into other Indian languages. Also, ICT tools and digital media should be designed accordingly for students with special needs. School education boards at the national and state levels should define guidelines for digital use for teaching and examinations, digital safety, and security. Students having problems related to ICT use and skills, such as digital anxiety and lack of motivation, should be referred to school counsellors. ICT training centres should be identified in each area so that students from schools lacking in infrastructure can regularly use them. ICT can be integrated with education through an integrated systems approach at the grassroots level.

References

Abdulkareem, Z., & Lennon, M. (2023). Impact of digital divide on students' performance in computerized UTME in Nigeria. *Information Development,* 1 – 16. https://doi.org/10.1177/02666669231191734

Baturay, M. H., & Demiray, U. (2018). The digital divide and its effects on education. *Journal of Education and Practice,9*(7), 1 – 8.

Bergdahl, N., Fors, U., Hernwall, P., & Knutsson, O. (2018). The use of learning technologies and student engagement in learning activities. *Nordic Journal of Digital Literacy, 13*(2), 113 – 130. https://doi.org/10.1826/ISSN.1891-943X-2018-02-04

Census of India (2011). *C-20: Disabled population by type of disability, age and sex (all), PC01_C20.* Government of India. https://censusindia.gov.in/census.website/data/census-tables

Centre for Catalysing Change & Digital Empowerment Foundation (2020). *Policy Brief: Bridging the digital divide for girls in India.*C3 & DEF.

Centre for Internet and Society (n.d.) https://cis-india.org/

Cetinkaya, L. (2017). The impact of WhatsApp use on success in education process. *International Review of Research in Open and Distributed Learning, 18*(7), 59 – 74. https://doi.org/10.19173/irrodl.v18i7.3279

Chinn, M. D., & Fairlie, R. W. (2010). ICT use in the developing world: An analysis of differences in computer and internet penetration. *Review of International Economics, 18*(1), 153 – 167. https://doi.org/10.1111/j.1467-9396.2009.00861.x

De-Sola Gutiérrez, J., Rodriguez de Fonseca, F., and Rubio, G. (2016). Cell-phone addiction: a review. *Frontiers in Psychiatry, 7,* 175. https://doi.org/10.3389/fpsyt.2016.00175

Dewan, S., & Riggins, F. J. (2005). The Digital Divide: Current and Future Research Directions. *Journal of the Association for Information Systems, 6*(12), 296-337. https://doi.org/10.17705/1jais.00074

Dobransky, K., & Hargittai, E. (2016). Unrealized potential: Exploring the digital disability divide. *Poetics, 58,* 18 – 28. http://dx.doi.org/10.1016/j.poetic.2016.08.003

Friedman, W. H. (2001). The digital divide. In *Proceeding of Seventh Americas Conference on Information Systems,* 2081 – 2086.

Government of India (2017). Saubhagya Dashboard. https://saubhagya.gov.in/

Government of India, Ministry of Education (2020-2021). Unified District Information System for Education Plus (UDISE+) 2019; Ministry of Education, Report on UDISE+, https://www.education.gov.in/sites/upload_files/mhrd/files/statistics-new/UDISE%2B2020_21_Booklet.pdf

Government of India, Ministry of Statistics and Programme Implementation & National Statistical Office (July 2017 – June 2018). *NSS Report No. 585: Household Social Consumption on Education in India.*www.mospi.gov.in

Joshi, B. M. (2023). Students in Economics facing a digital divide. *Molung Educational Frontier, 13,* 3 – 22. https://doi.org/10.3126/mef.v13i01.56021

Håkansson-Lindqvist, M. (2015). Exploring activities regarding technology enhanced learning in a one-to-one initiative. *Nordic Journal of Digital Literacy, 10*(4), 227-245. https://doi.org/10.18261/issn.1891-943X

Hobbs, R. (2010). *Digital and Media Literacy: A Plan of Action.* Washington DC: Aspen Institute.

Internet and Mobile Association of India (IAMAI) & Nielsen (2019). *Digital in India 2019 – Round 2 Report.* IAMAI.

International Telecommunication Union (2023). *Measuring digital development: Facts and Figures 2023.* ITU Publications.

KANTAR ICUBE & Internet and Mobile Association of India (2023). *Internet in India 2023. https://www.iamai.in/research/internet-india-2023*

Kirschner, P. A., & Karpinski, A. C. (2010). Facebook (R) and academic performance. *Computers in Human Behavior,26*(6), 1237 – 1245. https://doi.org/10.1016/j.chb.2010.03.024

Lui, K., and Wong, A. C. (2012). Does media multitasking always hurt? A positive correlation between multitasking and multisensory integration. *Psychonomic Bulletin Review, 19,* 647 – 653. https://doi.org/10.3758/s13423-012-0245-7

Mathrani, A., Sarvesh, T., & Umer, R. (2021). Digital Divide framework: online learning in developing countries during the COVID-19 lockdown. *Globalization, Societies and Education, 20,* 625 – 640. https://doi.org/10.1080/14767724.2021.1981253

Ministry of Rural Development. (2019). Mission Antyodaya. https://missionantyodaya.nic.in/ma2019/preloginStateRuralIElectricityReports2019.html

National Commission for Protection of Child Rights (2021). *Effects (Physical, Behavioral, and Psycho-social) of using mobile phones and other devices with internet accessibility by children.*

Reeve, J. (2012). A Self-determination Theory Perspective on Student Engagement. In *Handbook of Research on Student Engagement* (pp. 149 – 172). Boston, MA: Springer US. https://doi.org/10.1007/978-1-4614-2018-7_7

Schwab, K. (2016). "The Fourth Industrial Revolution: what It Means, how to respond." *World Economic Forum,* 14 January. https://www.weforum.org/stories/2016/01/the-fourth-industrial-revolution-what-it-means-and-how-to-respond/. Accessed 10 November 2024.

Tichenor, P. J., Donohu, G. A. & Olien, C. N. (1970). Mass media flow and differential growth in knowledge. *Public Opinion Quarterly, 34* (2), 159 – 170.

UNICEF (December, 2017). *The State of the World's Children 2017: Children in the Digital World.*www.unicef.org/media/48581/file/SOWC_2017_ENG.pdf. https://doi.org/10.1086/267786

UNICEF (2021, May). *Rapid assessment of learning during school closures in the context of COVID.* https://www.unicef.org/india/reports/rapid-assessment-learning-during-school-closures-context-covid-19

van Dijk J (2005). *The Deepening Divide: Inequality in the Information Society.* Thousand Oaks, CA: SAGE Publications, Inc.

Vygotsky, L. S. (1978). *Mind in Society: The development of higher psychological processes.* MA: Harvard University Press.

Warschauer, M. (2003). *Technology and social inclusion: Rethinking the digital divide.* Cambridge, MA: MIT Press. https://doi.org/10.7551/mitpress/6699.001.0001

Warschauer, M., & Matuchniak, T. (2010). New Technology and digital worlds: Analyzing evidence of equity in access, use, and outcomes. *Review of Research in Education, 34* (1), 179 -225. https://doi.org/10.3102/0091732X09349791

Wei, K.-K., Teo, H.-H., Chan, H. C. & Tan, B. C. Y. (2011). Conceptualizing and testing a Social Cognitive Model of the Digital Divide. *Information Systems Research, 22*(1), 170 – 187. https://www.doi.org/10.1287/isre.100.0273

Chapter Thirteen

Innovative Approaches Used in Diverse Educational Contexts

Soumya Thankam Varghese

O.P. Jindal Global University, Haryana, India

Abstract

The global educational landscape is transforming, emphasizing holistic well-being and technology-driven learning. Educators and stakeholders are exploring innovative strategies to create enriching, student-centred experiences. Understanding existing pedagogies in diverse contexts is crucial for this evolution. Technology-driven learning equips educators with personalized digital tools, fostering collaboration among stakeholders, educators, and learners. It facilitates personalized learning and adaptive assessments, adaptable to various educational contexts. This synergy of pedagogy and technology promotes engagement, equity, and lifelong learning, while addressing cultural nuances, socioeconomic disparities, and diverse learning needs. A scoping review on innovative learning strategies in diverse educational contexts aims to identify opportunities and challenges for educators. This research focuses on articles published from 2022 to 2024, sourced from the Scopus database using keywords like innovative pedagogy, technology, diverse needs, and future learning. A total of 63 articles were selected for thematic analysis. The findings will provide insights to empower stakeholders and educators in creating an innovative, holistic, well-being-oriented educational system. This research will help build education systems that nurture every child, fostering academic excellence and holistic development.

Keywords: innovative pedagogy, diverse needs, future learning, technology-induced learning.

***

Introduction

Educational perspectives are deeply rooted in educational philosophies, shaping pedagogical practices that offer a holistic understanding of teaching and learning processes across various contexts. However, the rapid changes in social and technological spheres significantly impact education, driving continuous and innovative pedagogical transformations. This ongoing process of redefining ourselves and discovering new ways of learning collectively is what we call "learning in a collective sense." The vision for future education by the Organization for Economic Co-operation and Development (OECD, 2018) emphasizes the significance of acquiring a complex skill set for the development and growth of a person as a holistic being. This kind of learning environment should be enhanced with innovative pedagogical approaches. Innovative pedagogical approaches are essential for adapting to the rapidly changing educational landscape. This will create an enriching educational experience for learning in a diversified context. Innovative pedagogical approaches are essential for adapting to the rapidly changing educational landscape (Peterson et al., 2018). These approaches will create an enriching educational experience for learning in diverse contexts. This integration of innovative pedagogical methods is crucial in today's dynamic educational environment as it can enrich learning experiences in diverse contexts (Suryani et al., 2024). Innovative pedagogical approaches are essential for adapting to the rapidly changing educational landscape. The significance of innovative pedagogy cannot be emphasized enough in the context of modern education (Hamzah et al., 2024). As the world continues to change at a rapid pace, majorly influenced by technological advancements and consequently evolving societal needs, the traditional ways of teaching are often insufficient to meet the concerns and demands of contemporary learners. Innovative pedagogy encompasses a variety of teaching strategies that are designed to engage students more deeply, foster critical thinking, and prepare them for the complexities of the modern world. For instance, experiential learning permits students to gain hands-on experience and apply theoretical knowledge in real-world scenarios, thereby enhancing their understanding and retention of information (Qian, 2023). Project-based learning encourages students to work on complex problems over extended periods, promoting deep learning and the development of problem-solving skills (Botes & Rossi, 2024; Zakaria et al., 2024).

Moreover, collaborative learning emphasizes the importance of teamwork and communication, skills that are essential in almost every professional field. The integration of technology in the classroom, such as through the use of digital tools and online resources, makes learning more interactive and accessible, catering to the diverse needs of students. Flipped classrooms, where instructional content is delivered outside of class and in-class time is used for interactive activities, allow for more personalized and engaging learning experiences (Chen, 2023; Gopalan, 2024).

Gamification, which incorporates game elements into the learning process, can significantly increase student motivation and engagement. Personalized learning is another critical aspect of innovative pedagogy, as it tailors educational experiences to meet the individual needs and learning styles of each student (Mourtzis et al., 2023). This approach not only helps in addressing the unique challenges faced by different learners but also ensures that each student can achieve their full potential. Furthermore, innovative pedagogical methods often emphasize the development of soft skills, such as critical thinking, creativity, and emotional intelligence, which are increasingly recognized as vital for success in the twenty-first century (Husanu & Ertekin, 2024).

Aim and Objective

Aim

The aim of the current scoping review is to explore and analyse innovative pedagogical approaches being implemented across diverse learning contexts. With the rapid advancement and integration of technology in education, traditional learning platforms have undergone significant transformation. This study seeks to identify and evaluate the effectiveness of these novel practices, focusing on how they enhance learning outcomes, foster inclusivity, and accommodate the varied needs of learners. By inspecting a range of technological tools and methodologies, this research aims to provide an inclusive understanding of the current landscape of educational innovation and its implications for future teaching and learning strategies.

Objective

To identify innovative pedagogical approaches in diverse learning contexts.

Method

The current scoping review was methodically conducted following the framework established by Arksey and O'Malley, which encompasses six essential steps (Arksey & O'Malley, 2005). First, the research question was clearly identified to guide the scope of the review. Next, relevant studies were identified through a comprehensive search of the SCOPUS database. Following this, a selection process was implemented to determine which studies would be included in the review, ensuring that only the most pertinent articles were considered. The fourth step involved charting the data, where key information from the selected studies was systematically extracted and organized. This was followed by collating, summarizing, and reporting the results, which provided a coherent synthesis of the findings. This structured approach ensured a thorough and reliable scoping review, mapping out the existing literature on innovative practices in learning and identifying key themes and gaps for future research.

Search strategy

Figure 13.1 The PRISMA Model Flow Chart.

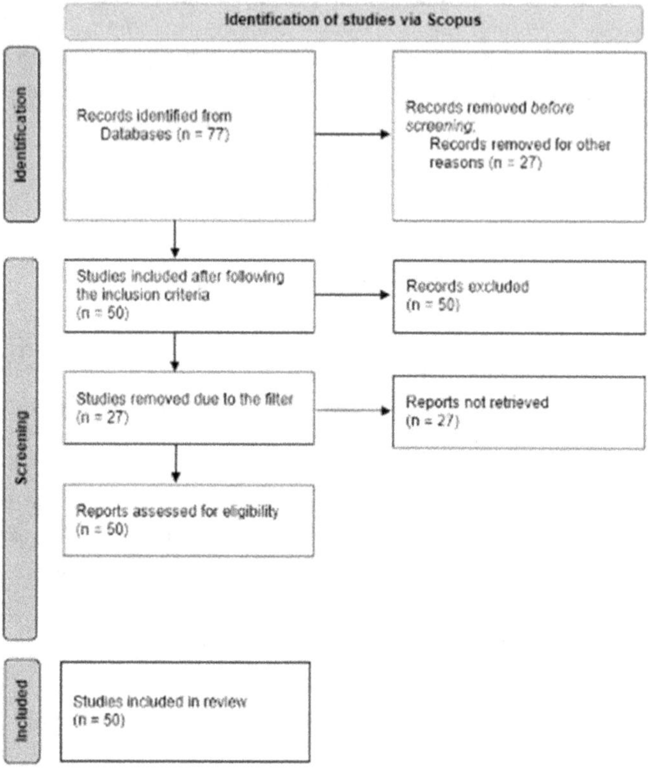

To address the research questions, a comprehensive literature review was conducted using recent articles available on the Scopus database. The authors identified several key keywords for the search, including "innovative pedagogy," "future learning," and "technology-based." The search was limited to journal articles retrieved from the Scopus database, published in English between the years 2022 and 2024, with specific keywords innovative pedagogy, technology, diverse needs, and future learning. A total of 77 articles were identified, and among them, 50 were further processed for the final research synthesis. The process is illustrated in Figure 13.1. The selection is done by excluding book chapters and articles in the press. The inclusion criteria were strictly adhered to, ensuring that only relevant journal articles were considered. Excluded from the study were book chapters, articles in press, and articles published in other languages or outside the specified years.

Procedure

After finalizing the research questions, the next step involved the systematic gathering of relevant literature from the Scopus database. Initially, a comprehensive search was conducted to download a wide array of articles. This initial collection included potential duplicates, which were meticulously screened and removed to ensure the uniqueness of each entry. Following this, a dual-stage review process was employed to refine the selection further. This process began with a preliminary review of the titles, allowing for the exclusion of those that were not directly relevant to the research questions. Subsequently, a more detailed examination of the full papers was conducted to shortlist the most pertinent articles for theme generation.

The selected articles, spanning the years 2022 to 2024, focused on innovative practices in learning. To analyse these articles, thematic analysis as outlined by Braun and Clarke (2017) was utilized. This method involved several key steps: familiarization with the data, generating initial codes, searching for themes, reviewing themes, defining and naming themes, and finally, producing the report. Initially, the researchers immersed themselves in the data to gain a thorough understanding. This was followed by the generation of initial codes, which involved identifying significant features of the data that were relevant to the research questions. These codes were then organized into potential themes, which were reviewed and refined to ensure they accurately captured the essence of the data. The final themes were clearly defined and named, providing a structured framework for the analysis. This rigorous process ensured that the themes retrieved were robust and provided meaningful insights into innovative practices in learning.

The Themes Emerged for Discussion

The thematic analysis for theme generation effectively categorized various innovative educational approaches. This process identified distinct themes such as interactive learning, AI-based learning, experiential-blended learning, 3D printing and design learning, and inquiry-based project learning (Chen, 2023). Each theme represents a unique facet of modern pedagogy, highlighting the diverse methods educators can employ to enhance student engagement and learning outcomes. By segregating these themes, the analysis provides a structured framework for integrating these approaches into educational practices, fostering a more dynamic and personalized learning environment.

Interactive learning

Interactive learning is a dynamic and engaging approach that significantly enhances the connection between students, teachers, and peers. One of the key advantages of interactive learning is its versatility, as it can be implemented both technologically and face-to-face. This flexibility allows educators to tailor their teaching methods to suit various learning environments and student needs.

Technological tools such as interactive whiteboards, educational software, and online platforms facilitate interactive learning by providing diverse and engaging content. These tools enable real-time feedback, instant access to resources, and the ability to track student progress. For instance, quizzes and polls can be conducted online, allowing for immediate assessment and adjustment of teaching strategies based on student performance (Quintana-Ordorika et al., 2024; Chen, 2023). Face-to-face interactive methods, such as debates and discussions, foster a collaborative learning environment where students can actively participate and express their ideas. These activities encourage critical thinking, enhance communication skills, and promote a deeper understanding of the subject matter. Debates, for example, require students to research and articulate their viewpoints, while discussions provide a platform for sharing diverse perspectives and reflecting on different ideas (Shivaram, 2024).

Reflective practices are an integral part of interactive learning. By engaging in activities that require reflection, such as group discussions and peer feedback, students develop the ability to critically analyse their learning experiences. This process not only helps them understand their strengths and areas for improvement but also fosters a growth mindset (Youhasan et al., 2022). The shift towards digitally oriented classrooms has significantly transformed educational practices, with technology playing a pivotal role in enhancing the learning experience. The COVID-19 pandemic underscored the importance of technology in education, as it became the primary mode of learning during that period (Triyanto, 2024). With schools and universities closed, digital platforms emerged as the saviour for children, ensuring continuity in education. This period of reliance on technology has led to a lasting integration of digital tools in classroom practices, making learning more interactive and engaging (Cowan & Farrell, 2023).

Today, classrooms are increasingly equipped with various technological tools that facilitate interactive learning. Teachers are leveraging digital platforms to conduct quizzes, gather student opinions, and facilitate reflective note writing and project work (Chen, 2023). These tools not only make learning more engaging but also provide real-time feedback and foster a collaborative learning environment. For instance, online quizzes allow for immediate assessment of student understanding, enabling teachers to adjust their teaching strategies accordingly. Opinion polls and discussion forums encourage students to express their thoughts and engage in meaningful dialogue with their peers.

Reflective note writing, facilitated by digital tools, allows students to critically analyse their learning experiences and articulate their thoughts more effectively. This practice helps in developing critical thinking skills and promotes a deeper understanding of the subject matter. Additionally, project work conducted through digital platforms enables students to collaborate seamlessly, share resources, and receive instant feedback from their teachers (Botes & Rossi, 2024).

The integration of technology in classroom practices has thus made learning more dynamic and interactive. It has bridged the gap between traditional and modern educational methods, creating a more inclusive and engaging learning environment (Chen, 2023). As technology continues to evolve, its role in education is likely to expand, offering new opportunities for enhancing the teaching and learning experience.

AI-based learning

Artificial intelligence (AI) is revolutionizing educational practices by offering innovative pedagogical approaches that enhance learning experiences and outcomes. One of the most significant contributions of AI in education is its ability to provide personalized learning experiences. AI algorithms can analyse individual student data to tailor educational content and activities to their unique needs, learning pace, and preferences. This personalized approach ensures that each student receives the support they need to succeed, making learning more effective and engaging (Alam & Asmawi, 2023).

Moreover, AI-powered tools such as virtual reality (VR) are proving to be invaluable, especially for children with various disabilities. VR can create immersive and interactive learning environments that cater to the specific needs of these students, presenting content in formats that are accessible and engaging. This technology helps overcome learning obstacles by providing alternative ways to experience and understand complex concepts.

Intelligent tutoring systems (ITS) are another remarkable application of AI in education. These systems can deliver personalized instruction and feedback, simulating one-on-one tutoring experiences. ITS can adapt to the learner's progress, offering additional support or advanced challenges as needed, thereby enhancing the learning process.

AI also plays a crucial role in automating administrative tasks such as grading and assessment. Automated grading systems can quickly and accurately evaluate student work, providing immediate feedback and freeing up valuable time for teachers to focus on more interactive and personalized teaching activities. This automation not only increases efficiency but also ensures consistency and objectivity in assessments (Bennett & Abusalem, 2024).

Experiential blended learning

Experiential blended learning is emerging as a transformative pedagogical approach that combines the strengths of traditional classroom methods with the dynamic capabilities of online educational materials and interactive activities. This hybrid learning style is particularly relevant in the contemporary educational landscape, where the integration of technology into teaching practices is not just beneficial but necessary. By merging face-to-face instruction with digital tools, experiential blended learning creates a more engaging and immersive learning environment that fosters deeper understanding and retention of knowledge. One of the key advantages of experiential blended learning is its ability to present content from textbooks in a more realistic and interactive manner (Chen, 2023). Digital tools such as simulations, virtual labs, and multimedia presentations can bring theoretical concepts to life, making them more tangible and easier to grasp. For instance, a biology lesson on cellular processes can be enhanced with 3D animations and interactive models, allowing students to visualize and manipulate the components of a cell. This not only makes the learning experience more engaging but also helps students develop a more comprehensive understanding of the subject matter.

Moreover, experiential learning, which emphasizes learning through experience and reflection, is inherently more engaging and can significantly induce curiosity among students. By participating in hands-on activities, real-world problem-solving, and collaborative projects, students are encouraged to explore, experiment, and reflect on their learning experiences. This active involvement in the learning process helps to cultivate critical thinking skills, creativity, and perseverance. When students are curious and motivated, they are more likely to persist in the face of challenges and develop a growth mindset (Botes & Rossi, 2024).

Blended learning environments also provide opportunities for personalized learning, where students can progress at their own pace and receive tailored support based on their individual needs. Online platforms can offer a variety of resources, such as video tutorials, interactive quizzes, and discussion forums, that cater to different learning styles and preferences. This flexibility allows students to take ownership of their learning journey and engage with the material in a way that suits them best. Experiential blended learning as an innovative pedagogy leverages the best of both traditional and digital educational practices to create a more interactive, engaging, and effective learning experience (Love, 2024). By combining hands-on, experiential activities with the vast resources and capabilities of digital tools, this approach not only enhances student engagement and curiosity but also fosters the development of essential skills such as critical thinking, creativity, and perseverance (Husanu & Ertekin, 2024). As technology continues to evolve, the potential for experiential blended learning to transform education and improve learning outcomes will only grow (Botes & Rossi, 2024).

3D Printing and Design

The integration of 3D printing technology with the design-thinking process represents a powerful and innovative pedagogical approach that provides students with hands-on experience and fosters creative problem-solving. This combination allows students to transform theoretical knowledge into practical applications, thereby enhancing their understanding of complex concepts and encouraging innovation. 3D printing technology enables students to create tangible models of their ideas, making abstract concepts more concrete and accessible. By visualizing and manipulating these models, students can better grasp intricate details and relationships within the subject matter. For instance, in a biology class, students can print 3D models of cellular structures, allowing them to explore and understand the spatial relationships and functions of different cell components. This hands-on experience not only reinforces theoretical knowledge but also makes learning more engaging and memorable (Xiao & Evans, 2022).

The design-thinking process complements 3D printing by encouraging students to tackle real-world problems through a structured approach that promotes creativity and innovation. This process typically involves several stages: empathizing with users, defining the problem, ideating solutions, prototyping, and testing. By following these steps, students learn to approach problems from multiple perspectives, generate innovative ideas, and iteratively refine their solutions based on feedback (Conlon, 2024). When combined with 3D printing, the design-thinking process allows students to rapidly prototype their ideas and test them in a tangible form. This iterative cycle of designing, printing, and testing helps students develop critical thinking and problem-solving skills. They learn to identify and address design flaws, optimize their solutions, and understand the practical implications of their theoretical knowledge (Zakaria et al., 2024).

Moreover, this innovative pedagogy empowers students to create their own products and contribute meaningfully to society. By working on projects that address real-world challenges, students can develop solutions that have a positive impact on their communities. For example, students might design and print assistive devices for individuals with disabilities, educational tools for underprivileged schools, or sustainable products that address environmental issues. These projects not only enhance students' learning experiences but also instil a sense of social responsibility and purpose (Qian, 2023).

The integration of 3D printing technology with the design-thinking process as an innovative pedagogy provides students with a comprehensive and immersive learning experience (Hamzah et al., 2024). It bridges the gap between theoretical knowledge and practical application, fosters creativity and innovation, and encourages students to develop solutions that benefit society. By engaging in this hands-on, problem-solving approach, students are better equipped to understand complex concepts, develop essential skills, and make meaningful contributions to their communities (Zakaria et al., 2024).

Inquiry-Driven Project-Based Learning

Inquiry-driven learning, combined with project-based learning, represents a powerful and innovative pedagogical approach that fosters deep understanding and practical application of knowledge (Hamzah et al., 2024). This model encourages students to take ownership of their learning journey by actively seeking answers to their questions and exploring topics that spark their curiosity. By engaging in inquiry-driven learning, students develop critical thinking skills, enhance their problem-solving abilities, and cultivate a lifelong love for learning.

In an inquiry-driven project model, students begin by identifying a question or problem that intrigues them (Zakaria et al., 2024). This initial stage is crucial as it taps into their natural curiosity and motivates them to delve deeper into the subject matter. As they investigate their chosen topic, students engage in research, gather information from various sources, and analyse their findings. This process not only helps them build a solid foundation of knowledge but also teaches them how to evaluate and synthesize information critically.

The project-based aspect of this model allows students to apply their theoretical knowledge in practical, real-world contexts. Over an extended period, they work on projects that require them to design, create, and implement solutions to the problems they have identified. This hands-on approach ensures that learning is not confined to the classroom but extends into real-life applications. For example, a student interested in environmental science might investigate the impact of plastic waste on local ecosystems and develop a project aimed at reducing plastic use in their community.

Throughout the inquiry-driven project, students are encouraged to think creatively and explore multiple perspectives. The design-thinking process, often integrated into this model, further enhances their ability to innovate and solve complex problems. By iterating on their ideas, testing prototypes, and refining their solutions, students learn the value of perseverance and resilience. This iterative process also fosters a growth mindset, as students understand that failure is a natural part of learning and an opportunity for improvement.

The final output of an inquiry-driven project is typically a tangible product or presentation that showcases the student's learning journey and the solutions they have developed. This could be a scientific model, a digital presentation, a community initiative, or any other form of project that demonstrates their understanding and creativity. By sharing their work with peers, teachers, and the broader community, students gain valuable feedback and recognition for their efforts. Inquiry-driven project model as an innovative pedagogy empowers students to take charge of their learning, encourages deep exploration of topics, and promotes the practical application of knowledge. By fostering curiosity, creativity, and critical thinking, this approach prepares students to tackle real-world challenges and contribute meaningfully to society (Husanu & Ertekin, 2024). It transforms the learning experience from passive reception of information to active, engaged, and purposeful inquiry.

Discussion

Organizing innovative pedagogies systematically and introducing them hierarchically in the learning process can significantly enhance educational outcomes. By categorizing these methodologies based on the nature of the content being explored and the level of understanding required, teachers can effectively tailor their instructional strategies to meet the diverse needs of their students (Hamzah et al., 2024).

At the foundational level, interactive learning serves as the initial step in engaging students. This approach includes methods such as quizzes, debates, and discussions, which foster active participation and immediate feedback. Interactive learning helps students build a strong connection with the material and their peers, laying the groundwork for deeper engagement (Chen, 2023). One of the primary benefits of interactive learning is that it transforms students from passive recipients of information into active participants in their education. This shift encourages students to think critically, ask questions, and engage with the material on a deeper level. For example, during a debate, students must research their positions, articulate their arguments, and respond to counterarguments, all of which promote critical thinking and effective communication skills.

Interactive learning also leverages the power of social interaction. Group activities and peer-to-peer learning opportunities allow students to collaborate, share diverse perspectives, and learn from one another. This collaborative approach mirrors real-world scenarios where teamwork and communication are essential, thereby preparing students for future professional environments. Moreover, interactive learning can be seamlessly integrated with technology to further enhance its effectiveness. Digital tools such as interactive whiteboards, educational software, and online platforms provide a variety of ways to engage students. These tools offer real-time feedback, instant access to resources, and the ability to track student progress, making the learning experience more personalized and responsive to individual needs (Triyanto, 2024).

Moving up the hierarchy, experiential learning comes into play. This method emphasizes learning through experience and reflection, making it particularly effective for subjects that benefit from hands-on activities. Blended learning, which combines traditional classroom methods with online educational materials, falls under this category. By integrating digital tools, experiential learning makes theoretical content more tangible and engaging, encouraging students to explore and experiment. The experiential learning cycle, as described by David A. Kolb, consists of four stages, namely concrete experience, reflective observation, abstract conceptualization, and active experimentation. In the first stage, students participate in a concrete experience, such as a lab experiment, a field trip, or a simulation. This hands-on involvement allows them to engage their senses and immerse themselves in the learning activity.

Following the experience, students move to the reflective observation stage, where they reflect on what they have done and observed. This reflection helps them to connect their feelings and thoughts with the experience, leading to deeper insights and understanding. Reflective observation is crucial as it encourages students to think critically about their actions and the outcomes, fostering a deeper level of learning. In the abstract conceptualization stage, students use their reflections to develop theories, concepts, or general principles. This stage involves thinking and reasoning, allowing students to form new ideas or modify existing ones based on their experiences. By abstracting their learning, students can apply their knowledge to different contexts and situations. The final stage, active experimentation, involves applying the newly acquired knowledge and concepts to new situations. Students test their theories and ideas through further experimentation, which leads to new experiences and the continuation of the learning cycle. This iterative process ensures that learning is continuous and evolving.

Experiential learning is particularly effective because it caters to different learning styles and preferences. It allows students to learn by doing, which can be more engaging and motivating than traditional lecture-based methods. Additionally, it helps students develop essential skills such as problem-solving, critical thinking, and collaboration, which are valuable in both academic and real-world settings.

AI-based learning can be seamlessly integrated at this stage to further personalize the learning experience. AI tools can adapt content to individual student needs, providing tailored support and enhancing engagement. Virtual reality (VR) can also be utilized to create immersive learning environments, especially beneficial for students with disabilities, by presenting content in accessible formats. AI-powered tools such as intelligent tutoring systems (ITS) are particularly noteworthy. These systems simulate one-on-one tutoring by delivering personalized instruction and feedback. ITS can adapt to the learner's progress, offering additional support or advanced challenges as needed, thereby enhancing the learning process. For example, an ITS might provide extra practice problems for a student struggling with a particular concept or introduce more complex topics for a student who has mastered the basics.

Moreover, AI can significantly aid students with disabilities by providing accessible learning environments. Virtual reality (VR) and augmented reality (AR) technologies, powered by AI, can create immersive and interactive experiences that cater to the specific needs of these students. For instance, VR can simulate real-world scenarios that help students with autism develop social skills in a controlled and safe environment.

AI also plays a crucial role in automating administrative tasks such as grading and assessment. Automated grading systems can quickly and accurately evaluate student work, providing immediate feedback and freeing up valuable time for teachers to focus on more interactive and personalized teaching activities (Chen, 2023). This automation not only increases efficiency but also ensures consistency and objectivity in assessments. Furthermore, AI can enhance collaborative learning by facilitating group projects and discussions. AI-driven platforms can analyse group dynamics and suggest optimal group compositions based on students' strengths and weaknesses. These platforms can also provide real-time feedback on group interactions, helping students develop better collaboration and communication skills.

In addition to these applications, AI can support teachers in developing and delivering more effective lesson plans. AI tools can analyse vast amounts of educational data to identify trends and best practices, helping educators design lessons that are more likely to engage and benefit their students. For instance, AI can suggest multimedia resources that align with the curriculum and cater to different learning styles. Overall, AI-induced pedagogy represents a significant advancement in educational practices. By providing personalized learning experiences, supporting students with disabilities, automating administrative tasks, enhancing collaborative learning, and aiding in lesson planning, AI is transforming the educational landscape.

As students' progress, the design-thinking process becomes crucial. This approach encourages creative problem-solving and innovation by having students tackle real-world problems. The integration of 3D printing technology allows students to visualize and create tangible models, bridging the gap between theoretical knowledge and practical application. This hands-on experience not only deepens understanding but also fosters creativity and critical thinking. Design thinking is an innovative, human-centred approach to problem-solving that emphasizes understanding users, challenging assumptions, redefining problems, and creating innovative solutions (Zakaria et al., 2024). This iterative process is effective for tackling complex scenarios where students need to gain clarity with respect to theory and application. The design thinking process typically involves five key stages: Empathize, Define, Ideate, Prototype, and Test. The first stage involves gaining a deep understanding of the users and their needs. This is achieved through observation, interviews, and other research methods. The goal is to set aside personal biases and truly understand the user's experiences and challenges. The insights gathered during the empathize phase are synthesized to define the core problem. This involves creating a clear and concise problem statement that guides the rest of the design process. A well-defined problem statement focuses on the user's needs and provides a clear direction for ideation. With a clear problem statement, the ideation phase involves generating a wide range of ideas and potential solutions. This stage encourages creativity and out-of-the-box thinking. Techniques such as brainstorming, mind mapping, and sketching are commonly used to explore various possibilities. The most promising ideas are then turned into prototypes. Prototyping involves creating simple, tangible representations of the ideas, which can range from sketches and models to interactive digital simulations. The goal is to quickly and cheaply test the feasibility of the ideas. The prototypes are tested with users to gather feedback and insights. This stage is crucial for understanding how well the solutions meet the users' needs and identifying areas for improvement. The feedback collected during testing is used to refine the prototypes and iterate on the solutions. Design thinking is not a linear process; it often involves moving back and forth between stages as new insights and ideas emerge.

The design thinking process fosters a culture of innovation and collaboration. Involving cross-functional teams and encouraging diverse perspectives helps generate creative solutions that might not emerge through traditional problem-solving methods. In educational settings, design thinking can be a powerful tool for fostering creativity, critical thinking, and problem-solving skills among students. By engaging in design thinking projects, students learn to empathize with others, define problems clearly, generate and prototype ideas, and test their solutions in real-world contexts.

At the pinnacle of the hierarchy is project-based learning. This method involves students working on projects over an extended period, promoting deep understanding and practical application of knowledge. Inquiry-driven learning, which encourages students to find answers to their own questions, complements project-based learning by fostering curiosity and exploratory thinking. The final output of these projects often involves students presenting their findings or creating models based on their explorations, demonstrating their comprehensive understanding and innovative solutions (Hamzah et al., 2024).

By categorizing and introducing these pedagogies hierarchically, teachers can create a structured and progressive learning environment. This approach ensures that students build a solid foundation through interactive and experiential learning, develop critical thinking and creativity through design-thinking processes, and achieve deep understanding and practical application through project-based learning. Such a systematic approach not only enhances student engagement and learning outcomes but also prepares them to tackle real-world challenges with confidence and innovation.

Project-based learning (PBL) is a dynamic and hands-on educational approach where students gain knowledge and skills by working on extended projects that investigate and respond to complex questions, problems, or challenges. This method transforms the traditional classroom into an active learning environment where students are deeply engaged in their education.

In PBL, students are not just passive recipients of information; they become active participants in their learning journey. They start by identifying a real-world problem or a significant question that interests them. This initial stage is crucial as it taps into their natural curiosity and motivation. Once the problem or question is defined, students embark on a process of inquiry and research, gathering information from various sources, analysing data, and synthesizing their findings.

Throughout the project, students apply critical thinking and problem-solving skills to develop solutions or answers to the questions they are investigating. This process often involves collaboration with peers, which enhances their ability to work as a team and communicate effectively. By working together, students learn to appreciate diverse perspectives and leverage each other's strengths to achieve common goals.

One of the key benefits of PBL is that it promotes deep understanding and retention of knowledge. Because students are actively involved in the learning process and can see the real-world applications of what they are studying, they are more likely to retain and understand the material. This approach also helps students develop essential life skills such as time management, organization, and self-directed learning. Moreover, PBL culminates in the creation of a tangible product or presentation that demonstrates the students' learning and achievements. This final output could be a scientific model, a digital presentation, a community project, or any other form of work that showcases their understanding and creativity. Presenting their projects to an audience, whether it be their classmates, teachers, or the broader community, provides students with valuable feedback and a sense of accomplishment.

Implication

The review analysis underscores the potential for integrating diverse, innovative educational approaches to create a more engaging and effective learning environment. By categorizing themes such as interactive learning, AI-based learning, experiential-blended learning, 3D printing and design learning, and inquiry-based project learning, educators can tailor their teaching strategies to meet the varied needs of students. This structured framework not only enhances student engagement but also promotes personalized learning experiences, ultimately leading to improved educational outcomes.

Conclusion

Interactive learning emphasizes the importance of active student participation and collaboration, fostering a more engaging and dynamic classroom environment. AI-based learning leverages advanced technologies to provide personalized learning experiences, adapting to individual student needs and promoting more effective learning outcomes. Experiential-blended learning combines traditional and hands-on learning methods, offering students practical, real-world experiences that enhance their understanding and retention of knowledge. The inclusion of 3D printing and design learning introduces students to cutting-edge technologies and creative problem-solving skills, preparing them for future careers in various fields. Inquiry-based project learning encourages students to take ownership of their learning through exploration and investigation, promoting critical thinking and independent learning skills.

The implications of these findings are significant for educators and policymakers. By recognizing and integrating these diverse educational approaches, we can create a more inclusive and effective learning environment that caters to the varied needs of students. This structured framework not only enhances student engagement but also promotes personalized learning experiences, ultimately leading to improved educational outcomes. Furthermore, the integration of these themes into educational practices can foster a culture of innovation and continuous improvement within educational institutions. By staying abreast of the latest pedagogical trends and technologies, educators can ensure that their teaching methods remain relevant and effective in an ever-evolving educational landscape.

In conclusion, the review has provided valuable insights into the potential of innovative educational approaches to transform the learning experience. By embracing these diverse methods, educators can create a more dynamic, engaging, and personalized learning environment that meets the needs of all students. This study underscores the importance of continuous exploration and adaptation in education, paving the way for a future where learning is more interactive, personalized, and effective.

References

Alam, M. S., & Asmawi, A. (2023). Chat GPT in Constructivist Language Pedagogy: Exploring Its Potential and Challenges in Legal Vocabulary Acquisition. *Applied Research on English Language, 12*(4), 131-148.

Arksey, H., & O'Malley, L. (2005). Scoping studies: towards a methodological framework. *International journal of social research methodology, 8*(1), 19-32. https://doi.org/10.1080/1364557032000119616

Bennett, L., & Abusalem, A. (2024). Building Academic Integrity and Capacity in Digital Assessment in Higher Education. *Athens Journal of Education, 11*(1), 71-94. https://doi.org/10.30958/aje.11-1-5

Botes, M., & Rossi, A. (2024). At the Verge of 'Is' and 'Could Be': Storytelling as Medium to Develop Critical Ethical Skills. In *Building Inclusive Ethical Cultures in STEM* (pp. 313-330). Cham: Springer International Publishing. https://doi.org/10.1007/978-3-031-51560-6_17

Chen, L. Y., Zhou, W. W., Hsieh, W. Z., & Chou, R. J. (2023). Establishing a professional learning community for cultivating future design talents using a 'peer coaching'mechanism. *Heliyon, 9*(10). https://doi.org/10.1016/j.heliyon.2023.e20906

Clarke, V., & Braun, V. (2017). Thematic analysis. *The journal of positive psychology, 12*(3), 297-298. https://doi.org/10.1080/17439760.2016.1262613

Conlon, J. (2024). Using product lifecycle management (PLM) to re-think fashion business education: an assessment of pedagogical practices and learning benefits. *Research Journal of Textile and Apparel, 28*(2), 282-298. https://doi.org/10.1108/RJTA-10-2021-0128

Cowan, P., & Farrell, R. (2023). Virtual Reality as the Catalyst for a Novel Partnership Model in Initial Teacher Education: ITE Subject Methods Tutors' Perspectives on the Island of Ireland. *Education Sciences, 13*(3), 228. https://doi.org/10.3390/educsci13030228

Gopalan, C., Halpin, P. A., Selvam, A. A. A., & Hung, W. C. (2024). Exploring educational transformations through the Innovative Flipped Learning Instruction Project Symposium. *Advances in Physiology Education, 48*(3), 505-511. https://doi.org/10.1152/advan.00027.2024

Hamzah, F., Abdullah, A. H., & Ma, W. (2024). Advancing Education through Technology Integration, Innovative Pedagogies and Emerging Trends: A Systematic Literature Review. *Journal of Advanced Research in Applied Sciences and Engineering Technology, 41*(1), 44-63. https://doi.org/10.37934/araset.41.1.4463

Husanu, I. N. C., & Ertekin, Y. (2024, June). Board 111: Transformative Approach of Engineering Technology Curricula Based on Sustainability, Systems Thinking, Creativity, and Alignment with Industry Needs. In *2024 ASEE Annual Conference & Exposition.*

Love, J. O. (2024, June). Engineering'STEAMs' Up Elementary Education: Impacts of the COVID-19 Pandemic (Fundamental). In *2024 ASEE Annual Conference & Exposition.*

Mourtzis, D., Panopoulos, N., & Angelopoulos, J. (2023). A hybrid teaching factory model towards personalized education 4.0. *International Journal of Computer Integrated Manufacturing, 36*(12), 1739-1759. https://doi.org/10.1080/0951192X.2022.2145025

OECD (2018). The Future of Education and Skills. Education 2030. Paris:OECD Publishing.

Olorunfemi, B. O., Longe, O. M., & Darsot, F. M. (2023). Survey Analysis on Engineering Students' Experience of Future-fit Classroom Learning Environment. *International Journal of Engineering Pedagogy, 13*(6). https://doi.org/10.3991/ijep.v13i6.38683

Peterson, A., Dumont, H., Lafuente, M., & Law, N. (2018). Understanding innovative pedagogies: Key themes to analyse new approaches to teaching and learning.

Qian, C. (2023). Research on Human-centered Design in College Music Education to Improve Student Experience of Artificial Intelligence-based Information Systems. *Journal of Information Systems Engineering and Management, 8*(3), 23761. https://doi.org/10.55267/iadt.07.13854

Quintana-Ordorika, A., Camino-Esturo, E., Portillo-Berasaluce, J., & Garay-Ruiz, U. (2024). Integrating the Maker pedagogical approach in teacher training: the acceptance level and motivational attitudes. *Education and Information Technologies, 29*(1), 815-841. https://doi.org/10.1007/s10639-023-12293-4

Shivaram, J. M. (2024). On Instilling Research Paradigm in PG Engineering Students through Guided Open Ended Lab Component Implementation. *Journal of Engineering Education Transformations, 37*(3). https://doi.org/10.16920/jeet/2024/v37i3/24001

Suryani, A., March, S., Fleer, M., & Rai, P. (2024). Contributing to SDGs through conceptual PlayWorlds: Changing the STEM story for children, families and teachers. In *Early childhood voices: Children, families, professionals* (pp. 59-70). Cham: Springer International Publishing. https://doi.org/10.1007/978-3-031-56484-0_5

Triyanto, T., Kholifah, N., Nurtanto, M., Nur, H. R., Saputro, I. N., Istanti, H. N., & Gadi, A. C. Z. (2024). Student e-learning effectiveness based on pedagogy, evaluation and technology dimensions (PET-D): Empirical studies in higher education in the COVID-19 epidemic. *Multidisciplinary Science Journal, 6*(12), 2024245-2024245. https://doi.org/10.31893/multiscience.2024245

Xiao, J., & Evans, D. J. (2022). Anatomy education beyond the Covid-19 pandemic: A changing pedagogy. *Anatomical Sciences Education, 15*(6), 1138-1144. https://doi.org/10.1002/ase.2222

Youhasan, Punithalingam, Yan Chen, Mataroria P. Lyndon, and Marcus A. Henning. "University teachers' perceptions of readiness for flipped classroom pedagogy in undergraduate nursing education: A qualitative study." *Journal of Professional Nursing* 41 (2022): 26-32. https://doi.org/10.1016/j.profnurs.2022.04.001

Zakaria, M. I., Nasran, N. A. H. N., Abdullah, A. H., Alhassora, N. S. A., Pairan, R., & Yanuarto, W. N. (2024). *Unlocking the Future: Mathematics Teachers' Insight into Combination of M-learning with Problem-Based Learning Teaching Activities.*

Chapter Fourteen
Navigating the Digital Classroom: Enhancing Educator Skills and Student Outcomes

Beenish Rasool

VIT-AP University, Amaravati, Vijayawada, Andhra Pradesh, India

Abdul Raffie Naik

VIT-AP University, Amaravati, Vijayawada, Andhra Pradesh, India

Abstract

In the context of today's rapidly advancing digital era, educators are confronted with the dual imperative of enhancing their digital competencies while adapting their instructional practices to meet the diverse needs of contemporary students. This chapter aims to investigate how digital technology influences teacher identity and roles, with a focus on identifying effective strategies to empower educators in improving their digital literacy and pedagogical approaches. Case studies will highlight a broad spectrum of teachers across different educational levels and geographical regions, with a notable eagerness among educators to incorporate digital tools into their teaching practices, albeit hindered by insufficient training and support systems. Effective empowerment strategies identified include ongoing professional development initiatives, fostering collaborative learning communities among educators, and ensuring access to resources that promote critical thinking and problem-solving skills. This thus underscores the necessity for a comprehensive approach that not only enhances technical proficiency but also encourages pedagogical innovation. By investing in continuous professional development and cultivating a collaborative culture, educators can adeptly navigate the challenges posed by the digital age, ultimately leading to enhanced student learning outcomes.

Keywords: digital competence, teacher identity, critical thinking, collaboration, lifelong learning, professional development, digital literacy.

***

Introduction

The concept of the digital classroom is an evolving feature in the delivery and experience of learning. With rapid changes in technology coupled with demands for flexibility, access, and interaction within learning spaces, digital classrooms have been implemented even more rapidly. The digital classroom employs online platforms, interactive tools, and multimedia resources for further enhancement of student and educator educational experiences. This chapter will, in effect, review the growth and development of digital classrooms, shifting pedagogical skill sets that are being demanded of educators for excellence in such learning environments, and case studies of successful integrations. This chapter also takes up the digital classroom as a global theme: how to measure impact, accountability strategies, and so on. It discusses active insights into practical strategies that educators and policymakers might use in light of the effective navigation of digital classrooms.

The Development of Digital Classrooms

The idea of digital classrooms has seen a lot of growth over the last few decades. This started with the integration of technology in education as straightforward technology tools that involved sending emails and discussion boards, which complemented traditional face-to-face instruction. However, as technology advanced, so did the potential for digital classrooms to transform education (Garrison & Kanuka, 2004). All of this changed with the rise of the internet in the 1990s and the early 2000s, when, for the first time, LMS systems such as Blackboard, Moodle, and Canvas could be developed. These allowed teachers to create a central online hub for course materials, communication, and tracking student performance (Wang, 2020).

The COVID-19 pandemic accelerated the adoption of the digital classroom across the world. During this pandemic, which locked the physical doors of schools and universities across the globe, educators swiftly transitioned to digital platforms. This marked push was, in turn, setting bright spotlights on the possibilities and difficulties that come with digital classrooms. Thus, while many have been able to transition their classes online, the hurdles ranged from technology access and digital literacy to student engagement (Selwyn, 2016). The pandemic certainly underlined the need for a more resilient and flexible education system that could adapt to changing circumstances to continue with quality education regardless of location.

Types of Digital Classrooms

There are several kinds of digital classrooms, and knowing all these is the key to choosing a suitable approach that will be fitting to the particular needs of educators and students. Each type of digital classroom comes with its own strengths and challenges; thus, the learning environment has to be tailored to suit the educational context.

1. Synchronous Digital Classrooms: Tutors and students interact in real time through synchronous classrooms using video conferencing tools like Zoom and Google Meet. This type of digital classroom largely resembles the traditional classroom experience in that feedback can be immediate, and discussions are dynamic (Martin & Parker, 2014). It has a limitation in that all participants must be online at the same time, something which is not easily done in diverse geographical settings or in time-constrained settings.

2. Asynchronous Digital Classrooms: Asynchronous classrooms allow students to log in at any time, at their convenience, view course materials, interactively participate in discussions, and submit assignments. This model is extremely helpful for learners whose schedules need to be accommodated due to job or personal commitments (Hrastinski, 2008). It may also tend to be less effective in engaging and motivating students because students can sometimes feel left out of the whole learning process due to a lack of real-time interaction.

3. Blended Digital Classrooms: Blended classrooms combine features of synchronous and asynchronous classes, offering a balanced model that allows learners to take full advantage of both models. For example, a blended course might consist of virtual, real-time lectures combined with self-directed online work. This way, it provides flexibility, like in the case of an asynchronous approach, without losing the interactive parts, as with synchronous instruction (Garrison & Kanuka, 2004).

4. In Flipped Classrooms, new content is introduced to the students outside of class time, usually through video lectures or readings, while class time itself is used for hands-on activities, discussions, and problem-solving exercises. This model stimulates active learning and enables educators to give more personalised support during class time (Bergmann & Sams, 2012).

5. Virtual Classrooms: The virtual classroom concept is totally online, and it can also include VR or AR to give an immersive feel to the learners. It will facilitate learning more in areas like STEM, where the learner will get to interact with three-dimensional models, conduct virtual experiments, or explore simulated environments (Merchant et al., 2014). While these tools make learning more interactive and engaging, VR and AR do require substantial technological infrastructure and resources.

6. Mobile Learning Classrooms: The mobile learning classroom makes use of portable devices such as smartphones and tablets to deliver learning. This model works best in areas where access to traditional classrooms is at a minimum, enabling students to learn at any time and from any location

(Traxler, 2007). Drawbacks associated with mobile learning include issues related to screen size, internet connectivity, and adaptation of content on the mobile platform.

7. Massive Open Online Courses: MOOCs provide free and low-cost access to a number of courses provided by various institutions across the globe. The population of students is huge, and these courses enable learning at scale. According to Yuan and Powell (2013), the major shortcoming is the lack of personalised teaching; similarly, limited interaction with the instructors can make MOOCs less effective for some students.

8. Gamified Digital Classrooms: Gamification embeds game elements within the learning context, such as points, badges, and leaderboards. It is meant to enhance engagement and motivation for learning. Gamified classrooms can make learning a little bit fun and encourage students to participate more actively in the educational experience. Indeed, "Gamification, in general, is about making something more engaging and fun" (Deterding et al., 2011). However, careful design must be considered for gamified activities so that they remain aligned with the educational objectives and do not interfere with the goals of learning.

Building Skills in Educators for the Digital Classroom

With the prevalence of digital classrooms, some of the hitherto required skills of educators are fast changing. While the traditional teaching skills would still be helpful, they need to be complemented with digital skills in making the educators effectively use technology in the lessons. The key competencies that an educator will possess are digital literacy, knowledge and actual use of the LMS platforms, and skills to produce relevant content in engaging ways through various available multimedia tools (Horn & Staker, 2015).

Professional development and continuous training have played a very significant role in the building of these skills among educators. Such training as in digital pedagogy, instructional design, and integration of technology into the curriculum has been found effective in enhancing readiness among educators within the digital classroom (Kim & Park, 2022). Examples include the following: One such in-service program, which focused on training teachers about the use of an interactive whiteboard and relevant educational software, ran in the United States and led to greater teacher confidence and better student outcomes. The program gave extensive time for hands-on practice, collaboration with peers, and follow-up support. For this reason, and many others, it succeeded (Johnson et al., 2016).

Despite these efforts, challenges remain. Most educators find it hard to adapt to new technologies because either they are unfamiliar with them, they do not have easy access to such resources, or they are just resistant to change (Mtebe & Raisamo, 2014). Despite these efforts, challenges remain. Most educators find it hard to adapt to new technologies because either they are unfamiliar with them, they do not have easy access to such resources, or they are just resistant to change (Mtebe et al., 2014). Overcoming this situation calls for more than mere technical training; it calls for an attitudinal change, a change of mindset. Educators need to be made to feel that technology will enhance and support traditional ways of teaching, but not replace them. This is what institutions can provide-institutional support in the form of access to technology, continuous professional development, and a culture that promotes innovation.

Case Studies on Successful Technology Integration

Successful integration of technology into the classroom seems to have an effect on the level of student engagement and learning outcomes. The section below gives an overview of three field-tested case studies showcasing effective uses of digital tools in diverse settings of education.

Case Study 1: Finland's Comprehensive Digital Learning Environment

Finland has always been in the slot of one of the best in education; the country has splendidly implemented digital classrooms in all three tiers of education. The concept of this nation shows that a well-structured digital learning environment comprises advanced technology, effective teacher training programs, and a curriculum that lays emphasis on digital literacy among students (Chen & Tsai, 2021).

Digital devices are used in Finnish schools for research, collaborations of all types, and creative projects. Teachers are trained to use these devices not as supplementing tools but to be integral to the process of learning. This has been leading to measurable positive changes in student engagement and academic success. For example, Johnson et al. (2016) reported that schools that have integrated the digital learning environment recorded a 20% increase in students' participation in online discussions and a 15% improvement in test scores after three years.

Case Study 2: Digital Classrooms in India's Rural Schools

While establishing digital classrooms in rural areas of India is not without its specific challenges related to access to technology and infrastructure, several initiatives have been highly successful in overcoming these challenges by bringing in innovative solutions fitted within the local contexts. Examples include the "Digital Village" projects that provided schools in remote areas with solar-powered tablets pre-loaded with offline digital libraries (Mtebe & Raisamo, 2014).

Teachers have been trained in the use of these facilities and how to deliver interactive lessons, involving students in active learning. This was also complemented by activities with local volunteers for technology maintenance and support of the teachers. There was a tremendous impact: student attendance rose to 30%, and important improvements in literacy and numeracy were registered among the students (Wang, 2020).

Case Study 3: The Use of Virtual Reality in the South Korean Classrooms

The implementation of virtual reality in South Korean classrooms has been a game-changing approach in teaching students in complex subjects, such as those in STEM. VR provides them with a virtual environment in which they will be able to practise the usage of three-dimensional models, running virtual experiments, and manipulation abstract ideas in a concrete manner (Kim & Park, 2022).

A pilot program initiated lessons in science classes in Seoul, using the basics of VR for learning, focusing on themes such as the human body, space travel, and chemical processes. The training also included how to effectively integrate virtual reality into lesson plans by teachers themselves. The results were amazing: Students in the VR group showed a better understanding of the subject matter and a 25% increase in exam scores over the traditional learning group (Chen & Tsai 2021).

Global Perspectives: Digital Classrooms Across Different Educational Settings

Because the market development differs largely within countries, regions, and continents, it is mainly driven by the income level, technological situation, and cultural background. The section shall give an overview of the implementation of digital classrooms in various regions, indicating successes as well as challenges.

In North America and Europe, the concept of the digital classroom has been well established, with good access to technology and immense institutional support. The focus in these countries has actually shifted toward the maximisation of the use of digital tools and studying other new technologies such as AI and VR for better learning outcomes (Johnson et al., 2016). For example, several European countries have integrated national policies for digital education, ensured funding for technologies in schools, and developed standardized curricula for digital skills (Garrison & Kanuka, 2004).

In developing countries, most of which are found in Africa and Asia, the establishment of such digital classrooms is way behind, due to a lack of infrastructure, funding, and equality in accessing technology. Yet, within this depressing reality, some inspiring examples of innovation and survival can be found. Recently, it was reported that the South African government had engaged in partnership with private companies to offer low-cost tablets, along with internet services in schools in some disadvantaged communities. The initiative has helped reduce the digital gap and supported thousands of students to improve their learning results (Mtebe & Raisamo, 2014).

Japanese education has integrated digital classrooms into the national curriculum, where digital literacy and critical thinking skills are honed. The Japanese system is designed in such a way that technology enhances collaborative learning and creativity: a preparation for twenty-first-century demands in the work environment. Kim and Park (2022) cite this as a reason for the high levels of student engagement in school, as well as good performance by Japanese students in international assessments.

Measuring the Impact of Technology Integration

Measuring the impact of technology integration is very important in ensuring that the digital classroom effectively serves to enhance educator performance and student outcomes. Herein, some key metrics, tools, and methods for assessing the successful integration of technology into education will be discussed.

Key Metrics and Indicators

The measurement of digital classroom impact might involve monitoring student engagement, academic performance, and educator effectiveness. Common metrics to track could include attendance rates, online activity participation, grades, and standardised test scores. Satisfaction of students and perceived effectiveness of the digital tools can be sourced from surveys and feedback forms (Horn & Staker, 2015).

Tools and Methods for Evaluation

This is where the digital classroom performance effectiveness is measured and analysed through various kinds of tools. Most LMSs (learning management systems) have built-in analytics features for educators to learn whether students are progressing properly and if extra help might be needed. For example, educators can track how much time students are spending on certain tasks, participation in discussions, and performance on assessments (Wang, 2020).

Analytics platforms can also be used to mine vast amounts of data and uncover trends that are time-based. By observing the performance and engagement trends of the students, the educators and administrators can make evidence-based decisions to bring effectiveness into the digital classrooms (Chen & Tsai, 2021).

Case Study: Using Data Analytics in a Blended Learning Environment

In the United States, one of the hybrid university courses used analytics with respect to measuring both the online and face-to-face components of a course involving student engagement and performance. Analytics showed that highly engaged students in the online discussions and activities fared better in the assessments and were more active in class participation (Johnson et al., 2016). This realisation thereby led to the course design inclusion of additional online interactive activities and providing targeted feedback to less-engaged students.

Strategies to Ensure Accountability in Digital Classrooms

Accountability in the digital classroom is crucial in ensuring that technology serves to improve learning outcomes for educational goals. The following are strategies in the development of accountability frameworks, monitoring progress, and ensuring transparency.

Accountability Framework Development

A good accountability framework requires well-set goals, along with their performance measures, and a system of periodic review and evaluation. Educators and administrators should work together and collaborate to set expectations about the use of technology, establish measurable goals, and develop a timeline for achieving these goals (Selwyn 2016).

Monitoring Progress

For digital classrooms to meet their goals, there is a need for ongoing monitoring. This is possible with regular checks, progress reports, and data analysis. Teachers need to have at their disposal tools and other resources necessary to keep track of the progress of their service delivery and improvements required (Mtebe & Raisamo, 2014).

Case Study: Accountability in a Digital Education Initiative in the UK.

In the UK, there was a digital education initiative at primary school levels, mainly for literacy and numeracy. Testing would be done on a regular basis, while continuous feedback loops and data harvesting for decision making were included in the program (Garrison & Kanuka, 2004). Teachers were held accountable for all students' performances, while support mechanisms were put in place to help those who may still be struggling with integrating technology into their teaching. The initiative saw a drastic rise in the performance of students, with a 15% increased rate of literacy over three years.

Overcoming Challenges in Digital Classrooms

With all the advantages of digital classrooms come a set of challenges that must be met in order to make the concept successful. The following section will address the common problems and explain how to overcome them.

Common Obstacles

The most salient challenge is the so-called digital divide, and that is the gap between those included and those excluded, basically from using technology. This can be reflected through issues in internet access, availability of digital devices, and digital literacy skills. Others include resistance to change: resistance among educators to adopt new technologies, and some students adapting to the digital learning environment (Johnson et al., 2016).

Strategies to Address Challenges

Equal access to technology and further support given to those who may be in need will help in minimising the digital divide. This would encompass buying affordable hardware and providing training programs that improve their digital skills, and sharing online resources for use by all students. Mtebe and Raisamo (2014) argue that, in addition to this, the spirit of innovation and openness to change in approach should be cultivated. The educators should be sufficiently motivated to attempt new gadgets or strategies; the students should be assisted in the acquisition of those sets of skills that enable them to succeed in the digital classrooms.

Case Study: Bridging the Digital Divide in Rural Regions of Canada

The Canadian program of bridging the digital gap in rural areas introduced students to the internet, digital devices, and training for students and educators alike. The community was engaged in a process that included parents and local leaders. This resulted in increased student online learning participation, while academic performance improved in all subjects (Chen & Tsai, 2021).

Future Directions and Trends in Digital Education

As technology is ever-changing, so will the digital classroom be. This section will now review a few current trends and their potential impact on education.

Emerging Technologies

New technologies, including artificial intelligence, augmented reality, and blockchain, among others, are bound to make digital classrooms even more vital in the future. These tools have the potential to allow personalised learning, increase engagement, and offer new methods for assessing the progress of students (Selwyn, 2016).

The Evolving Role of Educators

Educators themselves constitute another changing role in the digital classroom. As facilitators of learning, educators can no longer simply deliver content but have to guide students in how to use digital resources and work collaboratively with peers. That means new skills are acquired and change embraced (Horn & Staker, 2015).

Chapter Summary

The digital classroom is one of those present entities that will outlast COVID-19; it's the future of education. In this continuing evolution, educators and institutions have to invest more in enhancing educator skills, ensuring equity in technology access, and putting in place measures to ensure accountability. The flexibility, resilience, and further engagement add new dimensions to an education system better prepared to face the challenges well into the twenty-first century.

References

Bergmann, J., & Sams, A. (2012). *Flip your classroom: Reach every student in every class every day.* International Society for Technology in Education.

Chen, P.-Y., & Tsai, C.-C. (2021). Exploring the impact of virtual reality on student learning outcomes: A meta-analysis. *Educational Technology Research and Development, 69*(2), 217–239. https://doi.org/10.1007/s11423-02009874-2

Deterding, S., Dixon, D., Khaled, R., & Nacke, L. (2011). From game design elements to gamefulness: Defining "gamification." In *Proceedings of the 15th International Academic MindTrek Conference: Envisioning Future Media Environments* (pp. 9–15). ACM. https://doi.org/10.1145/2181037.2181040

Garrison, D. R., & Kanuka, H. (2004). Blended learning: Uncovering its transformative potential in higher education. *The Internet and Higher Education, 7*(2), 95–105. https://doi.org/10.1016/j.iheduc.2004.02.001

Horn, M. B., & Staker, H. (2015). *Blended: Using disruptive innovation to improve schools.* Jossey-Bass.

Hrastinski, S. (2008). Asynchronous and synchronous e-learning. *Educause Quarterly, 31*(4), 51–55.

Johnson, L., Adams Becker, S., Cummins, M., Estrada, V., Freeman, A., & Hall, C. (2016). *NMC horizon report: 2016 higher education edition.* The New Media Consortium.

Kim, K. J., & Park, Y. S. (2022). Enhancing student engagement through advanced digital tools: The South Korean experience. *Asia Pacific Education Review, 23*(1), 125–139. https://doi.org/10.1007/s12564-009-9007-2

Martin, F., & Parker, M. A. (2014). Use of synchronous virtual classrooms: Why, who, and how? *MERLOT Journal of Online Learning and Teaching, 10*(2), 192–210.

Merchant, Z., Goetz, E. T., Cifuentes, L., Keeney-Kennicutt, W., & Davis, T. J. (2014). Effectiveness of virtual reality-based instruction on students' learning outcomes in K-12 and higher education: A meta-analysis. *Computers & Education, 70,* 29–40. https://doi.org/10.1016/j.compedu.2013.07.033

Mtebe, J. S., & Raisamo, R. (2014). Challenges and opportunities of using e-learning in developing countries: A case of Tanzania. *Journal of Information Technology Education: Research, 13,* 31–53.

Selwyn, N. (2016). *Education and technology: Key issues and debates.* Routledge. https://doi.org/10.5040/9781474235952

Traxler, J. (2007). Defining, discussing, and evaluating mobile learning: The moving finger writes and having writ... *The International Review of Research in Open and Distributed Learning, 8*(2). https://doi.org/10.19173/irrodl.v8i2.346

Wang, M. (2020). The impact of learning management systems on student learning outcomes. *Journal of Educational Technology, 22*(4), 289–305.

Yuan, L., & Powell, S. (2013). MOOCs and open education: Implications for higher education. *Journal of Interactive Media in Education, 2013*(3). https://doi.org/10.5334/2013-18

PART IV: SPECIALISED APPLICATIONS OF TECHNOLOGY IN TEACHER EDUCATION

Chapter Fifteen
Technology Integrated Pre-Service Teacher Preparation in an Educational Psychology Course

Alpana Bhattacharya

Queens College, City University of New York, NY, USA

Abstract

This chapter explores the integration of technology in pre-service teacher preparation through an educational psychology course. It emphasizes alignment with key teaching standards, including CAEP, InTASC, and NYSTS, and utilizes the TPACK framework to interweave content, pedagogy, and technology knowledge. Activities such as collaborative group work, technology-integrated lesson design, and differentiated instruction highlight the development of pre-service teachers' competencies. Empirical studies validate the course's approach to building robust technological pedagogical content knowledge (TPACK) for secondary education across disciplines.

Keywords: technology integration, TPACK Framework, pre-service teacher preparation, educational psychology.

***

Introduction

Teaching Standards for Teacher Preparation

Most accredited teacher preparation programs in the United States require pre-service teachers to complete an educational psychology course for their teaching certification. Technology integrated pre-service teacher preparation showcased in this chapter is from a graduate course titled, *Educational Psychology,* offered at an accredited teacher preparation institution in New York. Therefore, pre-service teacher preparation in this educational psychology course is aligned with the *2022 Initial Level Standards for Educator Preparation,* specifically Standard 1 – *Content and Pedagogical Knowledge* (Council for the Accreditation of Educator Preparation [CAEP], 2022). Additionally, this educational psychology course also emphasizes the *Interstate Teacher Assessment and Support Consortium (InTASC) Model Core Teaching Standards* (2013) which require pre-service teachers to have essential knowledge, skills, and dispositions related to *Learners and Learning* (Standards 1-3), *Content* (Standards 4-5), *Instructional Practice* (Standards 6-8), and *Professional Responsibility* (Standards 9-10). Finally, the *New York State Teaching Standards* (NYSTS, New York State Education Department [NYSED], 2011), specifically *Knowledge of Students and Student Learning* (Standard 1), *Knowledge of Content and Instructional Planning* (Standard 2), *Instructional Practice* (Standard 3), *Learning Environment* (Standard 4), *Professional Responsibilities and Collaboration* (Standard 6), are also subsumed in this educational psychology course for preparation of pre-service teachers to teach secondary school students.

Given that a key component of pre-service teacher preparation in the educational psychology course is technology integration, this course emphasizes key elements of the *NYSTS* (NYSED, 2011). One key component of technology integrated teacher preparation emphasized within this educational psychology course is to promote pre-service teachers' "knowledge and understanding of technological and information literacy and how they affect student learning" (p. 2) an element of Standard 1 *Knowledge of Students and Student Learning* (NYSED, 2011). Another salient component of technology integrated teacher preparation emphasized within this educational psychology courses is to support pre-service teachers' knowledge, skills, and dispositions for engaging secondary school students in the "use of technology to solve real world problems" (p. 7), an element of Standard 3 *Instructional Practice* (NYSED, 2011). A final component of technology integrated teacher preparation emphasized in this educational psychology course is to train pre-service teachers to "organize and utilize available technology resources to create a safe and productive learning environment" and ensure that "students have equitable access to available resources and technologies" (p. 9), an element of the Standard 4 *Learning Environment* (NYSED, 2011).

The educational psychology course also ensures technology integrated pre-service teacher preparation by aligning course requirements with essential knowledge and critical dispositions stipulated within *InTASC* (2013) Standard 8 *Instructional Strategies.* One area of essential knowledge within *InTASC* Standard 8 is teacher understanding and using a wide variety of technological resources to engage students in learning. Another area of essential knowledge within *InTASC* Standard 8 is teacher understanding and using a variety of technologies to develop students' knowledge of content and skills to apply knowledge in meaningful ways. A final area of essential knowledge and critical dispositions within *InTASC* Standard 8 is teachers exploring use of new and emerging technologies to support and promote student learning and evaluating these resources for quality, accuracy, and effectiveness. Therefore, the educational psychology course requirements are aligned with the essential knowledge and critical dispositions of *InTASC* Standard 8 to ensure efficient technology integrated teacher preparation.

Educational Technology Framework for Teacher Preparation

Technological Pedagogical Content Knowledge (TPCK) by Mishra and Koehler (2006) is a salient conceptual framework that has informed technology integrated teacher preparation across the world. The three foundational elements of the TPCK framework are Content Knowledge (CK), Pedagogical Knowledge (PK), and Technology Knowledge (TK). CK requires an understanding of facts, concepts, theories, and policies pertaining to a specific subject area. PK involves processes, practices, and procedures related to teaching and learning of subject-area information. TK represents awareness of standard to advanced technologies such as smart phones, tablets, laptops, computers, email, Smartboard, Learning Management Systems, conferencing systems, and digital media.

Although CK, PK, and TK are foundational elements of the TPCK framework, which are integral part of CAEP (2022), INTASC (2013), and NYSED (2011) as well, Mishra and Koehler (2006) caution against considering them as unrelated elements and that attending to each as disjointed elements for teacher preparation would be an ineffectual approach. Consequently, these researchers advocate for a teacher preparation approach that integrates CK, PK, and TK as interrelated and overlapping elements. Based on the triangulation of CK, PK, and TK, Mishra and Koehler elicited four intersecting relationships leading to their comprehensive TPCK framework. The intersecting relationships emerging from the overlapping of content, pedagogical, and technology elements are Pedagogical Content Knowledge (PCK), Technological Content Knowledge (TCK), Technological Pedagogical Knowledge (TPK), and Technological Pedagogical Content Knowledge (TPCK).

According to Mishra and Koehler (2006), PCK is understanding how to select and apply specific teaching approaches in alignment with specific content. PCK also requires structuring elements of content for optimizing teaching and learning of subject matter information. TCK is knowing how to change subject matter information into appropriate representations with the application of varied technologies. TPK involves an awareness of the availability, features, functions, and performances of various technologies as instructional tools for augmenting specific processes, practices, or methods of teaching and learning in specific subject-area context. TPCK is an awareness that effective teaching and learning of subject matter information can emerge from interweaving of content whose representations have changed through application of varied technologies and that support context-specific approaches for teaching and learning of subject specific content.

Current studies have examined pre-service teacher preparation in relation to Mishra and Koehler (2006) TPCK framework (see Açıkgül & Aslaner, 2020; Izgi-Onbasili et al., 2022; Meroño et al., 2021). More specifically, studies have examined technology integrated pre-service teacher preparation in relation to PCK, TCK, TPK, and TPCK. One such study is the Haley-Mize and Walker (2014) investigation which surveyed pre-service teachers' knowledge of teaching and technology before and after their participation in a teacher preparation course involving face-to-face technology enhanced learning experience, fully online instruction, or face-to-face instruction. The survey (Schmidt et al., 2009) tapped into pre-service teachers' TPACK. Pre-post survey results indicated improvement in pre-service teachers' TPACK when they were in a course that provided technology enhanced learning experience through creation of a digital space for collaborative learning, activities requiring integration of technology, collaborative wiki creation, and modeling digital tools.

Similarly, Harvey and Caro (2017) also examined pre-service teachers' TPACK through a pretest-posttest study. One group of pre-service teachers was prepared through technology integration in alignment with the TPACK framework, whereas the other group was not. As with the Haley-Mize and Walker (2014) investigation, the Harvey and Caro (2017) study also ascertained pre-service teachers' TPACK through a survey (Schmidt et al., 2009) which represented the TPACK framework. Again, the pre-service teachers who were engaged in tasks requiring reflection on their fieldwork experience and course experience in alignment with the TPACK framework had high scores on the posttest than the pre-service teachers whose learning did not align with the TPACK framework.

Compared to the quasi-experimental studies summarized above, Beuno and Niess (2023) examined development of TPACK amongst pre-service mathematics teachers through a qualitative cross-case study. These researchers conducted observations of pre-service mathematics teachers' experiences in their teacher preparation course and teaching practices in a high school mathematics classroom, with follow-up semi-structured interviews to track their TPACK development over a semester within the context of a redesigned *Mathematics Teaching Practice IV* course. The redesigned course involved the use of Information Communication Technology (ICT) for pre-service teacher preparation. Seven pre-service mathematics teachers participated in this qualitative case study. The seven pre-service teachers had been oriented to instructional technologies during their earlier teacher preparation courses. In the *Mathematics Teaching Practice IV* course, as the study and learn phase, the pre-service teachers learned to use Geo Gebra, Google Presentation, Padlet, Pear Deck, Kahoot, Mentimeter, and other technologies for teaching mathematical concepts through course instructor demonstrations and participation in collaborative activities. Following the study and learn phase, pre-service teachers were organized into three groups and assigned to teach mathematical concepts (Trigonometry – sine and cosine functions; Algebra – linear and quadratic functions; and Algebra – linear functions) in a high school class. Results indicated that Group 1 pre-service teachers had formed a favourable attitude toward teaching and learning mathematics with appropriate technology, although they were hesitant to let the high school students explore technology for learning mathematical concepts. Comparatively, Group 2 and Group 3 used the technology in their lessons and recognized its alignment with the mathematics content. Collectively, the seven pre-service teachers showed improvement in their TPACK based on their course experiences and their teaching practices in the high school mathematics class.

In keeping with the qualitative methodology, Srisawasdi (2012) also adopted a case study approach for examining the role of TPACK in preparing pre-service teachers for teaching physics. Three pre-service physics teachers' TPACK development was examined, corresponding to their participation in a case-based ICT learning module, a teaching internship course, and teaching methods in a high school physics classroom using computerized laboratory environments. The ICT module integrated the three foundational elements of the TPACK framework (Mishra and Koehler, 2006) for the preparation of the pre-service physics teachers' CK, PK, and TK. The subject matter content integrated in the ICT module was from nanotechnology, mathematics, physics, chemistry, and biology. Inquiry-based learning, problem-based learning, constructivist learning, and web-based learning were teaching methods used as pedagogy. Computer-based laboratory, computer simulation, web-based learning environment, web-based Artificial Intelligence, and Geographic Information System (GIS) were integrated as technology in the case-based ICT module. Pre-service physics teachers had to design tasks for teaching in a high school, one before and another after participation in the case-based ICT learning module. Pre-service teachers' tasks were analyzed in alignment with the seven elements of TPACK. Results showed that the three pre-service teachers' TPACK had advanced following their participation in the case-based ICT module. Their second task for teaching in a high school reflected higher TPACK competencies than the first task. More specifically, their competencies related to the interactive elements of the TPACK framework, specifically pedagogical content knowledge, technological content knowledge, and technological pedagogical knowledge, had progressed from perception level to conception level and action level.

Piotrowski and Witte (2016) also examined pre-service English language arts teachers' construction of TPACK in their teacher preparation course through the case study approach. These researchers examined the effectiveness of pre-service teachers' application of TPACK in creating technology-integrated flipped lessons. Nine pre-service English language arts teachers participated in this investigation related to technology-integrated teacher preparation. An adapted version of the Schmidt et al. (2009) survey was administered to determine pre-service teachers' TPACK. Data was collected through pre-post interviews, weekly online posts related to prompts and questions from the course instructor, flipped lesson videos, and written reflections. Results indicated that pre-service teachers benefited from creating technology-integrated lesson plans. Pre-service teachers' preparation for teaching English language arts reflected their ability to align their lessons with the overlapping knowledge categories of the TPACK framework.

Educational Psychology and TPCK Framework Alignment

The educational psychology course assignments align with the seven elements (CK, PK, TK, PCK, TCK, TPK, and TPACK) of the Mishra and Koehler (2006) framework. The educational psychology course has four required course components: class participation (25%), fieldwork assignment (25%), technology-integrated curriculum unit presentation (25%), and case study analysis (25%). This course is conducted through online synchronous or face-to-face/in-person modalities. Irrespective of the course delivery format, the above-identified course components are integral for the preparation of pre-service teachers in the educational psychology course. This chapter focuses exclusively on pre-service teachers' engagement in activities assigned as a part of their weekly class participation in the educational psychology course.

The class participation component requires pre-service teachers to engage in individual and collaborative activities pertaining to teaching and learning in grades 7-12 subject-area classes for fifteen class sessions, 2.5 hours each. These class participation activities require pre-service teachers to analyse classroom teaching-learning vignettes through videos, evaluate educational methods, materials, and measures on websites, review New York State Learning Standards by Content-Area through digital media, research instructional technologies on the World Wide Web/Internet, and design teaching-learning tasks with computer-based technologies for grades 7-12 subject-area classes. The class participation activities are based on pre-service teachers' knowledge of educational psychology theories, principles, frameworks, and concepts (CK), subject-area facts, concepts, and rules (CK), teaching-learning processes, practices, and approaches (PK); and features, functions, and performances of standard technologies (TK). The class activities call for application of educational psychology information (CK) as a foundation for anchoring subject-specific instructional tasks (PCK), which can be implemented and completed through the pathway of teaching-learning processes (PK) and integration of technologies (TK) corresponding with educational psychology information (PCK), subject-area information (PCK), and teaching-learning approaches aligned with educational psychology information and subject-area information, thereby resulting in advancement of TPCK. The interweaving of pre-service teachers' CK, PK, TK, and the emerging intersecting relationships, PCK, TCK, TPK, and TPCK, are represented as robust knowledge (Richey & Nokes-Malach, 2015), which is deep, connected, and coherent knowledge of teaching and learning developed through class participation activities.

Table 15.1 outlines class participation assignments and corresponding CK, PK, and TK emphasized during weekly class sessions. The fifteen activities displayed in Table 15.1 are drawn from assigned readings that contain key educational psychology theories, models, and concepts. The processes, practices, and approaches listed as pedagogies are also drawn from educational psychology. Because pre-service teachers enrolled in this course are from different teacher preparation programs (i.e., art, English, family consumer science, music, mathematics, physical education, science, social studies, TESOL, and world languages), the class activities are completed in subject-area groups. The technology integration listed in Table 15.1 are modes of representing teaching-learning processes, practices, and methods; however, the outcomes from collaboratively completed activities are illustrated and presented within the context of specific subject matters, as are the pedagogies and technologies. The activity completed for Class 13 required pre-service teachers to find a computer-based instructional application on the World Wide Web for promoting self-regulated learning of high school students in their teacher certification subject area. For example, the pre-service English language teachers identified "AI for Teachers," a computer-based interactive tutoring application (TK), which they paired with teaching a novel, *Ponyboy's Journey*, for the curriculum unit, *Heroism* (CK). Furthermore, CK and TK were connected to the phases of self-regulated learning (i.e., forethought, performance, reflection), the PK. Specifically, the subject matter, heroism (CK), wherein pre-service English language teachers were to facilitate high school students' self-regulated learning processes (PK) through application of the computer-assisted instructional application (TK) required students to a) compare Ponyboy's journey with their own personal journey of heroism, and b) create a new heroism journey based on similarities and differences between their own and Ponyboy's journey. The computer-assisted instructional application, AI Interactive Tutor, was to help high school students plan (forethought phase) their comparison of Ponyboy's journey with their own and creation of a new heroism journey (TPACK). A comparison chart was to be developed (performance phase), with AI Interactive Tutor guidance (TPK), to help students identify similarities and differences (PCK) between their own and Ponyboy's journey and create a new journey (TCK). The comparison of Ponyboy and his own journey and the creation of a new journey (PCK) is confirmed through the use of an AI Interactive Tutor (TPK) to ascertain whether the identified similarities and differences between Ponyboy's and his own journey (TCK) are accurate or questionable. Comparison of AI Interactive Tutor feedback and self-generated revisions (self-reflection) to the comparison chart and creation of Ponyboy and own journey is to provide validation for self-regulated learning about Heroism (TPACK).

Empirical Support for Technology-Integrated Teacher Preparation

The educational psychology course referenced in this chapter was offered as a synchronous online course. The Blackboard Learning Management System (LMS) was the Virtual Learning Environment (VLE) for pre-service teacher preparation in this graduate-level educational psychology course. The weekly virtual class sessions were conducted through Blackboard Collaborate Ultra. The collaborative tasks listed in Table 15.1 were completed by pre-service teachers in the VLE of Blackboard, specifically, in breakout groups on Blackboard Collaborate Ultra. Another feature, the Discussion Board, was to facilitate peer-to-peer and pre-service teachers-to-course instructor communication in the VLE. Furthermore, multimedia resources were shared through the Interactive Whiteboard of the Blackboard Collaborate Ultra.

Teng and Allen (2005) have examined the effectiveness of Blackboard LMS, including its communication and interactive features, boosting pre-service teachers' confidence in integrating technology in their future classes. This study had a pretest-post-test methodology of examining pre-service teachers' confidence in integrating technology in their future teaching. Participants were pre-service teachers from various teacher preparation programs (elementary education, secondary education, communication disorders, art education, music education, and special education), enrolled in four sections of an undergraduate educational psychology course. The pretest survey ascertained pre-service teachers' perception of computer skills and their ability to integrate technology in their teaching. The post-test survey focused on determining pre-service teachers' perception of whether Blackboard activities helped develop their computer skills and confidence about integrating technology in their teaching, and which Blackboard activities developed their computer skills, technology integration confidence and knowledge of educational psychology principles. Course instructors for the four sections of the educational psychology course and pre-service teachers enrolled in these sections were interviewed at the end of the semester to elicit pre-service teachers and their professors' thoughts about teaching and learning with the use of Blackboard. Blackboard was extensively used for discussions of educational psychology topics, to turn in assignments, and to access course materials.

Repeated measures ANOVA comparison of pretest-post-test survey responses from Teng and Allen (2005) study indicated that there was a significant difference in pre-service teachers' perception of their own computer skills and their confidence about technology integration, before and after the course. Results also showed that pre-service teachers' computer skills, including their use of word processing, PowerPoint, e-mail, listserv, educational software, World Wide Web/Internet, to both locate and present information, were significantly different before and after the course. Furthermore, pre-service teachers indicated that Blackboard was helpful for developing their computer skills and their confidence in technology integration in teaching. With regards to Blackboard activities that were beneficial for advancing computer skills and confidence in integrating technology for teaching, 53% to 92% of pre-service teachers provided high ratings for Discussion, Lecture Notes, PowerPoint presentations, technology assignments, case studies, and external website links as beneficial. The benefits of Blackboard were further corroborated with interview responses from pre-service teachers and their course instructors, wherein Course Content (lecture notes, PowerPoint presentations), Assignments (technology assignments, case studies), External Links (resources or assignments), Discussion Board, Gradebook, E-mail, and Announcements were once again identified as useful for teaching.

Findings from Teng and Allen (2005) provide evidence-based support for integrating Blackboard LMS for pre-service teacher preparation in this chapter's educational psychology course. In addition to using Blackboard features such as Course Content, Assignments, External Links, Discussion Board, Gradebook, E-mail, and Announcements, which were identified as useful for teaching in the Teng and Allen study, the educational psychology course showcased in this chapter utilized a key function included within the Blackboard Collaborate Ultra, breakout groups. Because the focal educational psychology course is student-centred, all 15-week class sessions involve collaborative groupwork. Breakout groups are created, and pre-service teachers are assigned to subject-area groups for collegial discussions, debates, and designing of instructional activities, practices, and assessments. The course instructor visits each breakout group to moderate and facilitate their discussions, debates, and designs. Groupwork is illustrated on the Interactive Whiteboard of Blackboard Collaborate Ultra. Groups utilize technologies such as the World Wide Web/Internet, Multimedia Tools, Digital Databases, Computer-Based Applications, and Computer Application Software to produce and present their collaboratively generated subject matter activities, approaches, and assessments. Group-generated and presented products are posted on the Discussion Board as a repository of resources, wherein peers can post questions, comments, and evaluations, moderated by the course instructor. Technology-integrated student-centred preparation of pre-service teachers through the Breakout Group feature of the Blackboard Collaborate Ultra is part of all 15 class activities listed in Table 15.1.

To substantiate the narrative provided in this section about collaborative construction of instructional activities, approaches, and assessments in the Breakout Group of Blackboard Collaborate Ultra, the activity assigned and completed by pre-service teachers during Class 11 of the synchronous educational psychology course is represented herein. Class 11 was based on the educational psychology topic, Differentiated Instruction, a teaching approach for students with diverse learning needs. Reading materials and PowerPoint slides related to differentiated instruction and adaptive teaching were uploaded to the Blackboard under the menu item Course Content for Class 11. The collaborative activity illustrated in Figure 15.1 was posted on the Discussion Board of Blackboard. Prior to their collaborative groupwork, pre-service teachers were expected to watch a videotaped lesson from their subject area. The hyperlinks for the video-taped lessons for the secondary education subject areas were posted on the Course Content of Blackboard. For example, prior to their collaborative groupwork, pre-service art teachers were asked to watch this 11-minute video, *Figure Drawing in the Middle School Art Classroom* https://youtu.be/tM_YCcHzqqQ. Following the viewing of their subject-area videotaped lessons, pre-service teachers were instructed to join their breakout groups for collaborative work. Groups had 30 minutes to discuss and design their subject-area differentiated instruction activity. At the end of their collaborative work, pre-service teachers posted their product on the Discussion Board of Backboard and presented their product, including supplementary materials, using the Interactive Whiteboard of Blackboard Collaborate Ultra. Group presentations were followed up with class reaction to the presentations, and as per assigned clusters, groups posted their comments and questions corresponding to their assigned group's presentation on the Discussion Board. Pre-service teachers' responses to the peer group presentations indicated that they had benefited from this class activity. Specifically, they have a good understanding of the differentiated instruction approach for teaching high school students with different learning needs. Figure 15.2 illustrates Art Group's collaborative work for Differentiated Instruction.

Video integration has been pivotal in the educational psychology course examined in this chapter. As has been exemplified above, videotaped subject-area lessons served as a springboard for engaging pre-service teacher candidates in an activity on Differentiated Instruction. Similarly, other class activities also integrated video as a form of multimedia resource for pre-service teacher preparation in the educational psychology course. Video-based teacher preparation was also embedded in class activities aligned with Piaget's theory of cognitive development, Vygotsky's sociocultural theory, Intelligence, Baumrind's parenting, Banks' multicultural education, Skinner's theory of behaviourism, and Weiner's attribution theory. Video cases were implemented as a technology in teaching educational psychology to future teachers.

The effectiveness of video cases as an element of multimedia technologies has been examined in Anderson's (1998) study, specifically, the integration of multicultural multimedia materials in an undergraduate educational psychology course. Although the purpose of this study was to increase pre-service teachers' sensitivity towards people from cultures other than their own and to solicit their reaction to integration of computer software, videodiscs, videotape, and print media in a multicultural unit, students were to apply learning theories from educational psychology in teaching and model integration of technology in teaching. Multimedia was integrated through three technologies. Multimedia HyperCard Stack, which provided information about diverse cultures, was used in alignment with the Dimensions of Multiculture. It provided pre-service teachers with opportunities to add commentaries and create their own sequence of information. Decisions: Prejudice, a computerized simulation, was used to provide pre-service teachers the opportunity to collaboratively learn and talk about prejudice and discrimination and explore the processes that propagate their occurrences. GTV software and laserdisc enabled pre-service teachers to explore the diversity of American life and perspectives of different racial and ethnic groups. Analysis of pre-service teachers' responses to before and after questionnaires, as well as informal observation during the multicultural unit, indicated that their understanding of learning theories increased during the unit due to visualization of learning theories and participation in application activities. Pre-service teachers' responses about multimedia indicated that integration of varied technologies was effective in learning about the multicultural unit. Specifically, computer simulation was deemed effective because it made learning interesting and fun, enabled peer interaction, and required attention and active participation. Pre-service teachers also found videos helpful because they represented real schools and real classroom situations.

As with the Anderson study (1998), integration of videos in the educational psychology course highlighted in this chapter also provided pre-service teachers with opportunities to apply educational psychology concepts through peer interactions in relation to real classroom situations. Furthermore, as has been reported in current literature, digital media also serves as a platform for promoting meaningful peer interactions and enhancing educational outcomes. Full participation in digital forums has been noted to have resulted in the sharing of ideas and democratic discourse (Buckley-Marudas, 2016). By engaging pre-service teachers in planning, designing, and demonstrating educational psychology theories, models, and concepts in alignment with subject-area curriculum in this teacher preparation course, the anticipation is that pre-service teachers will sustain the TPACK for instructing diverse secondary school students.

In alignment with Anderson's (1998) findings, Myers and Rivero (2019) also contend that simulation in teacher education courses has the potential to re-create virtual classrooms to allow pre-service teachers to experience classroom dynamics and to experiment with teaching practices. Pre-service teachers also analyse simulated classroom situations in the educational psychology course through the application of relevant pedagogical approaches. The goal is to develop pre-service teachers' TPACK for teaching secondary school students in their subject-area classes. Classes 2, 3, and 8 of the educational psychology courses (see Table 15.1), for instance, engaged pre-service teachers in case analysis related to Piaget' logical thinking, Vygotsky's mediated learning, and Skinner's Operant Conditioning through integration of digital media (i.e., websites, computer software, e-books, databases) and media resources (i.e., satellite links and library collections of multimedia, film, and video). Thus, the technology-supported collaborative assignments provide a safe space for pre-service teachers to engage in discussion of educational psychology theories, principles, frameworks, and concepts as preparation for teaching secondary school students (Bhattacharya, 2019).

Technology Integrated Unit Presentation in Educational Psychology

The effectiveness of technology-integrated teacher preparation in the educational psychology course is determined with a transfer task. Towards the end of the semester, pre-service teachers present a collaboratively designed technology-integrated unit. The technology-integrated unit presentation requires triangulation of the subject-area curriculum unit (CK), educational psychology theme (PK), and digital media/multimedia resources (TK). Pre-service teachers post their technology-integrated unit presentation on the Discussion Board. The posted presentations are evaluated by the educational psychology class members and the course instructor with a grading rubric. Table 15.3 illustrates the grading rubric for the technology-integrated unit presentation. Peers and the instructor also provide written feedback for the presentations. The presentations posted on the Discussion Board are Microsoft PowerPoint files, with audio-video narration embedded in the slides. The presentations illustrate teaching, learning, and assessment processes within the context of a subject area unit of study, with integration of digital media (computer application, database, website) and multimedia (video, film, animation), and an educational psychology theme. Table 15.4 displays the alignment of educational psychology theme, subject-area unit, and technology application represented in the pre-service teachers' presentation. There are usually twenty-five pre-service teachers enrolled in each section of the educational psychology course. Usually, there are four sections of the educational psychology course, two sections conducted face-to-face/in-person, and two sections synchronously online. Pre-service teachers' presentations usually range from Excellent to Proficient as per rubric-based peer and course instructor evaluation. Therefore, it is safe to conclude that pre-service teachers can effectively integrate technology within their subject-area unit of instruction, with processes, practices, and approaches for teaching, learning, and assessment drawn from educational psychology. Pre-service teachers' technology integrated curricular units reflect appreciable alignment with CK, PK, PCK, TCK, TPK, TPACK of Mishra and Koehler (2006) framework.

Implications of Technology Integrated Pre-service Teacher Preparation

Table 15.4 outlines studies that have examined the relationship between teacher preparation and TPACK. Most studies listed in Table 15.4 that have investigated pre-service teachers' TPACK are qualitative case studies, 6 of 9 studies, with the number of participants ranging from 7 (i.e., Bueno & Niess, 2023) to 59 (i.e., Ersanli, 2016). Only three studies (i.e., Baran et al., 2023 [N = 215]; Lachner et al., 2021 [N = 208]; Max et al., 2023 [N = 495]), which integrated a quantitative methodology, have a large participant pool and experimental or quasi-experimental design. Furthermore, the measures used for collecting data have mostly been questionnaires, interviews, and observations. Finally, although all identified studies are about pre-service teacher preparation through technology integration, none have examined pre-service teacher preparation in educational psychology, in relation to subject-area teaching of secondary school students.

An analytical review of the studies listed in Table 15.4 suggests that future studies examining technology-integrated pre-service teacher preparation for teaching in secondary schools should consider an experimental or a quasi-experimental design, with a large number of participants, to ensure generalizability of findings. Another idea is to examine the intersecting relationships between educational psychology themes, subject-area content, and technology application for advancing pre-service teachers' TPACK and technology-integrated teaching in secondary schools. This is because searches of various digital databases (ScienceDirect, Google Scholar, PsycINFO, Web of Science, Scopus, and One Search) did not produce results, other than the two studies reviewed in this chapter (Anderson, 1998; Teng & Allen, 2005), that examined technology-integrated pre-service teacher preparation in an educational psychology course. Even the two reviewed studies did not connect educational psychology themes with secondary school subject areas for advancing technology-integrated pre-service teacher preparation.

Table 15.1 Technology Integrated Class Participation Activities for Pre-Service Teacher Preparation.

Sessions	Activity	Content	Pedagogy	Technology
Class 1	Create a task for Professional Practices	Danielson Framework for Teaching	Creative Thinking	World Wide Web Virtual Whiteboard Discussion Board
Class 2	Case Analysis of Logical Thinking	Piaget's Basic Assumptions	Critical Thinking	Multimedia Tools Virtual Whiteboard Discussion Board
Class 3	Case Analysis of Sociocultural Processes	Vygotsky's Basic Assumptions	Reasoning	Multimedia Tools Virtual Whiteboard Discussion Board
Class 4	Conduct Research on Brian Imaging Methods	Intelligences	Analytical Thinking	Multimedia Tools Digital Database Discussion Board
Class 5	Complete and Analyse Brain Processing Test	Information Processing Theory of Memory	Logical Thinking	World Wide Web Virtual Whiteboard Discussion Board
Class 6	Complete and Analyse Parenting Styles Survey	Baumrind's Parenting Styles Framework	Analytical Thinking	Multimedia Tools World Wide Web Discussion Board
Class 7	Complete and Analyse Identity Questionnaire	Marcia's Identity Statuses	Critical Thinking	World Wide Web Virtual Whiteboard Discussion Board
Class 8	Analysis of Classroom Behaviours	Skinner's Operant Conditioning	Analytical Thinking	Multimedia Tools Virtual Whiteboard Discussion Board
Class 9	Research and Demonstrate Technologies	Assistive Technology Digital Technology Programmed Learning	Discovery Learning	World Wide Web Digital Database Discussion Board
Class 10	Write Objectives Design Rubrics	Bloom's Taxonomy: Cognitive/ Psychomotor	Algorithmic Procedure	World Wide Web Virtual Whiteboard Discussion Board

Class 11	Illustration of Differentiated Instruction	Differentiation Adaptive Teaching	Problem Solving	Multimedia Tools Virtual Whiteboard Discussion Board
Class 12	Exemplar of Multicultural Education	Banks' Multicultural Education Dimensions	Analogical Thinking	Multimedia Tools Virtual Whiteboard Discussion Forum
Class 13	Research and Demonstrate Technologies	Zimmerman's Self-Regulated Learning Model	Discovery Learning	World Wide Web Digital database Discussion Board
Class 14	Complete/ Analyse Causal Dimension Scale II	Weiner's Attribution Theory of Motivation	Reasoning	World Wide Web Multimedia Tools Discussion Board
Class 15	Explore and Illustrate Digital Simulation/Game	Performance and Portfolio Assessment	Transfer of Learning	World Wide Web Digital Database Discussion Board

Table 15.2 Technology Integrated Curriculum Unit Presentation Themes, Topics, and Applications.

Art Education: Educational Psychology	**Art Education: Curriculum Unit**	**Art Education: Technology Application**
Goal Orientations	Shapes in art	Procreate
Culturally Relevant Pedagogy	Norton design	Drawing desk
Differentiated Instruction	Perspective	Monument Valley
Procedural Knowledge	Digital art	Fire Alpaca
Analytical Thinking	Life drawing	Magic Poser
Guided Participation	Figure drawing	Pose Croquis
Portfolio Assessment	Painting - landscapes	Terragaen
Multicultural Education	Comic books	World Anvil
English Education: Educational Psychology	**English Education: Curriculum Unit**	**English Education: Technology Application**
Culturally Relevant Pedagogy	Cultural concept: Heroism	Brainstorm
Portfolio Assessment	Story writing	StoriumEdu
Critical Thinking	Novel	Blooket
Transfer Learning	Argumentative essay	Animoto

Procedural Knowledge	Media literacy	Powtoon
Differentiated Instruction	Figurative language	Picktochart
Multicultural Education	Short stories	Jamboard
Guided Participation	Symbolism in poetry	Symbol-it
Math Education: Educational Psychology	**Math Education: Curriculum Unit**	**Math Education: Technology Application**
Analytical Thinking	Polynomials	DeltaMath
Problem-Solving	Solving equations	Cymath
Critical Thinking	Quadratic equations	Desmos
Goal Orientations	Line of best fit	Illuminations NCTM
Guided Participation	Factoring Trinomials	Phet Colorado
Differentiated Instruction	Angles	Sketchometry
Transfer Learning	Transformations	GeoGebra
Procedural Knowledge	Inverse functions	Pear Deck
Science Education: Educational Psychology	**Science Education: Curriculum Unit**	**Science Education: Technology Application**
Problem-Solving	Matter and Energy	CellEnergy Photosynthesis
Transfer Learning	Gene Expression	NOVA Labs
Critical Thinking	Biomolecules	Classcraft
Goal Orientations	Climate	Earth-Now
Differentiated Instruction	Landscapes	EarthViewer
Guided Participation	Minerals and Rocks	Gizmos
Spanish Education: Educational Psychology	**Spanish Education: Curriculum Unit**	**Spanish Education: Technology Application**
Culturally Relevant Pedagogy	Health	Wordwall
Guided Participation	Pronouns	Edpuzzle
Differentiated Instruction	News	Socrative
Discovery Learning	Supermarket	Quizziz
Transfer Learning	Clothing	Flippity
Multicultural Education	Seasons	Ling

Table 15.3 Peer and Instructor Grading Rubric for Technology Integrated Curricular Unit.

Presentation mechanics Presentation is of required duration and well-paced; style is engaging; illustrations are compatible; embedded audio-video narrative is viewable and coherent	**Digital resources:** Group presents digital resources that align with curricular content, educational psychology concepts, and technology application	**Assessment:** Group illustrates technology integrated assessment of student learning curricular content	**Instructional approaches and Activities:** Group explains pedagogical approaches and learning activities for technology integrated subject specific unit of instruction	**Content Connections:** Group explains alignment of subject area content with educational psychology concepts and technology tools	**Introduction** Group discusses the relevance of selected educational psychology theme and technology for curriculum unit	
Presentation mechanics are impressive	Resources are remarkable	Assessment is spectacular	Activities and approaches are impressively illustrated	Connections are extensive	Discussion is creditable	**Exemplary**
Presentation mechanics are good	Resources are valuable	Assessment is credible	Activities and approaches are efficiently illustrated	Connections are inclusive	Discussion is substantial	**Proficient**
Presentation mechanics are acceptable	Resources are appropriate	Assessment is adequate	Activities and approaches are adequately illustrated	Connections are adequate	Discussion is average	**Acceptable**
Presentation mechanics are marginal	Resources are basic	Assessment is limited	Activities and approaches are minimally illustrated	Connections are limited.	Discussion is sparse	**Emerging Proficiency**
Presentation mechanics are poor	Resources are trivial	Assessment lack clarity	Activities and approaches are inadequate	Connections are vague	Discussion lacks specificity	**Needs Improvement**
Not included	Not included	Not included	Not included	Not included	Not included	**Missing**

Table 15.4 Studies Investigating Pre-Service Teacher Preparation Aligned with TPACK Framework.

Studies	Number of Pre-service Teachers (grades 7-12)	Research Methodology	Data Source	Teacher Preparation Program (7-12 grades)	Technology Integration
Baran et al. (2023)	215	Correlational Research Design	Synthesis of Qualitative Evidence (SQD) Scale TPACK Practical Scale	English language Mathematics - Physical Education - Social Studies - Science	None
Bueno & Niess (2023)	7	Qualitative Case Study	Synthesis of Qualitative Evidence (SQD) Scale Observation Interview	Mathematics	- Pear Deck - Kahoot - Gartic - Geo Gebra
Caniglia & Meadows (2018)	51	Qualitative Coding (Open & Axial Coding)	Online Survey Lesson Plans	Mathematics	Websites
Chittleborough (2014)	28	Case Study	Unit Teaching Material Observation Survey Interview	Chemistry	Multimedia Whiteboard
Ersanli (2016)	59	Quantitative (T-test) Qualitative (Constant Comparison Method)	- TPACK Scale - Journal	- English language	None
Kohnke et al. (2024)	32	Qualitative Case Study	Questionnaire (N = 32) Interview (N = 12) Observation (N = 12)	English language	Mentimeter Podcast Moodle Canva Edpuzzle Quizlet
Lachner et al. (2021)	208	Quasi-Experimental Field Study	Self-efficacy Questionnaire Conceptual TPK-scale Situational TPK Test TPACK-tests SQD-Questionnaire	EFL Biology Mathematics Philosophy German	Digital Media Geo Gebra

Maeng et al. (2013)	27	Inductive Data Analysis	Interview Observation Lesson plan Artifacts Reflection	Science	Simulation
Max et al. (2023)	495	Structural Equation Modeling	Questionnaire	Science	Digital Media Makerspace

Figure 15.1 Interactive Activity Integrating Breakout Group Function of Blackboard Collaborate.

Class 11: November 13 **Topic: Differentiated Instruction**

<u>Group Activity</u>

This activity must be completed through collaborative discussion in your secondary education subject-area groups:

Group 1: Art
Group 2: English
Group 3: Mathematics
Group 4: Social Studies
Group 5: Science
Group 6: World Languages

1. Describe a high-level, a middle-level, and a low-level task to illustrate "Differentiated instruction" by adapting a) content, b) process, and c) product of instruction in your teacher certification subject-area for a 7-12 grade class.

2. Present your *Differentiated Instruction* activity to the class via the Virtual Whiteboard of Blackboard Collaborate Ultra. Each member of your group must present a segment of the Differentiated Instruction activity.

3. Post your group designed *Differentiated Instruction* activity, including all supplementary instructional materials, on the Discussion Board of Blackboard at the end of this class. Remember to include a) Activity Topic – Differentiated Instruction; b) Activity Date – November 13; and c) Group Member Names – First and Last Name of each person.

4. Post your group designed *Differentiated Instruction* activity, including all supplementary instructional materials, on the Discussion Board of Blackboard at the end of this class. Remember to include a) Activity Topic – Differentiated Instruction; b) Activity Date – November 13; and c) Group Member Names – First and Last Name of each person.

5. Respond to your peer groups with comments and questions as a discussion thread.
   a. Group 1 responds to Group 2/Group 2 counter responds to Group 1
   b. Group 2 responds to Group 3/Group 3 counter responds to Group 2
   c. Group 3 responds to Group 4/Group 4 counter responds to Group 3
   d. Group 4 responds to Group 5/Group 5 counter responds to Group 4
   e. Group 5 responds to Group 6/Group 6 counter responds to Group 5
   f. Group 6 responds to Group 1/Group 1 counter responds to Group 6

Figure 15.2 Art Education Pre-Service Teachers' Collaborative Work on Differentiated Instruction.

Activity: Intro to shading (7th grade)

Content: Shading an image of a pear

Assignment: Understanding value through shading

Preassessment: All students shade a gradient values worksheet to show their baseline understanding of shading

Level 1 Differentiation:
Content: A black and white image of a drawn pear is provided.
Process: Students must accurately copy the image of the drawn pear with a HB pencil.
Product: Students must reproduce pear image with one tome in 1 period.

Level 2 Differentiation:
Content: A black and white image of a real pear is provided.
Process: Students must accurately copy the image of a real pear with HB and 3B pencils.
Product: Students must reproduce pear image with two tones in 1 period.

Level 3 Differentiation:
Content: A colored image of a real pear is provided.
Process: Students must accurately copy the colored image of a real pear with multiple toned pencils.
Product: Students must reproduce pear image with multiple tones and highlights in 1 period.

References

Açıkgül, K., & Aslaner, R. (2020). Effects of GeoGebra supported micro teaching applications and technological pedagogical content knowledge (TPACK) game practices on the TPACK levels of prospective teachers. *Education and Information Technologies, 25*(3), 2023–2047. https://doi.org/10.1007/s10639-019-10044-y

Anderson, S. E. (1998). Integrating multimedia multicultural materials into an educational psychology course. By: *Journal of Technology and Teacher Education,* 6 (2-3),

Baran, E., Canbazoglu Bilici, S., Albayrak Sari, A., & Tondeur, J. (2019). Investigating the impact of teacher education strategies on pre-service teachers' TPACK. *British Journal of Educational Technology, 50*(1), 357–370. https://doi.org/10.1111/bjet.12565

Bhattacharya, A. (2019). Preparing pre-service teacher candidates for educating culturally and linguistically diverse learners. In A. Slapac & S. A.

Coopersmith (Eds.), *Beyond language learning instruction: Transformative supports for emergent bilinguals and educators* (pp. 76-103). IGI Global. https://doi.org/10.4018/978-1-7998-1962-2.ch004

Buckley-Marudas, M. F. (2016). "Truth" interrupted: Leveraging digital media for culturally sustaining education. *Multicultural Learning and Teaching, 12*(2), 1-19. https://doi.org/:10.1515/mlt-2015-0001

Bueno, R. W. da S., & Niess, M. L. (2023). Redesigning mathematics pre-service teachers' preparation for teaching with technology: A qualitative cross-case analysis using TPACK lenses. *Computers and Education, 205,* 104895-. https://doi.org/10.1016/j.compedu.2023.104895

Caniglia, J., & Meadows, M. (2018). Pre-service mathematics teachers' use of web resources. *International Journal for Technology in Mathematics Education,* 25 (3), 17–34. https://doi.org/10.1564/tme_v25.3.02

Chittleborough, G. (2014). Learning how to teach chemistry with technology: Pre-Service teachers' experiences with integrating technology into their learning and teaching. *Journal of Science Teacher Education, 25*(4), 373–393. https://doi.org/10.1007/s10972-014-9387-y

Council for Accreditation for Educator Preparation (2022). CAEP Revised 2022 Standards Workbook. Washington, DC: Author. https://caepnet.org/~/media /Files/caep/accreditation-resources/caep-2022-standards-workbook-final. pdf?la=en

Council of Chief State School Officers. (2013, April). Interstate Teacher Assessment and Support Consortium InTASC Model Core Teaching Standards and Learning Progressions for Teachers 1.0: A Resource for Ongoing Teacher Development. Washington, DC: Author. https://ccsso.org/sites/default/files/ 2017-12/2013_INTASC_Learning_Progressions_for_Teachers.pdf

Ersanli, C. Y. (2016). Improving Technological Pedagogical Content Knowledge (TPACK) of pre-service English language teachers. *International Education Studies, 9*(5), 18-27. https://doi.org/10.5539/ies.v9n5p18

Eutsler, L., & Long, C. S. (2021). Pre-service teachers' acceptance of virtual reality to plan science instruction.*Educational Technology & Society, 24*(2), 28–43. https://doi.org/10.30191/ETS.202104_24(2).0003

Haley-Mize, S., & Walker, D. The effect of instructional methodology on pre-service educators' technological, pedagogical, and content knowledge. *The International Journal of Learning in Higher Education, 20*(3), 13-25. https://doi.org/10.18848/1447-9494/CGP/v20i03/48697

Harvey, D. M., & Caro, R. (2017). Building TPACK in pre-service teachers through explicit course design. *TechTrends,61,* 106–114. https://doi.org/10.10 07/s11528-016-0120-x

Izgi-Onbasili, U., Avsar-Tuncay, A., Sezginsoy-Seker, B., & Kiray, S. A. (2022). An examination of pre-service teachers' experiences in creating a scientific digital story in the context of their self confidence in technological pedagogical content knowledge. *Journal of Baltic Science Education, 21*(2), 207-223. https://doi.org/10.33225/jbse/22.21.207

Kohnke, L., Foung, D., Zou, D., & Jiang, M. (2024). Creating the conditions for professional digital competence through microlearning. *Educational Technology and Society,* 27(1), 183-197. https://doi.org/10.30191/ETS.202401_27(1).SP05

Lachner, A., Fabian, A., Ulrike Franke, U., Preiß, J., Jacob, L., Führer, C., Küchler, U., Paravicini, W., Randler, C., & Thomas, P. (2021). Fostering pre-service teachers' Technological Pedagogical Content Knowledge (TPACK): A quasi-experimental field study. *Computers and Education, 174*, 1-14. https://doi.org/10.1016/j.compedu.2021.104304

Maeng, J. L., Mulvey, B. K., Smetana, L. K., & Bell, R. L. (2013). Pre-service teachers' TPACK: Using technology to support inquiry instruction. *Journal of Science Education and Technology, 22*(6), 838–857. https://doi.org/10.1007/s10956-013-9434-z

Max, A. L., Weitzel, H. & Lukas, S. (2023) Factors influencing the development of pre-service science teachers' technological pedagogical content knowledge in a pedagogical makerspace. *Frontiers in Education*, 8:1166018. https://doi.org/10.3389/feduc.2023.1166018

Mishra, P., & Koehler, M. J. (2006). Technological pedagogical content knowledge: A framework for teacher knowledge. *Teachers College Record (1970), 108*(6), 1017–1054. https://doi.org/10.1111/j.1467-9620.2006.00684.x

Meroño, L., Calderón, A., & Arias-Estero, J. L. (2021). Digital pedagogy and cooperative learning: Effect on the technological pedagogical content knowledge and academic achievement of pre-service teachers. *Revista de Psicodidáctica (English Ed.), 26*(1), 53–61. https://doi.org/10.1016/j.psicoe.2020.10.002

Myers, J. P., & Rivero, K. (2019). Preparing globally competent pre-service teachers: The development of content knowledge, disciplinary skills, and instructional design. *Teaching and Teacher Education 77*, 214-225. https://doi.org/10.1016/j.tate.2018.10.008

New York State Education Department (2011). New York State Teaching Standards. Office of Higher Education. Albany, NY. https://www.highered.nysed.gov/tcert/pdf/teachingstandards9122011.pdf

Piotrowski, A., & Witte. S. (2016). Flipped learning and TPACK construction in English education. *International Journal of Technology in Teaching and Learning, 12*(1), 33-46.

Richey, J. E., & Nokes-Malach, T. J. (2015). Comparing Four Instructional Techniques for Promoting Robust Knowledge. *Educational Psychology Review, 27*(1), 181–218. https://doi.org/10.1007/s10648-014-9268-0

Schmidt, D. A., Baran, E., Thompson, A. D., Mishra, P., Koehler, M. J., & Shin, T. S. (2009) Technological Pedagogical Content Knowledge (TPACK): The development and validation of an assessment instrument for pre-service teachers. *Journal of Research on Technology in Education, 42*(2), 123-14.

Srisawasdi, N. (2012). The role of TPACK in physics classroom: Case studies of pre-service physics teachers. *Procedia, Social and Behavioral Sciences, 46*, 3235–3243. https://doi.org/10.1016/j.sbspro.2012.06.043

Teng, Y., & Allen, J. (2005). Using Blackboard in an educational psychology course to increase pre-service teachers' skills and confidence in technology integration. *Journal of Interactive Online Learning, 3*(4), 1-12.

Chapter Sixteen

A World Apart or Same Difference? Reflections on Facilitating Online Global Creative Projects in Music Teacher Education

Pauline Black

University of the Highlands and Islands (UHI) Orkney, Scotland, UK

Emily Wilson

Faculty of Education, University of Melbourne, Australia

Abstract

Drawing on participatory culture in media education (Jenkins, 2006), collaborative co-construction of digital music projects using Green's (2002, 2008) informal music learning pedagogy can contribute to the development of student voice (Spruce, 2015). Similarly, digital music technologies provide a powerful means to support active, creative music making for students at all levels of prior musical experience (Kuhn & Hein, 2021). This chapter shares reflections based on collaborative self-study research (Samaras, 2011) that we undertook as music teacher-educators. We examine our experiences of facilitating two iterations of a global online creative music and video creation project and share insights that others might draw on. We discuss the following: implementation guidelines, benefits and challenges of the projects, pedagogy and frameworks used: Technological Pedagogical and Content Knowledge (TPACK) and Substitution, Augmentation, Modification, Redefinition (SAMR). We also unpack our choices in relation to the Digital Audio Workstation (DAW) that we chose, Soundtrap for Education. We argue for greater attention in teacher education to build confidence and competence in using online digital collaborative tools. Our experiences suggest that facilitating real-world creating projects may make teacher education meaningful for all through nurturing creativity and collaborative experiences using digital tools. Our research also highlighted the affordances of digital collaborative music technology tools to facilitate interdisciplinary, real-world creating projects for students at all levels of education, which can also make a potential contribution to sustainable development education more broadly.

Keywords: technology integration, TPACK Framework, Self-Study, teacher preparation.

***

Introduction

This chapter reports on lessons learned from two collaborative online international digital composing projects undertaken with music teacher education students in Melbourne, Australia and Aberdeen, Scotland. The projects were designed using Soundtrap for Education, a cloud-based digital audio workstation (DAW). 10,427 miles and 11 hours apart, students worked in groups with a mix of Australia and Scotland in each group to create interdisciplinary music and video artworks. In our work as music teacher-educators in Scotland and Australia, we facilitated the first iteration of this project during the 2021 COVID-19 lockdown entitled 'My Life in Isolation: A World Apart or Same Difference?' In 2022, we followed this with 'Caring for our planet: A World Apart or Same Difference?' responding to the UN Sustainable Development Goals (UNESCO, 2020), which represent a call to action for teacher-educators to embed climate consciousness in teaching and learning, amongst other areas of responsibility.

We begin by introducing the two iterations of our project. Our reflections as teacher-educators have led to some vibrant discussions on a wide range of topics, including attitudes, reactions, access, well-being, and ultimately, pedagogy and impact. We both very much enjoy designing and realising creative, real-world music projects, and these have been well received by our students and colleagues. We hope that by sharing some of our reflections that readers may gain some insight into the opportunities and challenges that arose, which may be helpful to design new projects across diverse contexts and disciplines.

Our Project

We begin by giving a brief overview of the process. In designing and facilitating these projects, we had several aims: to develop our students' skills and confidence to use creative as opposed to recreative music making in their teaching practice; to develop technological pedagogical skills; to support engagement and wellbeing, and, in the second iteration of the project, an additional aim was to promote climate awareness through music making.

The first iteration of our project in 2021 emerged in response to the extended lockdowns and prolonged remote learning that we, and our students, collectively experienced on opposite sides of the world. Using digital collaborative music and video creation was successful in addressing pedagogical aims. There were also outcomes related to well-being and engagement – the favourite part of the project for our students was making new friends on opposite sides of the world and seeing what their lives looked like. We designed the second iteration of the project in 2022 in response to the COP-27 UN Climate Change Conference in Sharm-el-Sheikh. We aimed to use music and video creation as activism to promote climate change awareness and embed the UN Sustainable Development Goals (UNESCO, 2020), education for sustainability (Østergaard, 2019) and beyond into regenerative practice (Armon, 2021) into our programs. From a teacher education perspective, we wished for our students to experience a real-world online collaborative creating project as learners so that they could then adapt it and facilitate a similar project with upper primary and secondary school students on placement as pre-service teachers or as graduate in-service teachers. To measure the impact of the projects, in our research (Wilson & Black, 2024; Black & Wilson, 2024), we investigated engagement (Fredricks et al., 2004), well-being outcomes (Seligman, 2018), knowledge of digital technology, collaboration and climate consciousness. We undertook this qualitatively using observation and thick description, but it could also be undertaken quantitatively.

Examples

We have two examples to share: First, we have a music and video creation from version one of the project, composed by Andy Turner, Calum Wood, Channelle Thomas, Kaitlin White, and Danna Yun.[1] Next, we present an example from the second iteration of the project composed by Holly Brow, Alastair Eddie, Hannah Rundell, Wenqi (Alex) Sun, and Ryan Hayward.[2] These examples serve to illustrate the themes that will be discussed in the rest of the chapter.

The Process

After identifying the overarching theme, we jointly devised a project brief. This contained a suggested process which was devised to try and pre-empt any technological and organisational challenges that we could foresee arising. Organisation and logistics were carefully considered and planned to avoid as many challenges as possible and ensure the successful completion of each

---

[1] Link to video file: https://youtu.be/h3zFNCL5Cew?list=PLgi37zowFl5X62fqrhDFbrZspVvmKj9YB

[2] Link to video file: https://youtu.be/O-QcdFaOF9I

group's project. Especially during the first project, we were concerned about time management aspects, knowing that students' lives were challenging, and we weren't necessarily aware of what was going on for them at home. While the students were not involved in conceiving the initial project, they had to subsequently plan their own group's project.

We began by meeting with the participants (85 and then 60) on Zoom to introduce the project and Soundtrap for Education via an online workshop. Finding a time period that aligned with our different time zones, different academic years, and when neither group of pre-service teachers was on placement was extremely challenging. In each version of the project, there was one cohort of students thrown 'in at the deep end' (Green, 2008, p. 25) and tackling this at the outset of their course when they did not really know each other, let alone know their international peers. We both used 3 hours of asynchronous class time to offer further support with Soundtrap for Education, along with composing and project design guidance. We had also prepared a bank of short tutorial videos so that students could work independently in between classes, to build confidence using the DAW. We also signposted our students to the online Soundtrap Educator Certification program, which most students completed, free of charge, to gain further professional learning.

After the initial workshop session, the students worked in groups of six, organised by us, with a mixture of Aberdeen and Melbourne students and accents in each group. Four is usually our preferred group size, but we deliberately made the groups slightly larger in case there were any reluctant participants – particularly during lockdown, it was difficult to know how students were managing at home; therefore, creating a supportive environment was important. Students had four weeks to complete their projects. They all gathered video of their local environment that was meaningful to them in response to the theme, using mobile phones. Then they edited their footage in whichever way they chose into 1-2 minutes to depict their collective experience of life in isolation (2021) and caring for the planet (2022). Some members of the group took more of a lead on video editing, and others fell more into organisational, compositional, or administrative roles. Once the videos were created, the students collaboratively created music to accompany their footage – this draws on the process that film composers use to create soundtracks.

The culmination of the project was a 'watch party' (Kuepper-Tetzel & Nordmann, 2021) in keeping with digital media culture (Cremata & Powell, 2017), five weeks after the project launch, where the final product was a YouTube playlist rather than an instrumental performance in a concert venue. At this sharing celebration, we sat in our respective classrooms projecting the videos and participants in their 'Zoom windows' simultaneously on the big screens and responding in real time via the chat function. Following the screening, student reflections were compiled digitally using Menti, an online polling tool.

The deadline of the watch party was important to support productive group learning. We deliberately kept the final artefact short at 1-2 minutes to ensure that everyone could produce a good-sounding product, but also allowed scope for students to create something musically satisfying if they chose to invest in that way. We suggested multiple channels of communication, including Zoom meetings and Messenger or WhatsApp chat groups, and students had to plan their own group's method(s) of communication while they planned for their group project. With the time difference that we were working with, we found that Messenger or WhatsApp was the easiest method of communication for our students. Soundtrap for Education also has a message function within it. Students tended to use what they were most familiar with.

Selecting a Platform – Why We Used Soundtrap for Education

We chose Soundtrap for Education because it works on any device, including mobile phones and is therefore useful when access to devices is limited, such as during remote learning, or when a school may have limited resources but may allow students to bring their own device (BYOD) to use in lessons. Online safety is of utmost importance when working in schools, and Soundtrap for Education addresses this concern by way of a 'walled garden' feature whereby the teacher oversees who can be invited into the group, restricting collaboration to authorised participants. In addition, Soundtrap for Education is accessible and intuitive to use. By positioning the pre-service teachers as learners and having them undertake a project, they had the opportunity to gain familiarity so that they could comfortably use the software with their own students.

Despite being an entry-level DAW, Soundtrap for Education is versatile, powerful and useful for more experienced musicians. It is widely used in education, including with primary school students, by generalist teachers and at senior secondary school level. In addition, it is used in a range of other subject areas including sciences, social subjects and languages. We have both used Soundtrap for Education extensively in our own teaching and music practice. In addition, Soundtrap for Education has built-in professional learning opportunities through their free educator certification programme as well as a range of exemplar resources and lesson plans.

Key considerations

We adopted the role of the teacher-as-facilitator, which was complex in this project. We are both experienced school music teachers and teacher-educators who have facilitated creative music-making projects with students for many years. This was the first time we had undertaken an international collaboration with our students. Working closely together with ongoing and frequent reflection was important. We drew on our previous experiences to design a project that was most likely to be successful; however, we felt like we were throwing ourselves into the deep end, not knowing if anyone would have anything to submit by the deadline that we had set for the watch party/sharing celebration.

Having a strong purpose, real-world context, and connection to place was important to the success of both iterations of the project. We carefully chose themes that were current and relevant to our students' lives at the time and would also be useful to explore with school students during teaching practicum placements or as graduate teachers. In particular, the ongoing climate change crisis is only becoming more prominent in all areas of education, and we were keen to continue to foreground the UN Sustainable Development Goals (UNESCO, 2020) in our teacher education courses.

We drew on the way that professionals work in the 'real-world,' in particular, filmmakers and composers (Keown, 2015). Although we are music educators and our students were all enrolled in music education subjects, the video aspect was equally important as the music. The whole is greater than the sum of the parts, which reflects research into participatory media cultures by Jenkins (2006).

Two of the challenges that we faced across both iterations of the project that are an ongoing focus of our facilitation approach relate to assessment and group learning skills. As is widely reported in the literature (Boud & Bearman, 2022), pre-service teachers, like most University students, are very assessment-driven. As is also widely reported in the music education literature (Fautley, 2021), assessment 'washback' can negatively impact creative projects, including ones such as these. This is something we have encountered in our teaching previously and have developed strategies to mitigate some of the challenges presented. Firstly, we have found that it is important to include the project as part of the assessment; otherwise, there is a risk that it won't be completed or will be completed with little investment. We have found that rather than assessing the artefact produced, which can negatively impact group dynamics and creativity, a valuable approach is to include a linked reflective task and assess that or to use the project as part of another task, such as a planning task. By including the project as part of the assessment, this also helps with managing class time, which is always a precious resource. The projects fed into a range of assessment tasks, such as designing a four-week scheme of work for school pupils to complete a composition project using Soundtrap for Education. Students had to devise plans and teaching materials for four lessons showing the development of skills. Another task was an essay where students critically reflected on their own personal and professional development throughout the course in relation to the SAMR and TPACK frameworks, uploading evidence of completed digital tasks (such as completed collaborative composition) into an e-portfolio.

Group learning challenges are widely reported in the school music education literature (see, for example, Wilson, 2019) and in our experience are also present to a lesser extent with adult pre-service teachers. To mitigate this, we tried to pre-empt problems connected with technology and organisation. In addition to carefully structuring the groups to have a reasonable number of students, slightly larger than our usual optimal group size of four, we ensured a mix of instruments and musical backgrounds of those who learn primarily through notation and those who primarily learn aurally. Given the timing of the projects and the associated global logistics, we felt that allocating groups was the most beneficial choice to make. Whilst this was not without its challenges, and removes some autonomy from students, we believe it was the most effective for the project, given that the students did not previously know each other. Despite implementing a range of strategies, we still received emails about some group members not responding to emails or not completing agreed-upon tasks in the agreed-upon timeline.

Pedagogy is Important

Across the two iterations of the project, we found that the pedagogical approach that we adopted was significant to the musical and extra-musical (collaboration, wellbeing and engagement) outcomes. Using Green's (2002, 2008) informal music learning pedagogy as a foundation, the project provided an opportunity for our students to create 'their' music which builds connections with everyday musical cultures, provides opportunities to make music in groups with their friends (and make new friends) and take ownership of the pedagogy (planning their own group's process) and curriculum (musical content) which impacts them. Spruce (2015) argues that student voice is supported when digital technologies, creating, and university/school classrooms are brought together, building bridges between institutional and everyday musical lives, which supports multiple and shifting musical identities (Hargreaves et al., 2017). This position is supported by Kuhn and Hein (2021), who contend that creating using digital technologies is best done using Green's (2002, 2008) pedagogical principles. Briefly, these principles are: students choose the musical content and set the direction of learning, music making is undertaken in peer groups with minimal adult involvement, and performing, composing and listening are integrated. Informal music learning pedagogy represents a reversal of the usual teacher and student roles, with the teacher acting as a facilitator rather than an instructor. Specifically, Green (2008) asks teachers to stand back and observe and then offer support in response to student-set objectives.

In our project, there were some interesting variations on these principles. As the work was completed largely online and asynchronously, as teacher-educators, we did not have a close overview of the student-led process in the way we usually would in a classroom situation. We were certainly standing back, but there was no observing until the final 'watch party.' Instead, students had to seek assistance if they were not able to solve problems themselves. This represents informal learning that is much closer to how it occurs in everyday music practice rather than in institutions like schools or even Universities. Rather than choosing who to work with in their friendship groups (Green, 2008), we had organised the groups, primarily to support musical skills, and students had to navigate making new friends with international and local peers. All the groups were able to navigate the group learning process and divide up the workload; there was some occasional low-level grumbling about uneven contributions, but nothing major. If we were facilitating the same project with school students, or if our students were facilitating the project with their own school students, there are some additional intervention strategies that we would use. Firstly, adopting a cooperative learning approach (Johnson & Johnson, 2009) and assigning group roles might have been useful. Secondly, formative peer- and self-assessment at the mid-point of the project to reflect and set goals for further refinement may also be useful for school students. Finally, providing additional and optional parameters for a composing toolkit may also have assisted some groups if they were paralysed by indecision. We also unpacked the project as a meta-pedagogical tool (Finney & Philpott, 2010) to help aid understanding of the process with our teacher-education students.

Frameworks We Used to Interrogate Practice: SAMR and TPACK

Frameworks are useful for music teachers and music teacher educators to understand the role of digital technologies in their classrooms. The SAMR and TPACK frameworks have been used extensively across education, not just in music education. We interrogated this project using these two frameworks and asked students to use them as tools for reflection.

SAMR

The SAMR model, developed by Puentedura in 2006, has four levels (substitution, augmentation, modification, and redefinition), and this provides a useful framework for music educators to think about regarding technology integration. The four levels are as follows:

Table 16.1 Sequences in Music Education (Puentedura, 2006).

Substitution	The technology provides a substitute for other learning activities without a functional change.
Augmentation	The technology provides a substitute for other learning activities, but with functional improvements.
Modification	The technology allows the learning activity to be redesigned.
Redefinition	The technology allows for the creation of tasks that could not have been done without the use of the technology.

Puentedura (2006) poses the following questions, which are useful to think about in a music education context when embedding digital technologies:

> What will I gain in the task by adding technology? Have I added an improvement to the task or process that could not be achieved without the technology? Does the new modification fundamentally depend on the technology? How is the task made uniquely possible by the technology?

We suggest that the Soundtrap for Education projects we undertook sit in the redefinition category, as the power of collaborating with international peers using cloud-based software and creating a collaborative video and music task demonstrated a task that would not have been possible without the technology.

TPACK

TPACK, or technological pedagogical content knowledge, was developed by Mishra and Koehler (2006), and it builds on Shulman's (1986) earlier PCK or pedagogical content knowledge framework. Music teachers can use the components and the intersecting components of this framework to examine their practice. Using music technology as a learner addresses the content and technology aspects of TPACK, whereas Dammers (2019) suggests that knowledge of how to facilitate student composition using a DAW or loop-based software is an example of all three areas of TPACK. He argues that a music teacher can ask questions such as 'Does my technology knowledge serve music learning effectively?' and that frequently the answer to this is 'no' and that a significant amount of technology usage by music teachers is not connected to facilitating student learning and lies outside the optimal TPACK space.

In both iterations of the project, we were impressed by the musical outcomes. We were surprised at the quality of the music and video creations produced, particularly in the short period of time allocated. We were reminded of the power of interdisciplinary arts practices and the affordances of digital tools to facilitate high-quality musical outcomes for experienced and inexperienced musicians alike, echoing Kuhn and Hein (2021), who highlight the affordances of DAWs alongside informal music pedagogy to promote participation. They state, "digital production is ideally suited to this goal [of producing "real" music], since student productions can sound quite legitimately "real." It feels wonderful to walk away from a class with a mixtape of your own great-sounding tracks!" (Kuhn & Hein, 2021, p. 12).

Together, the value of the music and video was greater than the sum of the parts, which reflects the redefinition level of the SAMR model. The music and video creations produced collaboratively would not have been possible without Soundtrap for Education. For example, a student in Aberdeen could record a bass part using loops, save it, then a student in Melbourne could listen to the bass part and record a guitar part on top, another student could add a vocal recording over the top of this and so on. Students could also leave messages for each other within the Soundtrap software and respond when they next logged on to the project. In the project, digital technologies were used as a tool for composing and not as an end in themselves, placing them in the technology and content areas of the TPACK framework.

Completing a project using a particular digital platform, Soundtrap for Education, was helpful for our students to have enough time to gain familiarity so that they could confidently use it in their teaching practice, which relates to all aspects of the TPACK model. We find that inexperienced music teachers usually underestimate how confident they need to be with technology before using it with their students. In particular, they need to be able to confidently troubleshoot; otherwise, lessons can get derailed. Several researchers have carried out studies into the concept of digital 'natives' (see, for example, Kirschner & De Bruckyere, 2017), and in line with this research, in our experience, the suggestion of digital natives is not actually a reality, and this is particularly relevant to using music technology pedagogically.

The students generally found Soundtrap for Education straightforward to use, and being browser-based and using cloud-storage made the logistics of working internationally much easier. Students commented on this project giving them more confidence to use DAWs in their teaching. The students in Aberdeen who were on the digital music practitioner course and not pre-service music teachers were also really enthusiastic about the potential of devising future community-based workshops using ideas gained from the project. This speaks particularly to the TPACK framework.

Reflecting the redefinition level of SAMR, our students commented on the benefits of the collaboration feature of Soundtrap for Education, which doesn't exist in other DAWs that they might have used before, such as Logic. Soundtrap for Education was also noted for its value as a digital composing tool that is accessible for school students and has value as a transformational tool. Students were able to reflect using the TPACK and SAMR frameworks and question whether their technology knowledge served music learning effectively as they progressed in their ongoing learning journey.

This project was suitable for students with varying levels of digital and compositional skills. Students were able to negotiate within their groups to select the role(s) and task(s) they undertook, with some students taking a lead on composing and songwriting, others on video editing, some taking more of an organisational and leadership role and many skill-sharing with their peers as the project unfolded. This exemplifies Lave and Wenger's (1991) theory of situated learning and is a model for showing how the more expert digital musician can support and inspire those who are less experienced.

Sustainable Development Goals

Students began to develop their practice as active global citizens, particularly through the second iteration of the project, with the focus on climate consciousness. Rich and powerful discussions evolved amongst groups arising from the initial video sharing. Place-based responses through video, lyrics, and music came together creatively in the international partnerships and highlighted our stimulus 'a world apart or same difference?' The students were very keen to highlight environmental issues through music and to find ways to incorporate this into their teaching. The Sustainable Development Goals (SDGs) (UNESCO, 2020) related to climate change consciousness, such as SDG 13 (take urgent action to combat climate change and its impacts), which is intrinsically linked to all the goals, are something we aim to continue to foreground in our teacher education courses.

Conclusion

The TPACK and SAMR frameworks have been useful for us to unpack our experiences integrating music technologies in this project, and the UN Sustainable Development Goals are valuable to give meaning and purpose to creative projects such as these. The project was helpful for our students to develop and practise much-needed skills in creating music and video using digital tools. However, what students commented most on was how much they enjoyed connecting with others and making new friends, reflecting clear well-being benefits. This was heightened by the global collaboration aspect; our students loved having new friends from across the globe, and seeing what the commonalities and differences in each other's lives looked like. The potential for global collaborations is only limited by the imagination of the project facilitator. For example, we are currently planning a further iteration of the project with primary school students and graduate teachers in countries in both the global north and south. In addition, this project reflects the way that musicians work, particularly in the post-COVID world. For example, songwriting is usually done in global teams, and the people involved frequently never meet. Their collaborative music-making process is done entirely online and asynchronously.

We undertook this project with secondary school music pre-service teachers; however, we believe it can be usefully implemented in a range of disciplines, both in secondary schools in cross-curricular projects and in primary generalist classrooms to explore cross- and interdisciplinary areas of inquiry. Soundtrap for Education has many pre-recorded sounds available and samples and loops are royalty-free therefore users do not need to be able to play a musical instrument to create music. Basic sound production and recording skills could be developed in collaborative projects such as podcasts, or with the addition of using moving image/film a short form video such as a TikTok, a TV advert or an audiobook could be devised. Teachers could consider designing multimedia projects to supplement written assignments, thus developing multiliteracies.

Particularly in the second iteration of this project, we found that music education and its potential for young people to make a positive contribution to the climate crisis was evident. From our perspective, our students engaged with their local environment in a different way when they were collecting their video footage. Exploring a topic such as climate change that can easily feel overwhelming through music and video was a way of approaching this issue in a way that students felt positive. Going forward, the UN Sustainable Development Goals will be much more prominent in our teaching and research, and we suggest that these are considered more widely.

If resources were limited, digital resources could be combined with acoustic musical instruments as well as found sounds and voices to devise creative compositions. Students could devise graphic scores (digitally or as an artwork) collaboratively responding to a theme such as climate change, and they could then respond to these musically. Performances could be recorded and shared, with different recorded interpretations of the graphic scores shared. Alternatively, a short moving image/film clip could be compiled, and students could respond by improvising live with acoustic musical instruments.

Real-world composing projects are important for music education in teacher education and schools, and can be facilitated effectively with digital technologies. We argue for greater attention to the affordances of digital collaborative music technology tools to facilitate real-world music and video creating projects for pre-service teachers and students at all levels of education.

References

Armon, C. (2021), "*Regenerative collaboration in higher education: a framework for surpassing sustainability and attaining regeneration*", Philosophies, Vol. 6 No. 4, pp. 1-11, https://doi.org/10.3390/philosophies6040082

Black, P. and Wilson, E., (2024). Digitally mediated collaboration and participation: composing 10,427 miles and 11 hours apart. *British Journal of Music Education,* pp.1-12. https://doi.org/10.1017/S0265051724000226

Cremata, R. and Powell, B. (2017), "*Online music collaboration project: digitally mediated, deterritorialized music education*", International Journal of Music Education, Vol. 35 No. 2, pp. 302-315, https://doi.org/10.1177/0255761415620225

Dammers, R.J., (2019). The role of technology in music teacher education. *The Oxford handbook of pre-service music teacher education in the United States,* pp.365-376. https://doi.org/10.1093/oxfordhb/9780190671402.013.17

Finney, J. and Philpott, C., (2010). Informal learning and meta-pedagogy in initial teacher education in England. *British Journal of Music Education, 27*(1), pp.7-19. https://doi.org/10.1017/S0265051709990167

Fredricks, J. A., Blumenfeld, P. C., & Paris, A. H. (2004). School Engagement: Potential of the Concept, State of the Evidence. *Review of Educational Research,* 74(1), 59-109. https://doi.org/10.3102/00346543074001059

Green, L (2002) *How Popular Musicians Learn: A Way Ahead for Music Education.* Aldershot: Ashgate.

Green, L. (2008), *Music, Informal Learning and the School: A New Classroom Pedagogy,* Ashgate, Farnham.

Hargreaves, D.J., Macdonald, R. and Miell, D., (2017). The changing identity of musical identities. In Handbook of musical identities, pp.3-23. MacDonald, R., Hargreaves, D.J. and Miell, D. eds., 2017. *Handbook of musical identities.* Oxford University Press. https://doi.org/10.1093/acprof:oso/9780199679485.003.0001

Jenkins, H., (2006). *Confronting the challenges of participatory culture: Media education for the 21st century. An occasional paper on digital media and learning.* John D. and Catherine T. MacArthur Foundation.

Johnson, D. W., & Johnson, R. T. (2009). An Educational Psychology Success Story: Social Interdependence Theory and Cooperative Learning. *Educational Researcher,* 38(5), 365-379. https://doi.org/10.3102/0013189X09339057

Kirschner, P.A. and De Bruyckere, P., 2017. The myths of the digital native and the multitasker. *Teaching and Teacher education,* 67, pp.135-142. https://doi.org/10.1016/j.tate.2017.06.001

Keown, D. J. (2015). Discovering the Lost Ark of Possibilities: Bringing Visibility to the Invisible Art Form of Film Music in Your Music Classroom. *Music Educators Journal,* 101(4), 70-76. https://doi.org/10.1177/002743211557387

Koehler, M. and Mishra, P., (2009). What is technological pedagogical content knowledge (TPACK)? *Contemporary issues in technology and teacher education,* 9(1), pp.60-70.

Kuepper-Tetzel, C. and Nordmann, E. (2021) "Watch party lectures: synchronous delivery of asynchronous material", *Journal of Learning Development in Higher Education,* (22). https://doi.org/10.47408/jldhe.vi22.696

Kuhn, W. and Hein, E. (2021), *Electronic Music School: A Contemporary Approach to Teaching Musical Creativity*, Oxford University Press, New York. https://doi.org/10.1093/oso/9780190076634.001.0001

Lave, J. and Wenger, E., 1991. *Situated learning: Legitimate peripheral participation.* Cambridge University Press. https://doi.org/10.1017/CBO9780511815355

Østergaard, E. (2019), "Music and sustainability education–a contradiction?", *Acta Didactica Norge*, Vol. 13 No. 2, pp. 2-20, https://doi.org/10.5617/adno.6452

Puentedura, R.R. (2020). SAMR - A research perspective. Retrieved on September 8, 2024 from http://hippasus.com/rrpweblog/archives/2020/01/SAMR_AResearchPerspective.pdf.

Samaras, A. (2011), *Self-study Teacher Research*, Sage Publications, Thousand Oaks, C.A.

Seligman, M. (2018) PERMA and the building blocks of well-being, *The Journal of Positive Psychology*, 13:4, 333-335, https://doi.org/10.1080/17439760.2018.1437466

Spruce, G. (2015), "Music education, social justice, and the "student voice": addressing student alienation through a dialogical conception of music education", in Benedict, C., Schmidt, P., Spruce, G. and Woodford, P. (Eds), *The Oxford Handbook for Social Justice in Music Education*, Oxford University Press, London, pp. 287-301. https://doi.org/10.1093/oxfordhb/9780199356157.013.20

UNESCO (2020), "Education for sustainable development: a roadmap", available at: https://unesdoc.unesco.org/ark:/48223/pf0000374802.locale=en (accessed 23 June 2023).

Wilson, E. and Black, P. (2024). Caring for our planet: a world apart or same difference? A global music collaboration. *Journal of Applied Research in Higher Education.* https://doi.org/10.1108/JARHE-08-2023-0360

Chapter Seventeen

Integrating Technology in Teacher Education for Medical and Veterinary Instruction: Perspectives on Advancing Practical Skills and Navigating Ethical Demands

Anna Tomańska

Wrocław University of Environmental and Life Sciences, Poland

Abstract

Modern challenges in medical and veterinary education, particularly regarding practical skills and ethics, have prompted the integration of advanced technologies like simulators, manikins, and AI. These tools replace live patients and models during early learning stages, ensuring better understanding while preparing students for real-world healthcare scenarios. Technologies such as robotic surgical systems and digital dissection tables provide realistic, interactive experiences, enhancing students' confidence and competence in a safe environment. Educators must adapt to effectively utilize these tools, requiring both technical proficiency and pedagogical expertise. Ethical concerns, especially stringent regulations on animal welfare in Europe, have accelerated the shift from live models to technological solutions. A balanced approach is needed to combine real specimens and technology, ensuring comprehensive training while upholding ethical standards. This study explores the use and challenges of integrating digital technologies in medical and veterinary education through online surveys, instructor interviews, and literature reviews. It identifies widely used technologies, evaluates their role in enhancing teaching, and examines educators' preparedness and gaps in training. Key issues include disparities in technology access, its impact on learning outcomes, and ethical and legal considerations shaping future advancements. The findings highlight the critical need for continuous professional development and the evolution of educational systems to align with technological advancements. The study provides actionable recommendations to improve technology integration, aiming to enhance educational practices and better prepare students for healthcare roles.

Keywords: veterinary, technology, animals, education.

***

Introduction

Modern challenges in medical and veterinary education, particularly those related to practical skills and ethical considerations, necessitate the integration of technology into the educational process. This is especially evident at the higher education level, where the maturity of students and the justification for using live models become more pronounced. Technologies such as simulators, mannequins, and artificial intelligence enable the replacement of live and experimental models during the stages when students do not yet possess a full theoretical foundation. This approach allows for better comprehension of the material and enhances student engagement. The integration of these technologies is especially valuable as it prepares students for real-world patient interactions, minimizing potential negative experiences for both students and patients during the learning process. Tools like robotic surgery systems and digital dissection tables provide realistic and interactive experiences that mimic real-world scenarios, allowing students to hone their skills in a safe and repeatable environment. This is crucial for the harmonious development of not just scientific acumen but also social, personal, and emotional skills. This practical approach enriches educational experiences and builds confidence before working with real patients. However, it also requires educators to have specialized knowledge to effectively utilize these tools and incorporate them into their teaching. As technology advances, some competencies among all participants in the educational process may prove insufficient.

The primary objective of this study is to explore the current use and challenges of integrating digital technologies in medical and veterinary education. By analysing data from an online survey and interviews with academic instructors, as well as conducting a literature review, the study aims to identify which medical and veterinary technologies are most utilized and valued in educational settings. It seeks to understand how these technologies are applied to enhance teaching and learning, assess the preparedness of educators to use these tools, and uncover gaps in training and support. Additionally, the research briefly addresses significant issues such as disparities in technology access, the impact of technology on educational outcomes, and provides foundational insights into the ethical and legal considerations associated with technology use, which are expected to drive future advancements. Ultimately, this study highlights the importance of developing insights into the effectiveness of current technological practices and offers recommendations for improving the integration of technology in medical and veterinary education.

The role of educators in integrating technology is critical, and their competencies must evolve accordingly. They need to be proficient not only in the technical use of these tools but also in their pedagogical application. Ethical issues surrounding the use of live models are particularly significant in Europe, where strict regulations regarding animal welfare exist. However, this legislative framework is becoming increasingly relevant on a global scale. These concerns drive the trend towards replacing live models with advanced technological solutions, necessitating a balance between using real specimens and technology to provide comprehensive training that aligns with ethical standards. Additionally, modern medicine evolves alongside technology, and its implementation requires various forms of education for both future doctors and their patients.

The demands placed on educators, along with the processes of technical upskilling, highlight the competencies of the future that will facilitate the integration of research devices across various areas of medicine and veterinary science. Studies conducted among academic teachers reveal the importance of contact with live patients for acquiring reliable medical knowledge and skills. They also point to the potential for the development of practical technological solutions in this regard, underscoring the need for continued evolution and adaptation of educational systems to keep pace with the dynamic advancement of technology.

Technological Shifts and the Digital Learner in Medical and Veterinary Education

Technological advancements have given rise to a new type of learner in the educational process: the digital student, which places new demands on educators regarding the use of digital technologies. Integrating practical skills with technology presents a significant challenge across various systems and countries, offering numerous benefits but also posing difficulties. Beyond individual challenges, systemic organization often fails to ensure effective use and equal access to new technologies, including opportunities for professional development for teachers in this field (Akram et al., 2022). Global trends toward providing free access to education recognize ethical standards and aim to bridge social disparities, ensuring the minimum necessary for societal and economic participation. From a legalistic perspective, public education as a personal interest also involves safeguarding the rights granted to citizens by the state (Shaw, 2022). While access to technology is a natural outcome of progress, it can also contribute to the creation of social inequalities (Tawfik, 2016).

Therefore, public and universal access to education is a crucial doctrinal consideration (Shaw, 2022).

The impact of technological changes on education has been a topic of discussion for some time; however, the shift within the teaching structure now sees Digital Natives—individuals born between 1980 and 1994—taking a leading role in the education sector. Interestingly, although younger generations, known as Digital Immigrants and Digital Learners, were raised in a virtual world, their competencies for conscious use of technology may be lower than those of Digital Natives (Creighton, 2018). Legislative changes and the development of international legal standards have also elevated health to one of the fundamental human rights (Toebes, 2001). However, without proper education, there can be no effective health protection or medical advancement. Societies with higher levels of education tend to enjoy better health and longer life expectancy; positive correlations are also seen with infant mortality, expected lifespan, and childhood vaccination rates. Educational interventions are crucial for levelling health disparities across different nations (Raghupathi, 2020). While this issue primarily addresses the technological education of teachers, it must also consider broader areas such as the overall development of technology and medicine, societal needs, and the intersections of public health and education. These interconnected elements highlight the complex landscape that educators must navigate as they integrate new technologies into medical and veterinary instruction, balancing the advancement of practical skills with the navigation of ethical demands.

New Technologies in Medical and Veterinary Education

In adult veterinary education, the concept of andragogy for educators might still be relatively new, although it has been observed that in some systems, the methodology of conducting classes has been completely restructured. In the United Kingdom, this includes radicalizing the curriculum by emphasizing practical skills and independent problem-solving over rote learning, a trend that the United States also follows. Since the 1960s, there has been an increasing emphasis on lifelong learning habits and skills, with a focus on problem-based learning and critical thinking. These changes represent a shift away from traditional teaching methods, despite resistance from some educators whose knowledge is rooted in the old system (Dale et al., 2008). These issues are often discussed, suggesting a gradual shift from lecture-based teaching to more engaging methods, and adapting lectures regularly to align with evolving technological capabilities (McLennan & Heath, 2008).

In veterinary medicine, various tools are adopted with some delay compared to human medicine, although there is a growing interest in developing e-learning tools that serve as multimedia repositories, including two-dimensional scans, three-dimensional models, and Digital Imaging and Communications in Medicine (DICOM) files. An example of this is the Interactive Veterinary Education Tool (IVET) platform (Xiberta & Boada, 2019). Veterinary education differs from human medicine as it integrates typical animal medical issues with agricultural sciences, including livestock breeding, food safety—especially of animal origin—and environmental hygiene (del Pino Palacios-Diaz & Mendoza-Grimón, 2023). It thus stands at the intersection of public health, serving both humans and animals.

Both disciplines face distinct challenges in their development and educational realities. Veterinary science is complex due to species diversity and numerous specializations, whereas human medicine treats each specialization in a highly focused manner, constantly striving to improve technologies and procedures for human health. A convergence of both fields is translational medicine, which aims to apply research findings across interdisciplinary domains. Advances in translational medicine include innovative diagnostic, clinical, and pharmacological solutions. Although translational medicine uses advanced biomedical technologies in laboratory settings, such as artificial microphysiological systems, 3D tissue models, synthetic tissues, implants, and smart materials, these are likely to progress from the experimental phase to clinical trials for use in humans or animals. Program concepts have been proposed to involve students in cooperation between researchers and their research results and patients, thereby bridging the communication gap between basic science researchers and clinicians (Valentijn et al., 2023).

Significant changes in medical education have been brought about by artificial intelligence (AI) and the development of large language models such as Chat Generative Pre-Trained Transformer (ChatGPT, OpenAI; Lee, 2023; Leng, 2023). A research gap has been noted in academia regarding the integration of AI technologies into medical education, with a lack of comprehensive guidelines and recommendations. AI tools enhance key skills in the healthcare sector as they significantly increase workforce efficiency, improve diagnostics, treatment planning, and overall patient care (Garcia et al., 2024). It has been observed that technological advancements are outpacing pedagogical strategies and the adaptation of medical school environments. However, this also highlights negative aspects associated with using ChatGPT. One major issue is the students' inability to distinguish false information generated by the tool. Tested in anatomy, ChatGPT achieved about 75-85% accuracy in this field, but it can also facilitate plagiarism and replace students in performing tasks, requiring careful consideration (Garcia et al., 2024).

Interesting observations have been made in low- and middle-income countries, where medical education faces certain barriers, and in China, where the large population is served by relatively few doctors and where rapid medical development reveals shortages in medical education resources. Additionally, body donation rates are low in China, limiting the available learning material for students, and access to detailed medical data is restricted; meanwhile, virtual patients lack personal data, which is therefore unprotected. In poorer countries, there is often a teacher-centred approach, and technology shortages are common, not only in terms of advanced equipment but also in access to and quality of internet connections, posing a general challenge for the future (Leng, 2023). For centuries, the use of cadavers has been considered beneficial in teaching at medical and veterinary schools. Alternative teaching methods support education with medical imaging, models, simulators, and virtual reality platforms. In surgical training, for instance, the Canine SyndaverTM is used. Another technology is computer-assisted learning (CAL) programs. Students from various nationalities indicate that real cadaver dissection is a fundamental method of learning anatomy, with other methods serving as supplementary. Educators emphasize that using cadavers is synergistic, allowing integration of basic sciences with clinical subjects (Varner et al., 2021), suggesting that cadavers should not be entirely replaced by technological methods.

The use of substitutes for living patients for training purposes is slightly different. In human medicine, simulation-based medical education (SBME) has been developed. It requires planning and tailored program development but offers many educational benefits (McGaghie et al., 2009). A significant aspect of its use is easing ethical tensions in medical education, where optimal patient treatment and the highest safety and well-being standards must be ensured. SBME training minimizes risk when students move on to real patients or actors prepared to fulfil this role. Thus, simulations can be used to enhance not only technical skills but also the behavioural and social development of future medical professionals. This provides instructors with the opportunity to set high standards for skill assessment, prepare students for unexpected scenarios in their future careers, and manage errors and patient safety – critical areas since students can explore and learn from mistakes in a safe environment where such errors would be intolerable in clinical practice (Amitai et al., 2023).

In SBME, the learner plays a central role, and with the significant growth of AI, simulations can become highly complex, including simulated environments for training various communication skills or rapid response capabilities of medical staff. Even dilemmas have emerged about situating technological education within the broader concept of lifelong learning. This would not only prepare students to learn using technology but also raise awareness of risks associated with diminished critical thinking and the establishment of safety systems for incidents involving AI. It is also crucial to emphasize that this technology should facilitate knowledge and skill acquisition during courses, considering both the students' and instructors' levels. This introduces a new role for the instructor, who must be proficient in both technical aspects and pedagogy (Komasawa & Yokohira, 2023).

Education can also take the form of games, designed for individual players or teams, often set in 3D environments. Multimedia games, however, have not yet been widely adopted; their quality and complexity require significant production costs. Nevertheless, they offer measurable outcomes, provide quick feedback, and engage learners. Given past perceptions of gamers and the often-controversial content of games, such as violence and sexual themes, educational games may be viewed sceptically as a time-wasting endeavour (de Bie et al., 2012), making them a surprising element in formal education.

In veterinary education, similar motivations apply, but animal patients are less predictable in their behavioural responses, making this an equally important skill to develop at some point in training. Simulators prepare students for interactions with live animals, and the effectiveness of teaching with such models has been confirmed (Noyes et al., 2022). Various simulators exist for both large and small animals, with rectal palpation simulators for pregnancy diagnostics in cattle being particularly common. There are also numerous devices available for teaching essential skills such as injections, biopsies, ophthalmic exams, anaesthesia monitoring, surgical procedures, intra-articular injections, castration, resuscitation, endoscopy, and laparoscopy. The trend of increasing simulator use has been noticeable since the 1990s, coinciding with legal advances emphasizing the need for alternative animal substitutes in education and research. Moreover, simulators can be used not only for preclinical preparation but also for skill enhancement and correction of errors in various medical procedures, such as surgical suturing (Braid, 2022).

An intriguing solution for both medicine and veterinary medicine is the use of interactive dissection tables. These tools develop a range of competencies, including cognitive-intellectual, diagnostic, prognostic, informational, analytical, and research skills. As with other tools, researchers highlight the need for scientific evaluation of their effectiveness. They have also noted correlations between changes in the medical education system and shifts in the information society, which is acquiring ever-greater digital skills. Regarding interactive tables, numerous benefits have been described that support cognitive processes, such as the development of logical and algorithmic thinking, allowing students to independently manage their learning, increased interest and motivation, and the development of experimental skills. A valuable interactive feature is the visualization of theoretical concepts, which accelerates knowledge acquisition (Abuvatfa et al., 2021).

Some skills have become essential competencies for physicians because medical practice necessitates the use of technological skills, such as surgical robots. There is even mention of a new era of medicine (Gosrisrikul et al., 2018), enhanced by robotic systems that eliminate or reduce common human errors. In human medicine, specialization divisions have been established, while in veterinary medicine, there is a recognized need to move away from the "universal veterinarian" concept, as future veterinarians will have to address increasingly diverse problems (Zemlijic, 2004).

Practical examples of the use of animals are shown in the photographs (Figures 17.1 to 17.4).

Figure 17.1 Specimens Prepared from Organs of Slaughtered Livestock for Dissection (on the Left: Cow Udder, Mammary Gland) And for Ultrasound Examination Exercises, Immersed in Gelatine (on the Right: Female Reproductive Tract and Ultrasound Transducer).

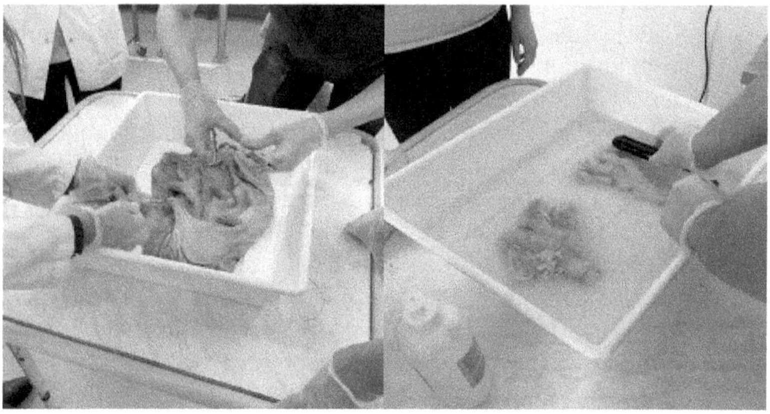

Note: Sessions Held at the Department of Reproduction and Clinic for Farm Animals, Wrocław University of Environmental and Life Sciences, Poland.

Figure 17.2 Educational Room Equipped with Veterinary Mannequins and Simulators.

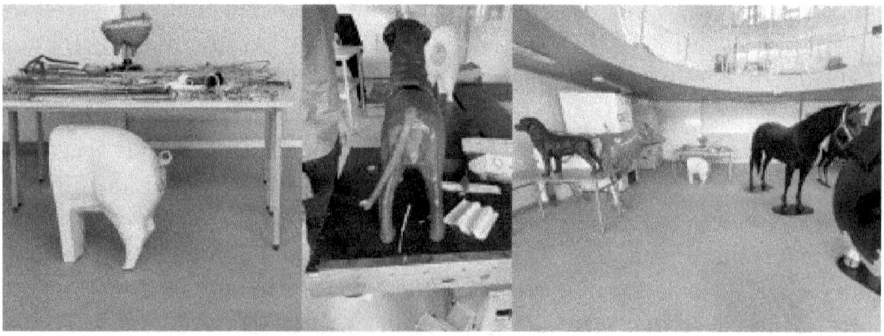

Note: Photographs were Taken during Sessions Conducted at the Department of Reproduction and Clinic for Farm Animals, Wrocław University of Environmental and Life Sciences, Poland.

Figure 17.3 Examples of Live Animals Interacting with Students During Veterinary Internships and Practical Training.

Note: Photographs were Taken during Clinical Rotations and Professional Training Sessions Conducted for Students by the Veterinary Clinics at Wrocław University of Environmental and Life Sciences, Poland.

Figure 17.4 Preserved Animal Specimens and their Tissues (on the Left: Histological Preparations; in the Centre: Animal Preserved in Formalin Solution; on the Right: Museum Collection Display).

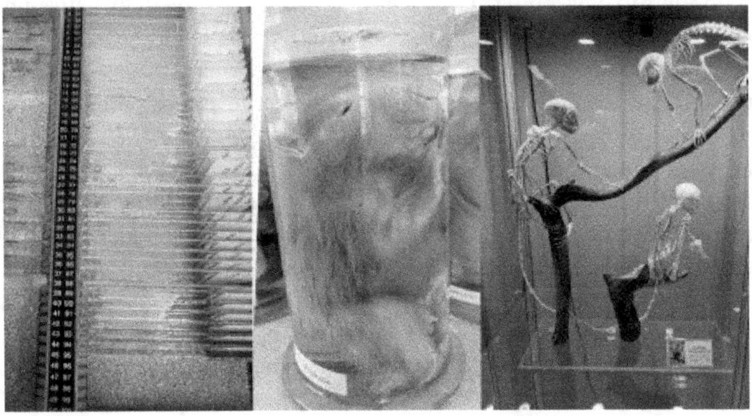

Note: Photographs were Taken at the Wrocław University of Environmental and Life Sciences, Poland.

The Drawbacks of Using Technology in Medical and Veterinary Education

Technology provides a wealth of benefits as an educational tool, but it also comes with notable drawbacks. As its use increases, we may see a decline in social interactions among students, heightened privacy concerns, and an increased potential for cheating (Kryeziu et al., 2021). Researchers emphasize the need for cohesive policies from educational administrations to not only integrate technology but also to connect various disciplines. This ensures that students maintain their problem-solving skills while also preventing social exclusion that can arise from the cost of additional equipment (Atta et al., 2020).

Not every technology suits specific educational content. For instance, VR 360° scenarios engage the senses of sight and sound and can elicit emotions that can be discussed during debriefing sessions, creating a space for reflection on values and attitudes. However, the limited opportunities for interaction can hinder the effectiveness of certain technologies compared to live models. Immersive scenarios, which deepen psychomotor experiences, often prove to be more effective in these contexts. Furthermore, students' excitement about virtual reality may divert their attention from understanding the underlying problems and finding solutions. Consequently, VR sessions often require repetition to help students recognize the cognitive benefits of using technology for learning rather than merely viewing it as a form of entertainment (Pfeifer et al., 2024).

This scenario can lead to future medical and veterinary professionals lacking a sense of responsibility for their decisions and failing to grasp the social implications of those decisions, posing challenges in their future careers. One approach to sensitizing students to these complex issues is to involve clinical practitioners in developing technological tools and pairing them with students to share their insights. To ensure that technology offers more than just a digital experience, it should be designed to support emotional processes involved in learning. This could involve exploring the connections between positive emotions, like joy and curiosity, and negative feelings such as fear and frustration (Loderer et al., 2020). The feeling of presence and making a tangible impact on their environment positively influences learning outcomes. However, it's essential to recognize that some students may experience cognitive overload as the use of technology expands their scope of exploration. This area warrants further research to investigate the relationship between learning styles and the effectiveness of various methods (Sari et al., 2021). For some, modern technologies may disrupt their perception and lead to inconsistencies in performance and assessment results, which may hinder the individualized needs of public education systems. When considering how to influence students' behaviours and thinking styles, examining empathic approaches is valuable (Samarasekera et al., 2023). In addition to empathy, compassion plays a vital role in providing quality patient care. These two factors can be collectively termed "impassion." Some scholars have developed scales to measure "impassion," underscoring that a lack of compassion alongside empathy may lead to manipulative behaviours. Thus, it's crucial to evaluate both elements together (Praszkier, 2023). The implementation of empathy scales presents a compelling opportunity to establish normative ranges that reflect appropriate values. One notable example is the Jefferson Scale of Empathy (JSE), which can assess not only the effectiveness of ethical education but also the suitability of individuals entering medical studies, helping to identify those who may require additional training in this domain (Kataoka et al., 2023).

The Role of Digital Competencies in Modern Education: Integrating Technology into Teaching Practices

Digital tools offer teachers a wide range of possibilities to enrich, and sometimes even replace, traditional teaching methods. Although some of these tools require scientific validation regarding their effectiveness and utility, experienced educators emphasize that educational tools should serve to harmonize teaching within defined frameworks and objectives, enhancing motivation and effectiveness while providing added value to the educational process. However, there are certain limitations associated with the competencies of teachers, who often need clear pedagogical support in integrating technology into education (Amhag et al., 2019).

Digital practices can alter interpersonal and social dynamics, complicating the learning process for teachers (Wood et al., 2020). Digital tools expand educational spaces by extending learning beyond the conventional classroom's time and place, prompting teachers to adapt to technology and critically engage with its potential (Coker, 2020). In pedagogy, these technologies are referred to as "design-supporting tools" (Masterman & Manton, 2011), underscoring the teacher's leading role in lesson management.

Interest in this topic is not new; the term "digital era" has been discussed in the context of teacher training, emphasizing that digital competencies, like others, should be acquired before entering the profession (Chin et al., 2019). The COVID-19 pandemic, which began in 2019, significantly impacted the global adoption of technology in education, including the growth of telemedicine. This period shifted the general trend toward remote work tools, resulting in many study programs or courses being offered entirely online, representing a future-oriented approach to education. Some experts speculate that transitioning to new systems involves significant work costs, including the burden of teacher self-education. Additionally, some foresee a phase of disruptive innovation for medicine (Park et al., 2021). Digital competencies also involve general skills, and these should be understood holistically as they pertain to various personal aspects of the teacher (Saaiq et al., 2024). Support for teachers in technological education is limited, yet technology itself offers numerous opportunities for self-improvement, such as bug-in-ear coaching, which enhances lesson design, assessment, and student surveying (McKenney & Visscher, 2019).

Research indicates that teachers' professional development in technology correlates with students' achievements, particularly in their sensitivity to scientific inquiry, though not necessarily with their own technological proficiency. This approach can have long-term benefits in improving student outcomes in response to the technological integration of their teachers (Lee et al., 2017). Integrated pedagogy also shows that overall experience and tenure influence technology use, with more experienced educators incorporating technology as a means of innovation more than novice teachers (Hughes, 2005).

In recent years, we have observed the integration of technology in schools and social environments, including private activities of schools and teachers on social media, which often evolve into building their personal brands. This development can be described as the "digital identity" of contemporary educators (Etedali & Finland, 2021). The technological landscape of the twenty-first century has evolved significantly (Urbani et al., 2017). Programs exist to help teachers integrate these skills, demonstrating a partnership responsible for global development (Sulaiman & Ismail, 2020). High-quality education aims to meet the demands of an evolving industry, not only by creating practical applications for computational tasks using educational robotics and programming but also by addressing and resolving moral dilemmas. Contextual competencies, such as ethical and cultural awareness, are crucial beyond basic skills. Teachers must link lessons to four dimensions of learning: skills, knowledge, character, and metacognition (González-Pérez & Ramirez-Montoya, 2022). However, there is no consensus among educators on which skills are crucial for the technological era, reflecting a broad range of concepts (Varas et al., 2023). Developing educational centres and multimedia databases that archive essential materials for student communication is also vital (Ganguly, 2010).

To effectively use technology, a teacher must understand its components (Cheffield et al., 2018), including basic operational principles and options. This can be particularly challenging for teachers who do not share enthusiasm for technology. Additionally, the average age of teachers, which in OECD countries is 44 and shows an aging trend, must be considered (Monteiro et al., 2020). In medicine and veterinary science, there are many specialization paths requiring deeper knowledge in areas such as medical imaging, surgical techniques, and diagnostic methods. Consequently, competency scope must be tailored to the specific needs of particular fields and consider the foundational knowledge levels of learners. Teachers and students should act as users, often prepared by clinical engineers or training personnel (Kanehira et al., 2015). Some medical devices, like robots supporting various clinical tasks, feature additional software that allows them to operate in a test phase, performing tasks in real-time but without real patients (Baumann et al., 2022).

The integration of technology has shifted from hardware and software to predominantly web-based digital content distribution (Ruggiero & Mong, 2015). Therefore, some advocate for teacher training in technology through programs that include key integration competencies like computational thinking, data sets, maker spaces, visual design, media editing, and other approaches (Luterbach, 2022). This evolving educational landscape requires a continuous adaptation of skills and knowledge by teachers, ensuring that they not only keep pace with technological advancements but also leverage them effectively to enhance learning outcomes.

Educational Challenges in the Innovation Sector for Medicine and Veterinary Science – Alternative Educational Models and Societal Pressure for Replacement

Medicine and veterinary science are rapidly evolving fields, with veterinary care for pets sometimes equalling or even surpassing human medical care. What is defined as progress positions veterinary medicine at a point where improvements in care quality are expected to become the norm in the near future. Advanced veterinary care can sometimes verge on experimentation, and it should be noted that regulations for animal experiments are not equivalent to those for human medical experiments. Advanced veterinary treatments can thus pose ethical dilemmas in everyday clinical practice (Quain et al., 2021). This is a consideration in the education of future veterinarians (Hernandez et al., 2018) as well as human doctors (Campbell et al., 2007).

While there is ongoing debate about the added value of alternative methods in medicine—such as replacing live patients with simulated ones—it has not yet been fully established how much these alternatives engage participants in the educational process (Spencer et al., 2001). In the context of animals, decisions to exclude them from the educational process can be more pronounced due to the lack of a consistent understanding of the effects of behavioural stress, fear, pain, and suffering in animals. Therefore, animal simulators hold significant potential and align with the 3Rs principle of Directive 2010/63EU (Humpenöder et al., 2021). This issue has been extensively explored by J. Balcombe, who notes that while computer simulations of animal dissections can be repeated and reversed multiple times, the tissues and organs of preserved organisms differ from those of freshly killed or living animals. Alternative models, such as plastinates, continue to allow the assessment of structures in three dimensions and colour, which is crucial for better understanding. These models are used in countries that have banned the killing of animals for practical training (e.g., the UK in 1876 and India in 1960).

While the value of hands-on learning is not negated, which favours an active learning process, alternative methods are considered capable of replacing most live-animal procedures. Discussions have also begun on objections from students who refuse to participate in animal-based classes due to conscientious or ethical reasons. Teachers' attitudes are crucial here; refusals should not be stigmatized as rebellion but seen as a demonstration of care and commitment to animal welfare. Animals were commonly used in pre-university schools worldwide up until the 1980s (Balcombe, 2000).

In the United States, the use of animals in research and education is regulated under the Animal Welfare Act (7 U.S.C. 2131). This law authorizes the United States Department of Agriculture to establish and enforce regulations detailed in 9 C.F.R. Parts 1-3, which provide standards for animal care and use. The rules for handling violations of the Act are specified in 9 C.F.R. Part 4. These regulations provide guidelines on standards for breeding, transport, euthanasia, and minimization of distress (United States Congress, 2018; U.S. Department of Agriculture, 2023a, U.S. Department of Agriculture, 2023b). In Europe, standards are based on Directive 2010/63/EU on the Protection of Animals Used for Scientific Purposes, which emphasizes the consideration of alternatives and the minimization of animal use. A crucial aspect is that animals should only be used in education if the educational goal cannot be achieved without using animal models (European Parliament and Council, 2010). The United Kingdom has also established that scientific progress must be demonstrated if animals are involved in the educational process, which must be transparent concerning established standards (UK Parliament, 1986). In India, the key piece of legislation is The Prevention of Cruelty to Animals Act, 1960, which underscores the importance of animal welfare and explicitly prohibits cruelty (Government of India, 1960). In the Republic of Korea, additional attention is paid to the social educational aspects, highlighting the significance of caregiver responsibility for animals (Republic of Korea, 1973). Similar regulations apply in China, where there is a strong focus on the organization and control of animal use, as well as the training necessary for researchers to participate in procedures involving animals. This legal system emphasizes the use of animals in both human and veterinary medicine education, and students are obliged to prepare ethically for classes. The principle of replacement is upheld (People's Republic of China, 1988). A similar legislative approach has been adopted in Brazil (Brazilian Government, 2008).

Overall, the legislation related to the experimental use of animals varies significantly across different countries in Asia and Oceania. It is evolving considerably in Latin America and the Middle East, as well as in Africa, but still lags behind the leaders in shaping standards for laboratory animals in Europe and North America (Baxter, 2020). Numerous literatures have emerged around the use of animals in science, including the need to specify certain terms that help regulate the scope of animal use globally. One example is the understanding of invasive procedures and humane euthanasia (Knight, 2022).

In a study by K.S. Baptista et al., medical and veterinary students support the need for enhancing animal procedures and their use in education to ensure the optimal acquisition of theoretical knowledge, attitudes, and technical skills. However, it is the teachers' responsibility to ensure the best methodology and implement more humane and efficient approaches (Baptista et al., 2024). The varied commitment of different countries to implementing a coherent animal policy can affect education and technological integration, as countries that require animal model users to adopt optimal or maximum replacement methods will more frequently and readily invest in various technological solutions.

The responsibility of using live models without the consent of an ethical committee and the considerations of medical and veterinary education in shaping knowledge and attitudes pose significant challenges. This has become one of the leading needs and reasons for changes in both social thinking and the individual needs of teachers and students. Teachers are tasked with managing and coordinating practical labs and sessions using such simulators, ensuring that students derive maximum benefit while upholding ethical standards, sometimes even bearing legal responsibility for the improper use of animals or unnecessarily causing them suffering or compromising their welfare. These challenges underscore the importance of continuous professional development for teachers and ongoing ethical reflections within medical and veterinary education.

Findings from Research

Preliminary research was conducted based on an online survey with 8 respondents and interviews with two academic instructors. The online survey was distributed to 30 academic institutions worldwide, but responses were received from only four institutions across four countries. Respondents were selected through voluntary participation via email, limited to medical and veterinary schools. The respondents represented a diverse group in terms of professional experience and origin, including one European country, two countries from the Middle East, Africa, and North America. The sample comprised 37.5% from veterinary disciplines, 37.5% from human medicine, and 25% from other related disciplines. The average work experience was 9.25 years (median 7.5 years, standard deviation 5.21 years).

Interviews were conducted via remote communication and covered questions from the survey while allowing for expanded discussion on various aspects. Among the respondents, 75% utilized technologies such as simulators, manikins, virtual reality, multimedia tables, or artificial intelligence. The most frequently chosen technologies in education included online learning platforms, interactive games, Internet communication tools for remote classes, and applications, with artificial intelligence also noted. Less frequently mentioned were 3D printing, scanning and modelling devices, robotic systems (e.g., surgical robots), digital dissection tables, and applications (Figure 17.5). Respondents indicated that technological tools are most beneficial for teaching surgery and clinical practice, as well as basic sciences such as anatomy. Technological potential was recognized in diagnostics education, client interaction, and management and organization. Less emphasis was placed on forensic medicine, pathology, pharmacology, or ethics, law, and administration, often attributed to their theoretical nature and the sufficiency of traditional lecture methods (Figure 17.6). It was noted that students sometimes or rarely bring their technological devices to class. Opinions on disparities in students' access to modern technologies like laptops, tablets, or relevant software were divided. In institutions facing shortages of mobile devices and network access, some technologies were avoided to prevent social exclusion or due to technical limitations. In several cases, respondents reported that students declined participation in tasks due to insufficient technical capabilities or technological knowledge.

Figure 17.5 The Respondents Indicated which Technologies, in their Opinions, are Most Commonly Used in the Education of Medical and Veterinary Students.

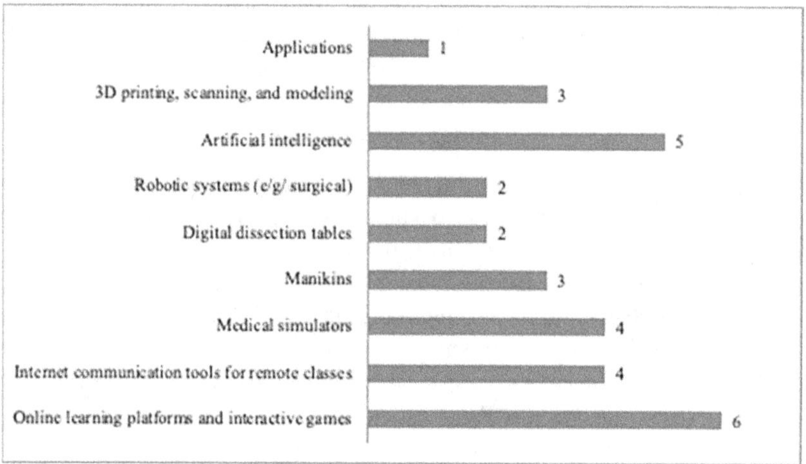

Figure 17.6 Distribution of Subjects Where Respondents Consider Innovative Technologies Most Useful in Education.

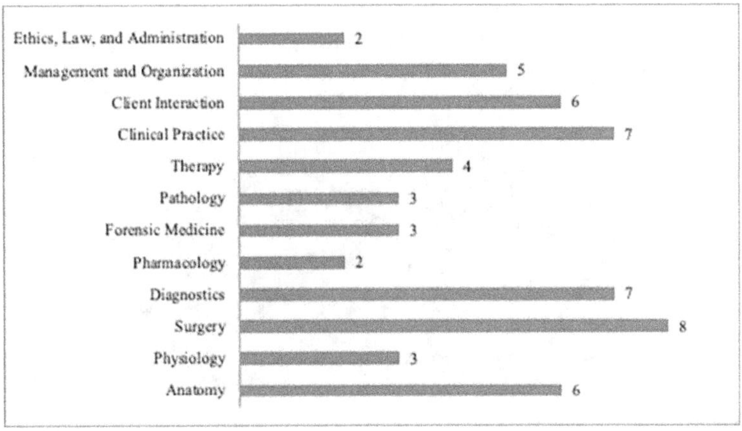

Figure 17.7 Respondents' Self-Assessment of Preparedness to Use Modern Technologies in the Educational Process.

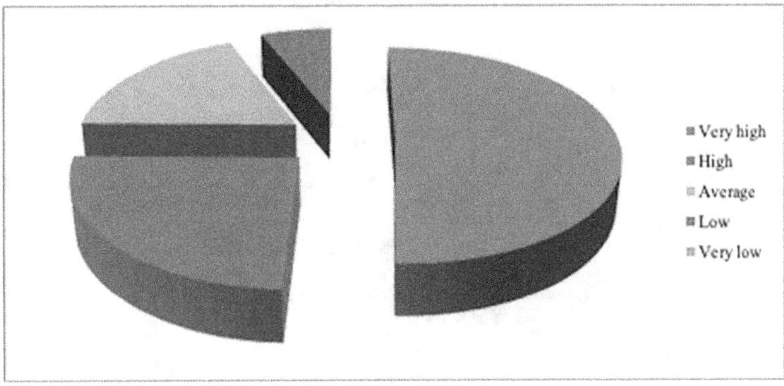

Certain simulators and the use of animals were deemed essential for planning educational processes. However, some activities could be performed on oneself or peers, which is useful for assessing reflexes, physiological dependencies, and anatomical functions. Only 37.5% of respondents reported receiving professional training specifically for utilizing technology in education or indirectly influencing their ability to operate devices. For more expensive educational tools using advanced solutions, basic training and support were often provided by the manufacturer. Many instructors stated that they adapted specific functions and advanced educational capabilities through organization and creativity based on their experience with technologies such as simulators. Half of the respondents rated their technological preparedness as high, but due to rapid technological advancements, this issue was not consistent, as highlighted during interviews. About 40% felt moderately prepared, and 10% felt poorly prepared, lacking confidence in modern technologies (Figure 17.7).

Figure 17.8 Availability of Regular Training and Support in the Use of New Technologies for Teaching at Respondents' Universities.

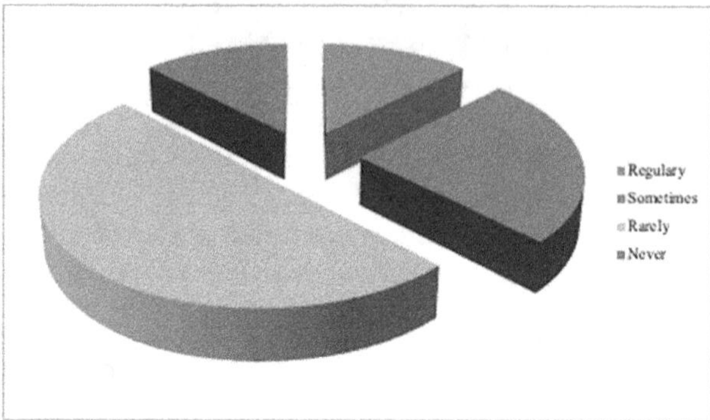

Figure 17.9 Respondents' Beliefs About the Adequacy of the Current Training System for Academic Teachers in Preparing Them to Use Modern Technologies in Education.

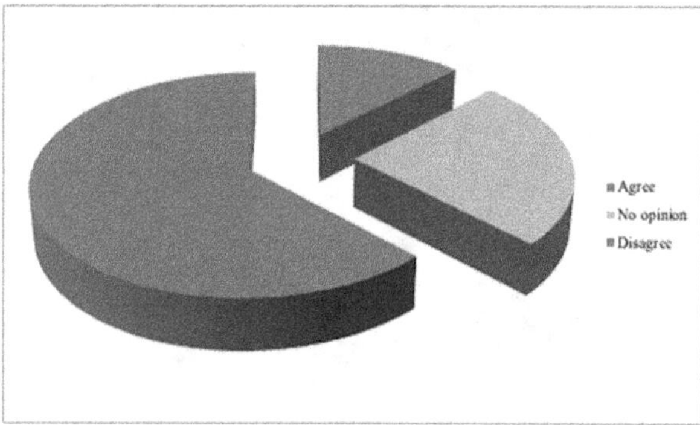

For academic instructors, valuable experiences could be gained from visiting better-equipped institutions. Relatively few universities offer integrated and regular training in technology (Figure 17.8). This data may suggest that the current system does not meet expectations for preparing teachers to fully utilize these technologies or that such training is not provided (Figure 17.9). Some respondents reported using their own resources for additional training in other disciplines, including acquiring technical skills. While they did not view spending their own money as a negative aspect, they noted competitive salary issues in technology sectors, which either discouraged academic careers or encouraged more effort towards re-skilling. Respondents identified the following areas for increasing their competencies to better use digital tools in their work: programming, 3D printing, virtual reality, application development, computer games, data analysis, robotics, teaching methodology, law on new technologies and intellectual property, with ethics and futurology research being less prioritized (Figure 17.10). Concerns were raised that legal regulations are lagging behind in protecting both animals and patients regarding technology use and personal data collection in education and medicine, potentially exposing gaps in compliance and emerging issues outside existing regulations (Figure 17.11).

Figure 17.10 Areas of Skill Enhancement Desired by Respondents if Time and Financial Constraints Were Removed.

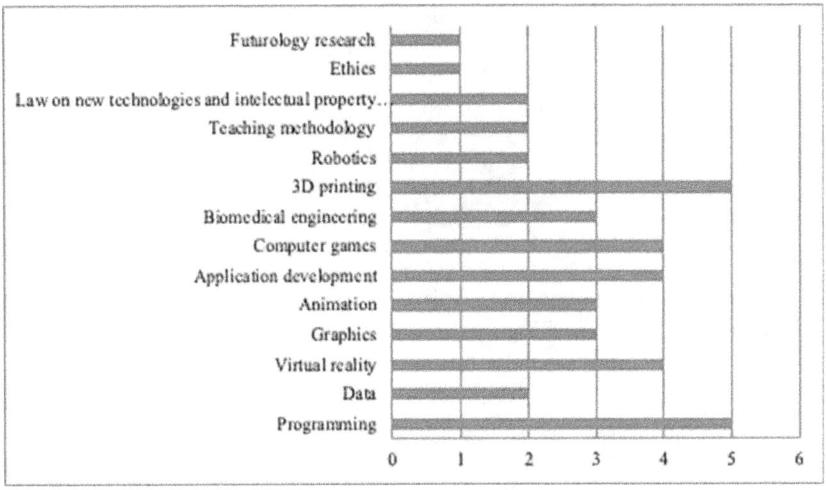

Figure 17.11 Incidence of Challenges in Complying with Data Protection Regulations in Medical/Veterinary Education Using Technology.

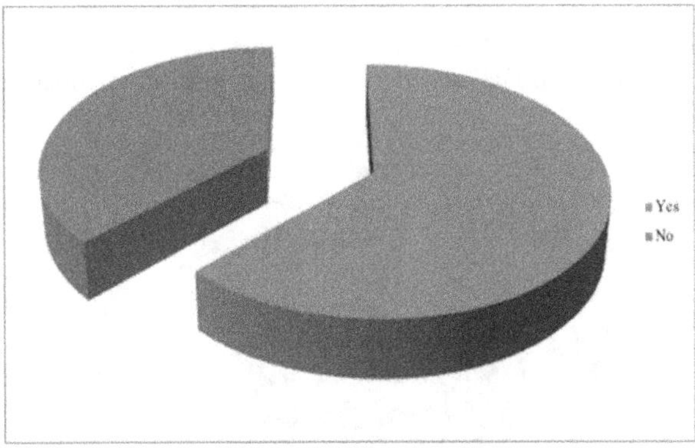

Interestingly, safety and procedures for preventing adverse effects in education were relatively rarely discussed. The most promising technologies for the next decade in medicine and veterinary education were identified as artificial intelligence and advanced simulators. Virtual and augmented reality (VR and AR) and the Internet of Things (IoT) also appeared as significant areas (Figure 17.12).

Figure 17.12 Technologies Most Promising for the Future of Medical/Veterinary Education in the Next 10 Years, According to Respondents.

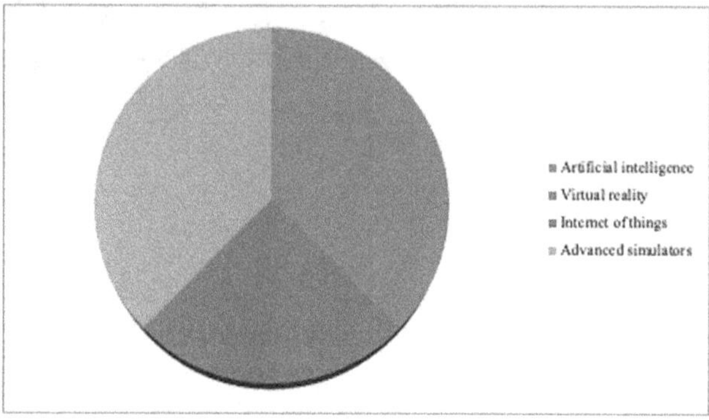

Results indicated a division in opinions on technologies like AI potentially replacing traditional teaching methods in medicine and veterinary education. While 12.5% strongly agreed, 50% agreed that AI could eventually replace traditional methods, and 37.5% disagreed (Figure 17.13). There was recognition of the potential for technology to gradually reduce the use of live-animal models, with 75% agreeing that certain activities with real models remain invaluable for learners and should not be fully eliminated. Fifty percent of respondents agreed that new technologies could entirely replace human and animal involvement in education, while 12.5% strongly disagreed (Figure 17.14).

Figure 17.13 Respondents' Opinions on Whether Technologies Such as AI Could Eventually Replace Traditional Teaching Methods in Medical/Veterinary Education.

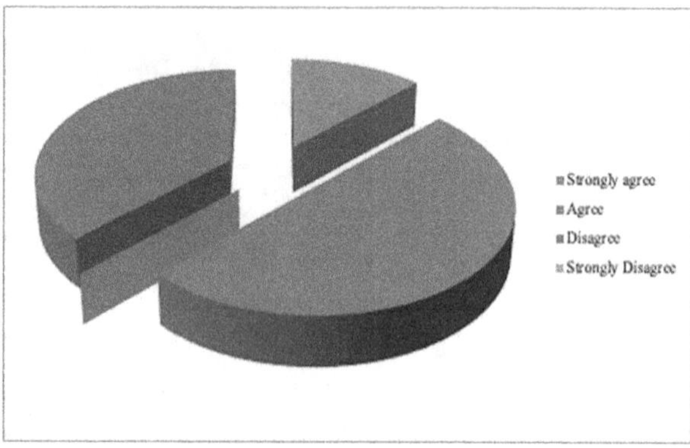

Figure 17.14 Respondents' Views on the Potential of Technology to Eliminate the Need for Live Models in Education.

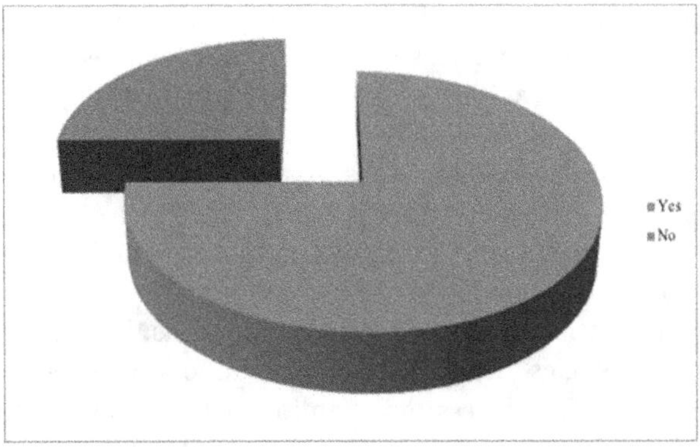

Figure 17.15 Respondents' Opinions on Whether New Technologies Can Fully Replace the Use of Human and Animal Cadavers in Education.

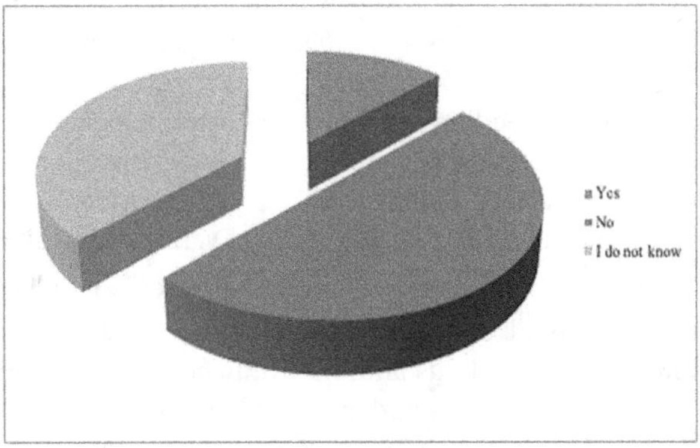

Summary and Future Directions

The advancement of digital technologies is transforming education, introducing a new type of learner—the digital learner—and presenting unique challenges for educators in integrating modern tools. Technologies such as artificial intelligence, simulators, and virtual reality offer significant benefits but also pose difficulties, including unequal access to resources and training, and systemic challenges in their implementation.

Globally, efforts to address social disparities in education must contend with the potential of technology to deepen existing inequalities. The contrast between digital natives and the limited technological skills of digital immigrants highlights the need for targeted efforts to ensure equitable access. In fields like medicine and veterinary science, the focus on developing practical skills and fostering lifelong learning is becoming increasingly important. In countries such as the UK and the USA, advanced technologies, including translational medicine and 3D modelling, are essential. However, the integration of these technologies requires further refinement of pedagogical guidelines. While artificial intelligence is a powerful innovation, it brings risks such as the spread of misinformation and plagiarism. In lower-income countries, challenges related to access to modern technologies are especially pronounced. Despite technological progress, traditional teaching methods, such as dissection, remain crucial for teaching anatomy and integrating clinical knowledge. Simulations provide valuable opportunities for safe learning and the development of technical and behavioural skills, but the integration of advanced technologies, such as surgical robots, necessitates ongoing evaluation of their effectiveness and attention to data safety and critical issues. Ethical considerations are increasingly calling for flexibility regarding the use of animals in education. Legal regulations, such as the EU Directive 2010/63/EU, aim to reduce cruelty and promote alternative methods, underscoring the importance of humane approaches and investments in substitute technologies.

The conducted research serves as a preliminary step in analysing the impact of digital technologies on medical and veterinary education. To gain a more comprehensive understanding of how new technologies integrate with traditional teaching methods, broader studies involving a larger number of respondents are necessary. Such research will provide a more precise evaluation of the effectiveness and efficiency of various educational methods and their associated costs in relation to actual student needs and outcomes.

Future research should encompass a global assessment of effectiveness, exploring how different methods and tools impact education and identifying the most beneficial technologies and their implementation costs. Additionally, analysing the costs and outcomes related to technology in medical and veterinary education will be crucial for understanding its impact on educational results and practical skills. While large-scale online surveys can be challenging, conducting localized studies that are later synthesized at a global level offers a viable approach for obtaining detailed data. Examining how new technologies integrate with traditional teaching methods and assessing their combined effectiveness will provide valuable insights. Furthermore, it is important to investigate how technological education is organized within medical and veterinary schools, including the presence of specialized roles and the consistency of teaching processes in terms of methodology and technology.

Finally, considering the impact of legal regulations and ethical standards on the use of technology in education will be vital. Future research should focus on creating a safe and ethical educational environment, balancing modern tools with traditional methods, and emphasizing continuous professional development and ethical reflection.

While technology has brought about a profound transformation in medical and veterinary education, it also [AT3] comes with notable challenges that must not be overlooked. Chief among these are the diminished opportunities for hands-on experience and the risk of students becoming overly dependent on technology. When students rely too heavily on digital tools, their ability to engage in critical thinking and solve complex problems can weaken over time. Additionally, there are ethical and technical constraints that must be considered, as well as the social inequalities exacerbated by the high cost and unequal access to advanced technologies. This reinforces the idea that technology should serve as an aid in the educator's toolkit, rather than being the primary means of instruction in medical and veterinary fields.

An essential element in both medical and veterinary education is the sensitization of students to the complexities of live patient care, which points to the need for an even stronger integration of ethical training. It is crucial to instill a sense of responsibility in students through consistent reminders of the real-world consequences of their decisions. Greater emphasis should also be placed on developing strong communication skills and fostering personal growth throughout their education. This is particularly important in cultivating empathy, emotional intelligence, and a deeper understanding of patient care—qualities that are vital for both human and veterinary practitioners.

Incorporating clinical dilemmas into case studies, especially through the lens of ethical decision-making, offers students the opportunity to engage with real-life scenarios in a controlled environment. This approach helps bridge the gap between theoretical knowledge and practical application, ensuring that students are not only technically competent but also equipped to make sound, compassionate decisions when dealing with living patients.

In conclusion, while digital technology enriches education, it necessitates further development of teacher competencies and adaptation of teaching methods. Addressing contemporary educational challenges requires a balance between innovative tools and traditional approaches, alongside ongoing professional growth and ethical considerations.

References

Abuvatfa, S.I., Bolilyi, V.O., Lunhol, O. and Sukhovirska, L.P. (2021) 'Use of Sectra virtual dissection table in the educational process', *Information Technologies and Learning Tools*, 83(3), pp. 49-59. https://doi.org/10.33407/itlt.v83i3.3798

Amitai, Z., Root, P., Stephen, S.D. and Shimon, G. (2006) 'Simulation-Based Medical Education: An Ethical Imperative', *Simulation in Healthcare: The*

Journal of the Society for Simulation in Healthcare, 1(4), pp. 252-256. https://doi.org/10.1097/01.SIH.0000242724.08501.63

Atta, I.S., El-Hag, M.A., Ihab Shafek, S., Al-Ghamdi, H.S., Al-Ghamdi, T.H. (2020) 'Drawbacks in the Implementation of an Integrated Medical Curriculum at Medical Schools and Their Potential Solutions', *Education in Medicine Journal*, 12(1), pp. 29-42. https://doi.org/10.21315/eimj2020.12.1.4

Balcombe, J. (2000) The Use of Animals in Higher Education: Problems, Alternatives, & Recommendations. Humane Society Press.

Baptista, C.S., Oliveira, P. and Ribeiro, L. (2024) 'An Exploratory Study of Students' Perception on the Use of Animals in Medical and Veterinary Medical Undergraduate Education', *Journal of Academic Ethics*, 22, pp. 115-136. https://doi.org/10.1007/s10805-023-09495-3

Baumann, O., Lenz, A., Hartl, J., Bernhard, L. and Knoll, A.C. (2022) 'Intuitive teaching of medical device operation to clinical assistance robots', *International Journal of Computer Assisted Radiology and Surgery*, 18, pp. 865-870. https://doi.org/10.1007/s11548-022-02802-0

Baxter, V.K. (2020) 'The Use of Animals in Research', in *Learning from Disease in Pets*, 1st ed. CRC Press, pp. 1-12. https://doi.org/10.1201/9780429056178-7

Braid, H.R. (2022) 'The Use of Simulators for Teaching Practical Clinical Skills to Veterinary Students – A Review', *Alternatives to Laboratory Animals*, 50(3), pp. 184-194. https://doi.org/10.1177/02611929221098138

Brazilian Government (2008) Law No. 11,794 of 2008 (Arouca Law). Brasília: Brazilian Government.

Campbell, A.V., Chin, J. and Voo, T.C. (2007) 'How can we know that ethics education produces ethical doctors?', *Medical Teacher*, 29(5), pp. 431-436. https://doi.org/10.1080/01421590701504077

Cheffield, R., Blackley, S. and Moro, P. (2018) 'A Professional learning model supporting teachers to integrate Digital Technologies', *Issues in Educational Research*, 28(2), pp. 487-510.

Chin, C.K., Munip, H., Miyadera, R., Thoe, N.K., Ch'ng, Y.S. and Promsing, N. (2019) 'Promoting Education for Sustainable Development in Teacher Education integrating Blended Learning and Digital Tools: An Evaluation with Exemplary Cases', *Eurasia Journal of Mathematics, Science and Technology Education*, 15(1), em1653. https://doi.org/10.29333/ejmste/99513

Coker, H. (2020) 'Why Does Digital Learning Matter? Digital Competencies, Social Justice, and Critical Pedagogy in Initial Teacher Education', *Journal of Teaching and Learning*, 14(1), pp. 133-141. https://doi.org/10.22329/jtl.v14i1.6259

Creigton, T.B. (2018) 'Digital Natives, Digital Immigrants, Digital Learners: An International Empirical Integrative Review of the Literature', *ICPEL Educational Leadership Review*, 19(1), pp. 132-140.

Dale, V.H.M., Sullivan, M. and May, S.A. (2008) 'Adult Learning in Veterinary Education: Theory to Practice', *Journal of Veterinary Medical Education*, 35(4), pp. 581-588. https://doi.org/10.3138/jvme.35.4.581

de Bie, M.H. and Lipman, L.J.A. (2012) 'Simulation as a Teaching Aid: The Use of Digital Games and Simulators in Veterinary Education: An Overview with

Examples', *Journal of Veterinary Medical Education*, 39(1), pp. 13-20. https://doi.org/10.3138/jvme.0511.055R

del Pino Palacios-Diaz, M. and Mendoza-Grimón, V. (2023) 'Environmental in Veterinary Education', *Veterinary Sciences*, 10(2), pp. 146. https://doi.org/10.3390/vetsci10020146

European Parliament and Council (2010) Directive 2010/63/EU of the European Parliament and of the Council of 22 September 2010 on the protection of animals used for scientific purposes. Official Journal of the European Union, L 276, pp. 33-79.

Ganguly, P.K. (2010) 'Teaching and Learning of Anatomy in the 21st Century: Directions and Strategies', *The Open Medical Education Learning*, 7(3), pp. 5-10.

García, M.B., Arif, Y.M., Khlaif, Z.N., Zhu, M., Pedro Pereira de Almeida, R., Simões de Almeida, R. and Masters, K. (2024) 'Effective Integration of Artificial Intelligence in Medical Education: Practical Tips and Actionable Insights', in *Transformative Approaches to Patient Literacy and Healthcare Innovation*, pp. 19. https://doi.org/10.4018/979-8-3693-3661-8.ch001

Hernández, E., Fawcett, A., Brouwer, E., Rau, J. and Turner, P.V. (2018) 'Speaking Up: Veterinary Ethical Responsibilities and Animal Welfare Issues in Everyday Practice', *Animals*, 8(1), pp. 15. https://doi.org/10.3390/ani8010015

Hughes, J. (2005) 'The Role of Teacher Knowledge and Learning Experiences in Forming Technology-Integrated Pedagogy', *Journal of Technology and Teacher Education*, 13(2).

Humpenōder, M., Corte, G.M., Pfützner, M., Wiegard, M., Merle, R., Hohlbaum, K., Erickson, N.A., Plendl, J. and Thöne-Reineke, C. (2021) 'Alternatives in Education – Rat and Mouse Simulators Evaluated from Course Trainers' and Supervisors' Perspective', *Animals*, 11(7), pp. 1848. https://doi.org/10.3390/ani11071848

Humpenōder, M., Corte, G.M., Pfützner, M., Wiegard, M., Merle, R., Hohlbaum, K., Erickson, N.A., Plendl, J. and Thöne-Reineke, C. (2021) 'Alternatives in Education-Rat and Mouse Simulators Evaluated from Course Trainers' and Supervisors' Perspective', *Animals*, 11(7), p. 1848. https://doi.org/10.3390/ani11071848

Issenberg, S.B., Barry, W.C., Petrusa, E.R. and Scalese, R.J. (2009) 'A Critical Review of Simulation-Based Education Research: 2003-2009', *Medical Education*, 44(1), pp. 50-63. https://doi.org/10.1111/j.1365-2923.2009.03547.x

Kanehira, R., Kawaguchi, K. and Fujimoto, H. (2015) 'Learning-Training System for Medical Equipment Operation', in *HCI International 2015 – Posters' Extended Abstracts*, pp. 321-327. https://doi.org/10.1007/978-3-319-21383-5_54

Kataoka, H.U., Tokinobu, A., Fujii, C., Watanabe, M., Obika, M. (2023) 'Eleven years of data on the Jefferson Scale of Empathy – medical student version: Japanese norm data and tentative cutoff scores', *BMC Medical Education*, 23(81). https://doi.org/10.1186/s12909-022-03977-5

Knight, A. (2022) 'Scientific and Educational Animal Use', in *Routledge Handbook of Animal Welfare*, 1st ed. Routledge, pp. 1-15. https://doi.org/10.4324/9781003182351-16

Komasawa, N. and Yokohira, M. (2023) 'Simulation-Based Education in the Artificial Intelligence Era', *Cureus*, 15(6), e40940. https://doi.org/10.7759/cureus.40940

Kryeziu, S.A., Avdiu, T.A., Avdiu, A. (2021) 'Examining the Teachers, Adnimistrators and Parents' View on Drawbacks of Technology Use in Education', *Illkogretim Online – Elementary Education Online*, 20(2), pp. 206-215. https://doi.org/10.17051/ilkonline.2021.02.26

Lee, H. (2023) 'The Rise of ChatGPT: Exploring Its Potential in Medical Education', *Anatomical Sciences Education*, 17(5), pp. 926-931. https://doi.org/10.1002/ase.2270

Lee, H., Longhurst, M. and Campbell, T. (2017) 'Teacher Learning in Technology Professional Development and Its Impact on Student Achievement in Science', *International Journal of Science Education*, 39(10), pp. 1282-1303. https://doi.org/10.1080/09500693.2017.1327733

Leng, L. (2023) 'Challenge, Integration, and Change: ChatGPT and Future Anatomical Education', *Medical Education Online*, 29(1). https://doi.org/10.1080/10872981.2024.2304973

Lenz, A., Baumann, O., Hartl, J., Bernhard, L. and Knoll, A.C. (2022) 'Intuitive Teaching of Medical Device Operation to Clinical Assistance Robots', *International Journal of Computer Assisted Radiology and Surgery*, 18, pp. 865-870. https://doi.org/10.1007/s11548-022-02802-0

Loderer, K., Perkun, R., Lester, J.C. (2020) 'Beyond cold technology: A systematic review and metaanalysis on emotions in technology-based learning environments', *Learning and Instruction*, 70, pp. 101162. https://doi.org/10.1016/j.learninstruc.2018.08.002

Luterbach, K.J. (2022) Creating Stellar Lessons with Digital Tools: From *Integration to Innovation in Technology-Enhanced Teaching*, 1st ed. pp. 1-276.

Masterman, E. and Manton, M. (2011) 'Teachers' Perspective on Digital Tools for Pedagogic Planning and Design', *Technology, Pedagogy and Education*, 20(2), pp. 227-246. https://doi.org/10.1080/1475939X.2011.588414

McGaghie, W.C., Issenberg, S.B., Petrusa, E.R. and Scalese, R.J. (2009) 'A Critical Review of Simulation-Based Education Research: 2003-2009', *Medical Education*, 44(1), pp. 50-63. https://doi.org/10.1111/j.1365-2923.2009.03547.x

McKenney, S. and Visscher, A.J. (2019) 'Technology for Teacher Learning and Performance', *Technology, Pedagogy and Education*, 28(2), pp. 129-132. https://doi.org/10.1080/1475939X.2019.1600859

McLennan, M.W. and Heath, T.J. (2008) 'The Role of Lectures in Veterinary Education', *Australian Veterinary Journal*, 78(10), pp. 702-709. https://doi.org/10.1111/j.1751-0813.2000.tb10412.x

Mendoza-Grimón, V. and Palacios-Diaz, M.D.P. (2023) 'Environmental in Veterinary Education', *Veterinary Sciences*, 10(2), pp. 146. https://doi.org/10.3390/vetsci10020146

Monteiro, A., Mouraz, A. and Dotta, L.T. (2020) 'Veteran Teachers and Digital Technologies: Myths, Beliefs and Professional Development', *Teachers and Teaching*, 26(7-8), pp. 577-587. https://doi.org/10.1080/13540602.2021.1900809

Noyes, J.A., Carbonneau, K.J. and Matthew, S.M. (2022) 'Comparative Effectiveness of Training with Simulators Versus Traditional Instruction in Veterinary Education: Meta-Analysis and Systematic Review', *Journal of Veterinary Medical Education*, 49(1), pp. 25-38. https://doi.org/10.3138/jvme-2020-0026

Park, J.C., Kwon, H.J.E. and Chung, C.W. (2021) 'Innovative Digital Tools for New Trends in Teaching and Assessment Methods in Medical and Dental Education', *Journal of Educational Evaluation for Health Professions*, 18, pp. 13. https://doi.org/10.3352/jeehp.2021.18.13

People's Republic of China (1988) Regulations for the Administration of Laboratory Animals. Beijing: State Council of the People's Republic of China.

Pfeifer, L., Fries, S., Striner, A., Nagel, L., Cohen, C., Aschentrup, L., Schönbeck, M., Naureth, A., Raschper, P., Herzig, T., Wrona, K.J. (2024) 'Positive Aspects and Potential Drawbacks of Implementing Digital Teaching/Learning Scenarios in Health Professions Using Nursing Education as an Example: A Research Report from Germany', *Nursing Reports*, 14(1), pp. 486-481. https://doi.org/10.3390/nursrep14010036

Praszkier, R. (2023) 'Skala empatii: wprowadzenie, walidacja i zastosowanie'. Polskie Forum Psychologiczne, XXVIII, 1, pp. 37-55.

Quain, A., Ward, M.P. and Mullan, S. (2021) 'Ethical Challenges Posed by Advanced Veterinary Care in Companion Animal Veterinary Practice', *Animals*, 11(11), pp. 3010. https://doi.org/10.3390/ani11113010

Raghupathi, V. and Raghupathi, W. (2020) 'The Influence of Education on Health: An Empirical Assessment of OECS Countries for the Period 1995-2015', *Archives of Public Health*, 78(20). https://doi.org/10.1186/s13690-020-00402-5

Republic of Korea (1973, amended 2013) Act on Welfare and Management of Animals. Seoul: Government of South Korea.

Ruggiero, D. and Mong, C.J. (2015) 'The Teacher Technology Integration Experience: Practice and Reflection in the Classroom', *Journal of Information Technology Education: Research*, 14, pp. 161-178. https://doi.org/10.28945/2227

Saaiq, M., Khan, R.A. and Yasmeen, R. (2024) 'Digital Teaching: Developing a Structured Digital Teaching Competency Framework for Medical Teachers', *Medical Teacher*, pp. 1-7. https://doi.org/10.1080/0142159X.2024.2308782

Samarasekera, D.D., Lee, S.S., Yeo, J.H.T., Yeo, S.P., Ponnamperuma, G. (2023) 'Empathy in health professions education: What works, gaps and areas for improvement'. *Medical Education*, 57(1), pp. 86-101. https://doi.org/10.1111/medu.14865

Sari, R.C., Warsono, S., Ratmono, D., Zuhrohtun, Z., Hermawan, H.D. (2023) 'The effectiveness of teaching virtual reality-based business ethics: is it really suitable for all learning styles?', *Interactive Technology and Smart Education* 20(1), pp. 19-35. https://doi.org/10.1108/ITSE-05-2021-0084

Shaw, M.P. (2022) 'The Public Right to Education', *The University of Chicago Law Review*, 89(5), pp. 1179-1244.

Spencer, J., Blackmore, D., Heard, S., McCrorie, P., McHaffie, D., Scherpbier, A., Gupta, T.S., Singh, K. and Southgate, L. (2001) 'Patient-oriented Learning: A Review of the Role of the Patient in the Education of Medical Students', *Medical Education*, 34(10), pp. 851-857. https://doi.org/10.1046/j.1365-2923.2000.00779.x

Sulaiman, J. and Ismail, S.N. (2020) 'Teacher Competence and 21st Century Skills in Transformation Schools 2025 (TS25)', *Universal Journal of Educational Research*, 8(8), pp. 3536-3544. https://doi.org/10.13189/ujer.2020.080829

Tawfik, A.A., Reeves, T.D. and Stich, A. (2016) 'Intended and Unintended Consequences of Educational Technology on Social Inequality', *TechTrends*, 60, pp. 598-605. https://doi.org/10.1007/s11528-016-0109-5

Toebes, B. (2001) 'The Right to Health as a Human Right in International Law', *Refugee Survey Quarterly*, 20(3), pp. 180. https://doi.org/10.1093/rsq/20.2.180

U.S. Department of Agriculture (2023) Animal Welfare Regulations. Code of Federal Regulations, Title 9, Parts 1-3. Washington, D.C.: U.S. Government Publishing Office.

U.S. Department of Agriculture (2023) Rules of Practice Governing Proceedings under the Animal Welfare Act. Code of Federal Regulations, Title 9, Part 4. Washington, D.C.: U.S. Government Publishing Office.

UK Parliament (1986, amended 2012) Animals (Scientific Procedures) Act 1986. London: The Stationery Office.

United States Congress (2018) Animal Welfare Act. United States Code, Title 7, Section 2131. Washington, D.C.: U.S. Government Publishing Office.

Urbani, J.M., Roshandel, S., Michaels, R. and Truesdell, E. (2017) 'Developing and Modeling 21st-Century Skills with Pre-service Teachers', *Teacher Education Quarterly*, 44(4), pp. 27-50.

Valentijn, F.A., Schakelaar, M.Y., Hegeman, M.A., Schot, W.D., Dictus, W.J.A.G., Crnko, S., ten Broeke, T. and Bovenschen, N. (2023) 'A Challenge-Based Interdisciplinary Undergraduate Concept Fostering Translational Medicine', *Biochemistry and Molecular Biology Education*, 52(2), pp. 198-209. https://doi.org/10.1002/bmb.21804

Varas, D., Santana, M., Nussbaum, M., Claro, S. and Imbarack, P. (2023) 'Teachers' Strategies and Challenges in Teaching 21st Century Skills: Little Common Understanding', *Thinking Skills and Creativity*, 48, 101289. https://doi.org/10.1016/j.tsc.2023.101289

Varner, C., Dixon, L. and Simons, M.C. (2021) 'The Past, Present, and Future: A Discussion of Cadaver Use in Medical and Veterinary Education', *Frontiers in Veterinary Science*, 8, 720740, pp. 1-5. https://doi.org/10.3389/fvets.2021.720740

Wood, N., Beavis, C., Cloonan, A., Hutchison, K., Pangrozio, L. and Sefton-Green, J. (2020) 'Teacher Learning and the Everyday Digital', *The Australian Educational Researcher*, 47, pp. 183-197. https://doi.org/10.1007/s13384-019-00326-2

Xiberta, P., Boada, I. (2019) 'IVET, an Interactive Veterinary Education Tool', *Journal of Animal Science*, 97(2), pp. 932-944. https://doi.org/10.1093/jas/sky471

Zemljic, B. (2004) 'Standards of Veterinary Education for the Future: Preparing the Profession for the New Century', Global Perspectives of Veterinary Education, *Journal of Veterinary Medical Education*, 31(1), pp. 13-14. https://doi.org/10.3138/jvme.31.1.13

Zhang, Y. and Xu, Y. (2024) 'Developing a Structured Digital Teaching Competency Framework for Medical Teachers', *Medical Teacher*, pp. 1-7.

About the Contributors

Bhaskar Gurramkonda is a research scholar in the Department of Education, Central University of Kerala. He has his Master of Education and Mathematics. He qualified for national-level tests, NET(Education), and CTET. He also qualified for APSET in both Mathematics & Education. His area of research work is AI Integrated Realistic Mathematics Education (AIREM), particularly focusing on digital aspects of assessment and teaching, Education Technology, and Teaching of Mathematics. Related to his research area, he has published research papers in reputable journals. He has presented papers at national and international conferences.

Vanitha, C. is an Assistant Professor in the Department of Education, Central University of Kerala. She has obtained a Ph.D from the University of Mysore on "Life Centred Career Education for Students with Mental Retardation in Karnataka State- An Exploratory Study." She has 11 years of teaching experience. Her research interests include Inclusive Education, Social Science Education, and Qualitative research. She has published research papers in the reputed Journals. She has presented papers at the National and International conferences and seminars. She has conducted webinars and Workshops at the National and International levels on Inclusive Education.

Shraddha Verma. Dr. Shraddha Verma is a distinguished educator and researcher with over 22 years of experience in the field of Education. She holds a Ph.D. in Education, an M.Phil., an M.Ed. (where she achieved a top 10 merit position), and an M.Sc. in Botany, specializing in Microbiology. Dr. Shraddha Verma is also UGC NET, CTET, and CGTET certified, reinforcing her academic rigor and dedication to educational excellence. Her contributions include authoring educational materials for SCERT and publishing numerous research papers in both national and international journals. With two patents, two sole-authored books, and four edited volumes to her credit, she remains a prominent voice in educational research. Dr. Shraddha Verma's active participation in conferences and seminars further enriches her perspective on contemporary educational practices. In addition to her academic and research pursuits, she has taken on several key administrative roles. She has served as Center Superintendent, a member of the Board of Studies, and has contributed to committees such as IQAC, NAAC, Academic Audit, Feedback, Scholarship, and Admissions. She is also entrusted with setting and evaluating examination papers, demonstrating her commitment to upholding academic standards. Since December 2022, Dr. Shraddha Verma has held the position of Dean,

Faculty of Education, at Kalinga University, where she leads with a vision of fostering growth and excellence in the field. Recognized for her contributions, she has received both the Innovative Educationist Award and the Academic Excellence Award, highlighting her dedication to advancing education and nurturing future educators.

Vijayalakshmi, A. is an Associate Professor in the Department of Statistics and Data Science at Christ University, with a keen focus on integrating emerging technologies into education. Dr. Vijayalakshmi A has an extensive background in artificial intelligence, machine learning, and IoT, with several published research articles and book chapters that showcase her contributions to these cutting-edge fields. With over 18 years of teaching experience, Dr. Vijayalakshmi A has significantly advanced classroom methodologies by integrating innovative tools and technologies, fostering enhanced student engagement and improved learning outcomes. As a passionate educator and researcher, Dr. Vijayalakshmi A has participated in workshops on topics such as AI in education and the role of data science in personalised learning in academia. Their work emphasises the transformative potential of technology in creating interactive, inclusive, and adaptive learning environments. This commitment to teaching excellence and innovation underpins their contributions to this book chapter on emerging technologies in classroom teaching.

Yashi Goyal. Ms Yashi Goyal is a post-graduate scholar and researcher enrolled in the Master of Education programme at the Department of Education, University of Delhi. She also holds a master's degree in Chemistry with a specialisation in Physical Chemistry from Miranda House, University of Delhi. Her research work is in the field of inclusive education with a focus on the transformative role of technology, particularly in science education. She has also presented her work at the 21st Annual Kaleidoscope Conference at the Faculty of Education, University of Cambridge, held in May 2024.

Amit Nitinrao Kolhe. is a visionary educationist and dedicated philanthropist committed to the economic and social development of rural communities through education and industrialization. With a degree in Computer Engineering and an MBA in Marketing, he blends technological expertise with strategic insight. Recognized as an ICON of Education by Lokmat newspaper and honoured by Maharashtra's Education Minister, Amit is the Managing Trustee of the Sanjivani Group of Institutes, a 38-year-old institution renowned for its academic excellence and infrastructure. He has extensive experience in managing educational institutions in rural India and serves as the Treasurer of the Association of Management of Unaided Engineering and Polytechnic Colleges for Maharashtra State. Additionally, he has contributed to the Maharashtra Government's Committee on Professional Education Policies and was elected to the CII Western Region Education Sub Committee for 2018-19. Through his leadership and the establishment of the Sanjivani Foundation, Amit aims to empower individuals and uplift society by making education a catalyst for transformative change.

Smita Tukaram Kumbhar is an accomplished Associate Professor and Head of the Department of Pharmaceutical Regulatory Affairs at Sanjivani College of Pharmaceutical Education and Research in Kopargaon, Maharashtra. With a PhD in Pharmacy and 16 years of research and academic experience, she is dedicated to advancing education in pharmacy and enhancing the field through innovative research and a strong commitment to mentoring the next generation of pharmacists. Dr. Kumbhar's contributions to the pharmaceutical sciences are marked by her extensive publications and her active role in guiding students toward academic and professional excellence.

Dhananjay Arun Kumbhar is a dedicated professional in the Department of Training and Placement at Sanjivani University in Kopargaon, Maharashtra. With a strong focus on bridging the gap between education and employment, he plays a pivotal role in facilitating career opportunities for students. Dhananjay is committed to enhancing students' employability through various training programs, workshops, and placement initiatives, ensuring they are well-prepared to enter the competitive job market. His efforts contribute significantly to the university's mission of fostering holistic development and success for its graduates.

Grace Okubo is a user experience designer boasting over a decade of experience in crafting user-centred products and leading the design and development of software products across various sectors such as real estate, fintech, and e-commerce. Her expertise lies in seamlessly blending Design Thinking and Agile Methodologies. At the same time, her role as an educator has seen her impart essential skills to a global audience in design and product management. Trained in Ethics in Design, Design Thinking, and Foresight Methodologies, Grace consistently proposes innovative solutions to complex challenges, driven by her goal of promoting equity, inclusivity, and sustainability.

Dr. Marina Bantiou is an Adjunct Lecturer of History Didactics and Oral History at the Department of Early Childhood Education at the University of Thessaly in Greece. She also serves as teaching staff in the Master's Program "Methodology of Criticism and Publication of Historical Sources" at the Department of History of the Ionian University, where she teaches the documentation of oral testimonies and the use of press sources in late modern and contemporary history. She graduated from the Department of Political Science and History of Panteion University and then obtained a Master's Degree in Historical Research, Didactics and New Technologies from the

Department of History of the Ionian University. She holds a PhD in Late-Modern and Contemporary History from the same department. Her first monograph in Greek, titled "The Battle of Crete in the Greek and International Press of the Second World War: Censorship and Propaganda," was published in 2024 by Radamanthys Press. Her research interests primarily revolve around Late-Modern and Contemporary Greek and European history, focusing on the intersection of history and media, public history, oral history and history education. She is also a Visiting Research Fellow at the Department of Political Science and International Relations at the University of Peloponnese.

Bonshri Gogoi is a postgraduate research scholar in the Department of Education at Tezpur Central University, Assam, India. Her research interests encompass Educational Psychology and Educational Technology. She has published three book chapters and presented research papers at an ICSSR-sponsored national seminar.

Ivy Shen is an Associate Professor of Political Science at Southeast Missouri State University. She holds an M.A. in Mass Media and Journalism, an M.A. in Political Science, and a Ph.D. in Political Science from the University of Oklahoma. Dr. Shen teaches American Politics at the undergraduate level and Public Administration courses at the graduate level. Her research has centred around the application of Artificial Intelligence in education, education policy, nonprofit management, performance management, and state politics.

Jana Gerard is the Coordinator of The EDvolution Center and adjunct faculty at the College of Education, Health, and Human Studies at Southeast Missouri State University. As part of her work with The EDvolution Centre, Jana ensures that teacher candidates at Southeast Missouri State are prepared to effectively and appropriately implement educational technology in their future classrooms. This is accomplished through providing instruction on digital literacy, digital competency, and digital citizenship through the lens of the ISTE Standards for Educators. Jana also provides students, faculty, and staff professional development on design thinking, coding, 3D printing, and virtual and augmented reality. Jana serves as the chair of the college technology and interprofessional education committees and a member of the educator preparation assessment committee, the equity in education committee, and the global citizenship committee. Jana is a member of the Missouri Regional Professional Development Committee's Technology Consortium, provides professional development on educational technology to local school districts, and works with local districts to maintain collaborative educational technology efforts. Jana presents at national, state, and local conferences and is an ISTE Certified Educator, Level 1 & 2 Google Certified Educator, Google Certified Trainer, Google Coach, Apple Coach, Microsoft Certified Educator, Microsoft Innovative Educator, Discovery Education

Network Star, Ozobot Certified Educator, Diifit ambassador, and Bloxels Ambassador. Jana has been married for over 30 years and has four adult daughters, four sons-in-law, and two giant dogs.

Elhadj Moussa BenMoussa is an experienced Academic English Instructor and PGR Adviser with over a decade of teaching and mentoring at universities around the world. He has worked in diverse roles, including Senior Lecturer, Academic Progress Coach, and TESOL Specialist, helping students succeed in their academic and professional journeys. With a PhD in TESOL and International Education, ElHadj combines deep expertise with a passion for creating positive, inclusive learning environments. Throughout his career, he has supported students in developing essential skills, from critical thinking to academic writing, and has guided many to achieve outstanding results. His international experience, including teaching at institutions in the UK, China, and beyond, reflects his adaptability and commitment to cross-cultural education. Outside the classroom, he is actively involved in research and academic events, fostering collaboration and growth within the academic community.

Abirlal Mukherjee is a digital humanist. He is a doctoral researcher in English (Digital Popular Culture) at GITAM Deemed to be University. Abirlal taught at Amity University as an Assistant Professor of English. He has been an InSIG 2024 Fellow at India School on Internet Governance and IIMB, a Teach for India Fellow - Delhi, batch 2023, a Research Associate Intern at the Indian National Law Review, The Law Learner and an Intern at National Human Rights Commission of India. Volunteered with UNDP. His research interests focus on the intersection of digital culture, memetics, content virality, digital pandemics and interdisciplinary education. Abirlal's current research delves into the impact of internet memes on cultural discourse, examining how they shape and reflect societal values.

Sunita Maharaj-Landaeta has been a Global Goodwill Ambassador since 2018, is an accomplished academic with extensive expertise in educational management, leadership, and inclusion and globalization. She holds a Postgraduate Certificate and M.Ed. from the University of Sheffield, UK, and a PhD in Education from the University of Trinidad and Tobago. Her diverse professional background spans human resources, forensic psychology, language and cultural exchange, and gender equality, bolstered by advanced training from the FLACSO Latin America Faculty of Social Sciences in Argentina. As a Cambridge English Teacher and former IELTS Examiner, Dr. Maharaj-Landaeta's career has bridged the intersections of migration, human trafficking, and education in the Caribbean and Latin America. With more than 20 years of experience, her work has focused on trauma-impacted students, curriculum development, and educational leadership, particularly within emergency and humanitarian contexts. From 2017 to 2019, she managed education projects for children on the move with the support of UN agencies, and since 2020, she has worked with the International Organization for Migration (IOM) on the Response for Venezuela (R4V) platform. Dr. Maharaj-Landaeta's scholarly endeavours include exploring the implications of emerging technologies on education, particularly their potential to enhance access and equity for vulnerable populations. Her critical analysis of artificial intelligence (AI) focuses on its impacts within educational contexts, emphasizing the importance of ethical practices and equity. Her research aligns with her PhD work at the University of Trinidad and Tobago, which examines educational responses for displaced children. Dr. Maharaj-Landaeta's commitment to innovation and leadership is further reflected in her completion of a UNHCR Innovation Fellowship and a Business Management and Leadership certificate at Harvard Business School.

Alfredo J. Landaeta is a seasoned procurement and supply chain professional with more than 25 years' experience in project management, policy implementation, and organizational development. With a focus on efficiency and innovation, Alfredo has worked across public and private sectors, contributing to the successful delivery of key initiatives. Since 2017, Alfredo has been actively involved in educational projects supporting migrants and refugees, offering his expertise and volunteering his time to improve access to learning and resources for underserved communities. He has a deep passion for initiatives that prioritize children's welfare and education, recognizing their critical role in shaping future generations. Recently, Alfredo has taken an active interest in the application of new technologies, including artificial intelligence (AI), related to everything we do daily, but especially in the educational arena. He is particularly focused on exploring how AI can streamline administrative processes, enhance personalized learning, and improve accessibility in educational systems. With a pragmatic approach to technology adoption, Alfredo emphasizes responsible implementation practices that balance innovation with ethical considerations and stakeholder engagement. His work is informed by a global perspective, developed through collaborations with diverse organizations at a local, regional and international level. Alfredo's commitment to advancing the integration of technology into education, combined with his dedication to supporting vulnerable communities, reflects his broader interest in fostering sustainable, practical solutions to modern challenges.

Alka Ranjan. Dr. Alka Ranjan is an Assistant Professor of Psychology at JAIN (Deemed-to-be University) specializing in Counselling Psychology and Public Health. Her research areas are positive psychology, family relationships, neuropsychology, Indian psychology, assessments, therapeutic processes, and qualitative research. She has experience teaching undergraduate, postgraduate, and PhD courses in psychology. She has published book chapters and research articles in national and international journals indexed in UGC -Care and SCOPUS. She supervises PhD scholars and postgraduate dissertations and has guided undergraduate research projects. She has organized National and International Conferences and conducted seminars, workshops, and community outreach programs. She is a practicing consultant psychologist with cross-cultural experience in Singapore, the U.S.A., and India.

Soumya Thankam Varghese is an Assistant Professor at the Jindal Institute of Behavioural Sciences (JIBS), O. P. Jindal Global University. She received her PhD in Psychology from the Vellore Institute of Technology, Chennai. Her primary area of interest revolves around academic performance, focusing on identifying and facilitating factors that contribute to it. NCERT, Govt. of India, funded her doctoral research. Throughout her career, she has taught various subjects, including research methods, educational psychology, organisational psychology and mental health at both undergraduate and postgraduate levels.

Beenish Rasool is a first-year PhD Scholar at VIT-AP University. She completed her M.A. in Applied Psychology from Jamia Millia Islamia and her Bachelor's degree in psychology from Kashmir University. Beenish also worked at Mount Talent Consulting & Aspire Human Capital Management Pvt. Ltd Gurgaon. She has presented three conference papers. Her research areas include Clinical, Positive, Counselling and Educational Psychology.

Abdul Raffie Naik is an assistant professor at VIT-AP, previously held positions at Acharya Institutes, Bangalore and Amity University, Mumbai. He worked as a Psychologist at Apollo Hospital, Hyderabad, for three years. He completed his Ph.D. from Aligarh Muslim University in 2019 and his M.Sc. from Central University of Karnataka. With 23 paper publications, six book chapters, and numerous conference presentations, he is also an active journal reviewer and editor. His research interests include Clinical and Counselling Psychology and Positive Psychology.

Alpana Bhattacharya is an Associate Professor of Educational Psychology in the Secondary Education and Youth Services department at Queens College and a Ph.D. Program in Educational Psychology at the Graduate Center, the City University of New York (CUNY). Her research interests include literacy acquisition, language development, reading and writing processes, metacognitive strategy instruction, specific learning disabilities, culturally and linguistically diverse learners, and teacher preparation for differentiated instruction. Her research contributions include journal articles, book chapters, and conference presentations in the areas of learning disabilities, literacy education, and teacher preparation. She also serves as a reviewer of journal articles and conference proposals for the American Educational Research Association

(AERA), American Psychological Association (APA), American Association of Colleges for Teacher Education (AACTE), Learning Disabilities Worldwide (LDW), Taylor & Francis, and Elsevier. As a tenured faculty at Queens College and Graduate Center, CUNY, she teaches undergraduate, graduate, post-masters, and doctoral courses in Educational Psychology including, *Development and Learning in Middle Childhood and Adolescence; Cognition, Technology, and Instruction for Diverse Learners; Classroom Management; Educational Psychology: Foundations and Contemporary Issues; Understanding Group Behavior and Cultural Differences in Schools;* and *Social-Emotional and Cultural Factors in Development and Education.* She also supervises doctoral dissertations in the areas of literacy education and language development from early childhood through adulthood.

Emily Wilson. Emily Wilson is Head of Music Education and Senior Lecturer at the Melbourne Graduate School of Education, University of Melbourne, Australia. Her PhD thesis (2019) is titled 'It's music and we came to play instruments': teaching for engagement in classroom music. Emily lectures in music education and teacher education at undergraduate and postgraduate levels, and for early childhood, primary and secondary pre-service teachers. She maintains a close relationship with primary and secondary schools and early childhood settings. She has over 20 years' experience as a music educator. As a school teacher, Emily has taught classroom and instrumental music at primary and secondary schools in Australia and the UK and has held Head of Department posts. Her research interests include student engagement, engaging teacher practices, decolonising music education, showcasing Indigenous knowledges through music and formal and informal learning in music education. Emily has published in national and international journals and presents regularly at national and international conferences.

Pauline Black. Pauline Black is a musician, educator, examiner and researcher based in Orkney, Scotland, UK. She is Deputy Principal of the University of the Highlands and Islands (UHI) Orkney, where she is also a lecturer in music teacher education. With thirty years of experience across secondary school, community settings, higher education and the music industry, Pauline has held a range of musical, academic and management roles. She was formerly Head of Music and Senior Lecturer in Music Education at the University of Aberdeen, leading the music education programme at undergraduate and postgraduate levels for a considerable time. Pauline has several years of experience in fulfilling board roles in the music industry and is the Past President of the Independent Society of Musicians (ISM). A trumpet player and singer, her practice bridges teaching, performance, creative practice and research. Pauline's interests include teaching and learning in jazz and improvised music; informal learning; collaborative composing with technology; creative learning;

participatory music; gender equity in music and identities in music. She publishes in national and international journals and presents regularly at conferences both nationally and internationally.

Anna Tomańska is a veterinarian and PhD candidate at the Department of Animal Biostructure and Physiology at Wrocław University of Environmental and Life Sciences. She is also a participant in the doctoral seminar on legal sciences at the Polish Academy of Sciences. Her research focuses on the use of animals in translational medicine and education, with an emphasis on developing new therapies and biomedical technologies. She is pursuing additional interests in forensic veterinary medicine. With six years of teaching experience across various educational levels, from preschool to university, she brings a diverse skill set to her work. During her studies, she contributed to an implementation project by developing an innovative hive for beekeeping, which sparked her interest in applying modern technology to animal observation and care. Innovative solutions continue to play a central role in her research and educational endeavours.

About the Editors

Harshith B Nair is a budding teacher educator at the Regional Institute of Education (NCERT), Mysuru. He specialises in mixed methods research and critical analysis. He serves on various editorial and reviewer boards of international organisations and journals. He has published various book chapters, papers and books in the area of education, technology and psychology. He wishes to pursue higher education in the area of teacher education and education technology and is currently looking for doctoral opportunities. He is an avid reader, theatre artist, and polyglot, and he believes that this helps him to be a better educator. He continues to work in the area of edtech and teacher education.

Prof. (Dr) G. Viswanathappa earned an M.Sc. in Statistics from Sri Venkateswara University, Tirupathi (1986), an M.Ed. from the Regional College of Education, Mysore (1988), and a PhD in Education from the University of Mysore (1992). After working with the Government of Andhra Pradesh, he joined NCERT on 16th October 1996 as faculty at the Regional Institute of Education (RIE), Mysore, later serving as Head of Extension Education and in charge of the Educational Technology Cell. He was Principal of RIE Ajmer (June 2017–June 2019) and Director of SIEMAT, Amaravati (June 2019–March 2020). He rejoined RIE Mysore in March 2020 as Professor of Education, teaching educational research, pedagogy, assessment, and ICT to graduate and postgraduate students. He has led state, regional, and national studies on quality interventions under SSA and RMSA, developed an online Certificate Programme in Educational Data Processing (CEDP), published 13 books, numerous journal articles, and guided 15 PhD scholars. A specialist in educational data analysis, he conducted capacity-building programs for researchers and developed accreditation guidelines for schools. He represented NCERT at UNICEF's School Leadership Programme in Bangkok (2018) and is a member of school boards at three central universities. He also contributed to developing the syllabus for NCTE's 4-year Integrated Teacher Education Programme.

Dr. Vasileios Paliktzoglou is an academic and industry professional and holds a Ph.D. in Computer Science from the University of Eastern Finland. He is an active contributor to academia, serving as a guest and associate editor for prestigious journals like the Journal of Information Systems Education (JISE) and the IAFOR Journal of Education. He has been a keynote speaker at several academic and industry events, sharing his insights on social computing, collaborative learning, and emerging technologies in higher education. Dr.

Paliktzoglou also actively participates in international research projects, highlighting his commitment to advancing knowledge and innovation as the Book Series Editor for the "Emerging Technologies in Education" at Vernon Press and "AI and Education" at STAR Scholars Press. His expertise continues to shape the academic and research landscape at Bahrain Polytechnic.

Index

www.ingramcontent.com/pod-product-compliance
Lightning Source LLC
LaVergne TN
LVHW020413220926
848484LV00018B/2564/J
* 9 7 9 8 8 8 1 9 0 3 7 1 8 *